CRC Handbook of Eicosanoids: Prostaglandins and Related Lipids

Volume I
Chemical and Biochemical Aspects

Part B

Editor

Anthony L. Willis, Ph.D.
Principal Scientist
Atherosclerosis and Thrombosis Section
Institute of Experimental Pharmacology
Syntex Research
Palo Alto, California

CRC Press
Taylor & Francis Group
Boca Raton London New York

CRC Press is an imprint of the
Taylor & Francis Group, an informa business

First published 1987 by CRC Press
Taylor & Francis Group
6000 Broken Sound Parkway NW, Suite 300
Boca Raton, FL 33487-2742

Reissued 2018 by CRC Press

A Library of Congress record exists under LC control number: 86018779

Publisher's Note
The publisher has gone to great lengths to ensure the quality of this reprint but points out that some imperfections in the original copies may be apparent.

Disclaimer
The publisher has made every effort to trace copyright holders and welcomes correspondence from those they have been unable to contact.

ISBN 13: 978-1-138-10573-7 (hbk)
ISBN 13: 978-0-203-71263-4 (ebk)

Visit the Taylor & Francis Web site at http://www.taylorandfrancis.com and the
CRC Press Web site at http://www.crcpress.com

DOI: 10.1201/9780203712634

PREFACE

I am greatly honored that Professors Ulf S. von Euler and George O. Burr, originators of the eicosanoid area, have written preface contributions to this handbook. This *CRC Handbook of Eicosanoids* is dedicated to these two great men and to the memory of Prof. Ulf von Euler, Nobel Laureate, who died in 1983.

To adequately cover the literature in this enormous and rapidly expanding field has proved to be a monumental task. Thus, I gratefully acknowledge the dedication and fortitude of the various authors who have contributed to this handbook. I also thank the publishers and our coordinating editor, Ms. Amy Skallerup, for their great patience.

Finally, I thank the members of our editorial board and many others for reviewing the manuscripts and Ms. Joan M. Fisher and Ms. Becky Nagel who have given so freely of their time in acting as editorial assistants.

Anthony L. Willis

EICOSANOIDS

U. S. von Euler*

When the high biological significance of certain unsaturated fatty acids was revealed through the discovery of the essential fatty acids in the 1930s by Prof. George Burr this opened the door for studies on a new class of physiological important compounds.

Polyunsaturated fats in the diet later moved into the center of interest of nutritionists, based on their alleged role in preventing the development of atherosclerosis and related conditions.

One of the essential fatty acids of the early studies, arachidonic acid, was later to become the cornerstone in a large chemical complex, comprising numerous pharmacologically active eicosanoids. In 1935, however, when the strong pharmacological actions of extracts of the sheep vesicular gland and of human seminal fluid was found to be due to fatty acids, apparently unsaturated, virtually nothing was known about the pharmacology of such compounds. In retrospect it seems that this finding should have prompted a contact with Dr. Burr which might well have contributed to the identification of the new class of pharmacologically active substances at an earlier date. Very likely it would have revealed the biosynthesis of the active compounds in the vesicular gland, using arachidonic acid as precursor.

At any rate the findings in due time added to the significance of the eicosanoids which in an unexpected way have proven to comprise compounds possessing biological actions of the most varying kind. Who could have foreseen 25 years ago that the main function of our most widely used remedy was to block synthesis of prostaglandins and thromboxanes? Or who could then have believed that nanogram doses of such eicosanoids could provoke asthma or send forth inflammatory signs or the local defense forces of the organism?

Through intensive chemical, physiological, and pharmacological work it has become clear that the chemical backbone for a large series of bioregulators is an eicosanoid. A question which is almost unavoidable in this context is why this class of substances have been chosen by nature to perform such a variety of functions. When more knowledge has accumulated, in the relatively new field involving the laws determining the interaction of agonists and receptors at the molecular level, we may find the answer to this question and may understand better why eicosanoids have been selected for so many important tasks.

Ulf S. von Euler
August 23, 1982

---

* Prof. von Euler died in 1983.

ESSENTIAL FATTY ACIDS

While Prof. von Euler and I were chatting at the Golden Jubilee International Congress in 1980, we found that our paths had crossed at the laboratory of Prof. Thunberg in Lund, Sweden, in 1933. But we did not meet. If we had met it is likely that we would have discussed our then current work on essential fatty acids and prostaglandins without noting any significant relationship between the two. Now, 50 years later, these compounds are being treated as a unit for the solution of their biological functions.

The presentations at the Golden Jubilee Congress came as a revelation to me who viewed them with the perspective gained by an absence of more than 35 years from the center of activity.

> "Perspective is a pleasant thing!
> It keeps the windows back of sills
> And puts the sky behind the hills."
>
> David McCord

It also pans the sand and gravel from the gold.

If the 165 papers presented at the 1980 Golden Jubilee Congress (on essential fatty acids and prostaglandins) are a harbinger of what is to come in this complex field of study, there is clearly a need for a handbook to collate the mass of data.

George O. Burr
March 27, 1982

THE EDITOR

Anthony L. Willis, Ph.D., is a Principal Scientist and Head of the Atherosclerosis and Thrombosis Section, Institute of Experimental Pharmacology, Syntex Research, Palo Alto, California.

Born in Penzance, Cornwall, England, Dr. Willis obtained degrees in pharmacology at Chelsea College, London, and at the Royal College of Surgeons of England, London. Dr. Willis is a member of several learned societies, including the British Pharmacology Society (of which he is a Sandoz Prizewinner), and a member of the Council on Atherosclerosis of the American Heart Association.

With the exception of brief sojourns at Stanford University, California, and Leeds University, England, Dr. Willis has spent his entire research career in the pharmaceutical industry: at Lilly Research (England), Hoffman-La Roche (U.S. and England) and now Syntex (U.S.)

Dr. Willis has made many fundamental and applied contributions of the area of prostaglandins and related substances now collectively termed the *eicosanoids*. He was among the first to delineate the role of prostaglandin as mediators of inflammation, including the first description of their presence in inflammatory exudate. Later, in work done alone and in collaboration, he shared in the discovery that platelets of human individuals synthesize and release prostaglandins and labile endoperoxides that induce platelet aggregation and that the mode of action of aspirin in inflammation, fever, and platelet aggregation was via inhibition of prostaglandin synthesis. His work included establishing isolation procedures of labile PG endoperoxides and description of deficiency prostaglandins and endoperoxide responsiveness in hemostatic disorders.

Later, Dr. Willis pursued the now very topical idea that thrombosis and other disorders may be preventable by redirecting eicosanoid biosynthesis addition to the diet of pure biochemical precursors of certain prostaglandins. This work led to the conclusion that there was considerable species variation in the enzymatic desaturation of unsaturated essential fatty acids and that metabolic pools of eicosanoid precursors may be of importance in basal production of prostaglandins by most tissues.

Most recently, Dr. Willis has developed several novel methods of thrombotic and atherosclerotic processes that allow rapid and predictive evaluation of test compounds.

This work has led to the development of potentially antiatherosclerotic prostacyclin analogs and a new description of this potential activity via inhibition of mitogen release and cholesterol metabolism.

Dr. Willis is author or co-author of almost 100 scientific articles and review articles, including a previous compendium of the properties of prostaglandins, which served as the starting point for this handbook.

ADVISORY BOARD

CONTRIBUTORS

Joseph G. Atkinson, Ph.D.
Senior Research Fellow
Merck Frosst Canada, Inc.
Kirkland, Quebec, Canada

Joseph M. Muchowski, Ph.D.
Assistant Director
Institute of Organic Chemistry
Syntex Research
Palo Alto, California

K. C. Nicolaou, Ph.D.
Professor of Chemistry
Department of Chemistry
University of Pennsylvania
Philadelphia, Pennsylvania

N. A. Petasis, Ph.D.
Research Associate
Department of Chemistry
University of Pennsylvania
Philadelphia, Pennsylvania

Joshua Rokach, Ph.D.
Executive Director of Research
Merck Frosst Canada, Inc.
Kirkland, Quebec, Canada

K. John Stone, Ph.D.
Group Head
Preclinical Development
Roche Products Ltd.
Hertsfordshire, England

Anthony L. Willis, Ph.D.
Principal Scientist
Atherosclerosis and Thrombosis Section
Institute of Experimental Pharmacology
Syntex Research
Palo Alto, California

TABLE OF CONTENTS

Volume I

Volume II

Chemistry of Eicosanoids

SYNTHESIS OF EICOSANOIDS

K. C. Nicolaou and N. A. Petasis

INTRODUCTION

The first major advances in the chemistry of prostaglandins (PGs) were made in the 1960s, almost 3 decades after their discovery in the early 1930s (see first chapter by Willis). During the 1970s, several oxygenated metabolites of arachidonic acid (AA) other than the PGs were also discovered. These include the thromboxanes, prostacyclin, the leukotrienes, and numerous other lipoxygenation products.

To describe all these various metabolites of AA and related C-20 fatty acids, Corey et al.[1] introduced the term eicosanoids to include the prostaglandins, the prostaglandin endoperoxides, the thromboxanes, the prostacyclins, and the leukotrienes. The biosynthesis of various eicosanoids from AA is outlined in Figure 1. Almost all natural eicosanoids have now been chemically synthesized and a large number of analogs produced.

Many excellent reviews dealing with the synthesis of eicosanoids have been published and the reader is referred to them for a more detailed account, especially of the earlier work.[2-9] The chemistry of the leukotrienes is described in detail in the contribution to this volume by Atkinson and Rokach. Here, we discuss several examples of natural and biologically interesting structural analogs of prostaglandins, prostaglandin endoperoxides, thromboxanes, prostacyclins, and leukotrienes. Biological properties of such synthetic analogs are described in the contributions to this volume of Willis and Stone and of Muchowski.

PROSTAGLANDINS

Figure 2 shows the naturally occurring prostaglandins of the 1-, 2-, and 3-series. All of these PGs can now be made synthetically by a variety of methods. Modifications are possible and a plethora of synthetic analogs has been designed. The ring system may be modified by the removal of existing functionalities, or by the addition of new substituents, and/or heteroatoms (Figure 3). The two side chains of the prostaglandin skeleton have also been modified in various ways. Indeed, this approach is very common (Figure 4). This has resulted in compounds with desirable properties of solubility, increased stability against biological degradation, and (particularly) selective biological activity.

In general, construction of the prostaglandin skeleton has been achieved either by cyclization of acyclic precursors, or by the proper substitution of a functionalized cyclopentane ring. However, the most efficient approaches are those utilizing a bicyclic precursor to introduce the various functionalities with stereochemical control. A great deal of synthetic manipulations were required for successful synthesis in this area, which resulted in the development of many elegant and novel approaches utilizing new synthetic methods.[10] The principles involved in most synthetic efforts in the area are illustrated in the ingenious bicycloheptane approach of Corey et al.,[11,12] as outlined in Figure 5.

The important features of Corey's approach are the cheap and readily available starting materials (cyclopentadiene), excellent stereocontrol, versatility, flexibility in analog synthesis, and high overall yield (23 to 27% from cyclopentadiene). The initial synthetic route, which has been varied and optimized extensively, has been used not only for the preparation of PGE and $PGF_{2\alpha}$, but it was also modified for the synthesis of PGs A, C, and D, as well as numerous analogs. Noteworthy is the use of a Diels-Alder reaction to construct a bicycloheptane intermediate which has the appropriate of four of the five asymmetric centers of the PGs. Also, the selective functionalization of the different sites of the molecule allows

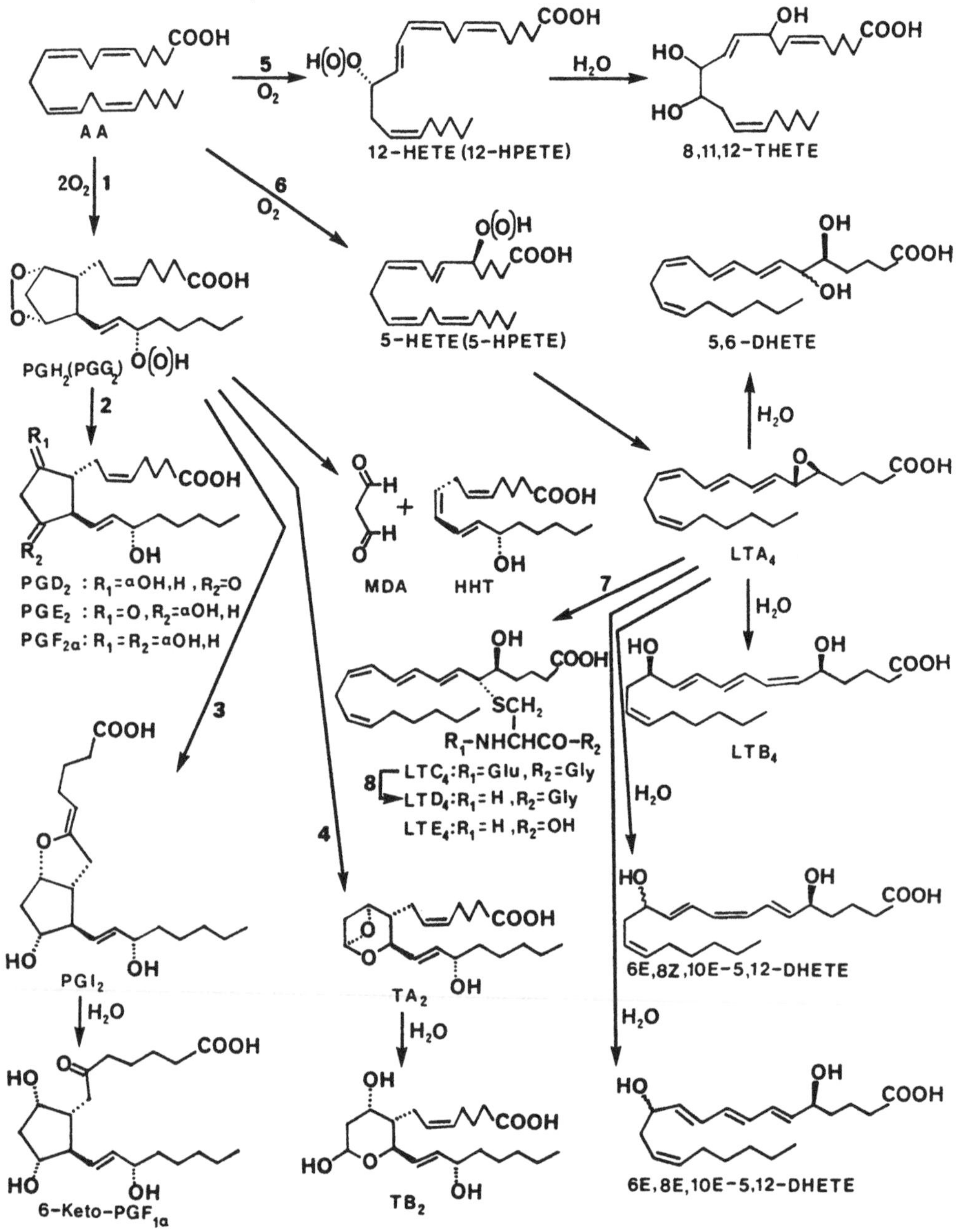

FIGURE 1. Biosynthesis of eicosanoids from AA Enzymes. (1) AA cyclooxygenase, (2) PG synthetase, (3) PGI$_2$ synthetase, (4) TXA$_2$ synthetase, (5) Δ^{12} lipoxygenase, (6) Δ^5 lipoxygenase, (7) glutathione S-transferase, (8) γ-glutamyl transpeptidase.

the introduction of not only the natural, but also various "unnatural" substituents. Finally, the optically active forms of these molecules can be obtained by resolution of an acid derivative at an early stage of the synthesis. For the reasons described above, Corey's approach has been widely used in the pharmaceutical industry.

PROSTAGLANDIN ENDOPEROXIDES

Both PG endoperoxides H$_2$ (PGH$_2$) and G$_2$ (PGG$_2$) have been synthesized in the labora-

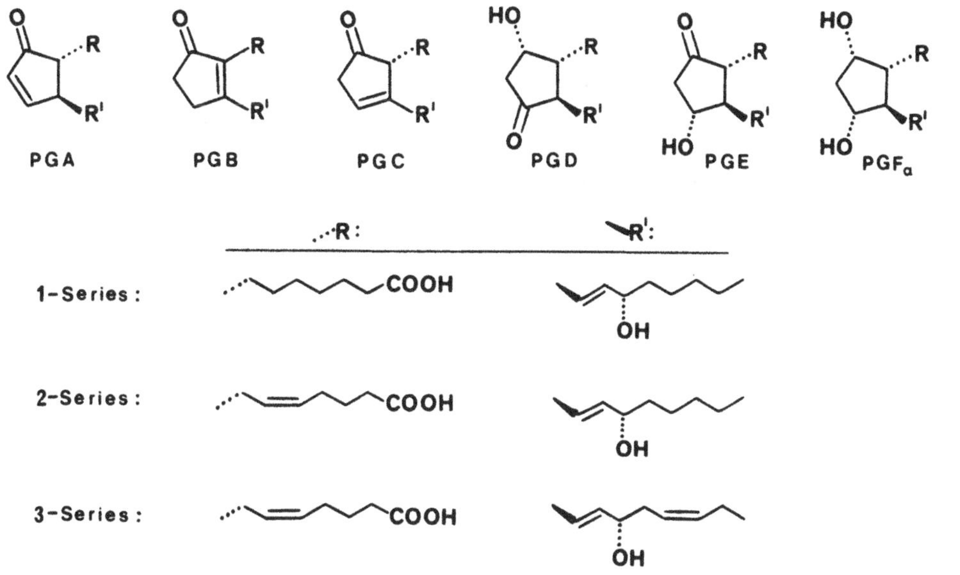

FIGURE 2.   Naturally occurring prostaglandins.

tory.[13-16] The most efficient methodology for these syntheses involves displacing halogens by hydrogen peroxide in the presence of silver ions to facilitate the departure of the halogens. Due to the relative instability of these natural endoperoxides a variety of stable analogs were synthesized. Figure 6 depicts some of the most important endoperoxide analogs with useful biological properties.

Figure 7 outlines a chemical synthesis by the Corey group,[17] of 9,11-azo-PGH$_2$, one of the first endoperoxide analogs to be synthesized and one of the most potent PGH$_2$ analogs. This synthesis starts with the readily available and optically active 9-epi,11-epi,15-acetoxyPGF$_2$ methylester and introduces the azo linkage via a double displacement with hydrazine followed by oxidation. 9,11-Azo-PGH$_2$ is substantially more active than PGH$_2$ (the natural endoperoxide) in stimulating blood platelets to aggregate and smooth muscle to contract.

THROMBOXANES

Despite the fact that the structure of thromboxane A$_2$ (TX$_2$) was proposed as early as 1975,[18] only recently synthesis of this important but unstable biomolecule has yet been reported;[206] the biological half life of TX$_2$ is only 32 sec at pH 7.4 (at 37°C). However, thromboxane B$_2$ (TXB$_2$), the stable metabolite of TXA$_2$ has been synthesized by several groups.[19-28] Several synthetic analogs of TX$_2$ with enhanced stability have been synthesized (Figure 8) and some of them were found to possess potent biological actions. A short and efficient total synthesis of the carbocyclic thromboxane A$_2$ (TXA$_2$) is outlined in Figure 9. In this synthesis by Nicolaou et al.,[29] bicyclo 3.1.1-heptan-2-one was utilized as the starting material containing the required thromboxane-like nucleus. The attachment of the bottom chain was achieved by a mixed cuprate reaction on and α,β-unsaturated aldehyde, whereas the upper chain was introduced by a standard Witting reaction after a simple aldehyde homologation. Finally, chromatographic separation led to pure TXA$_2$ which is available by this route either as the racemate or in its optically active form. Lefer and colleagues[30] found this compound to possess an interesting biological profile, being a potent constrictor of smooth muscle and selective thromboxane synthetase inhibitor.

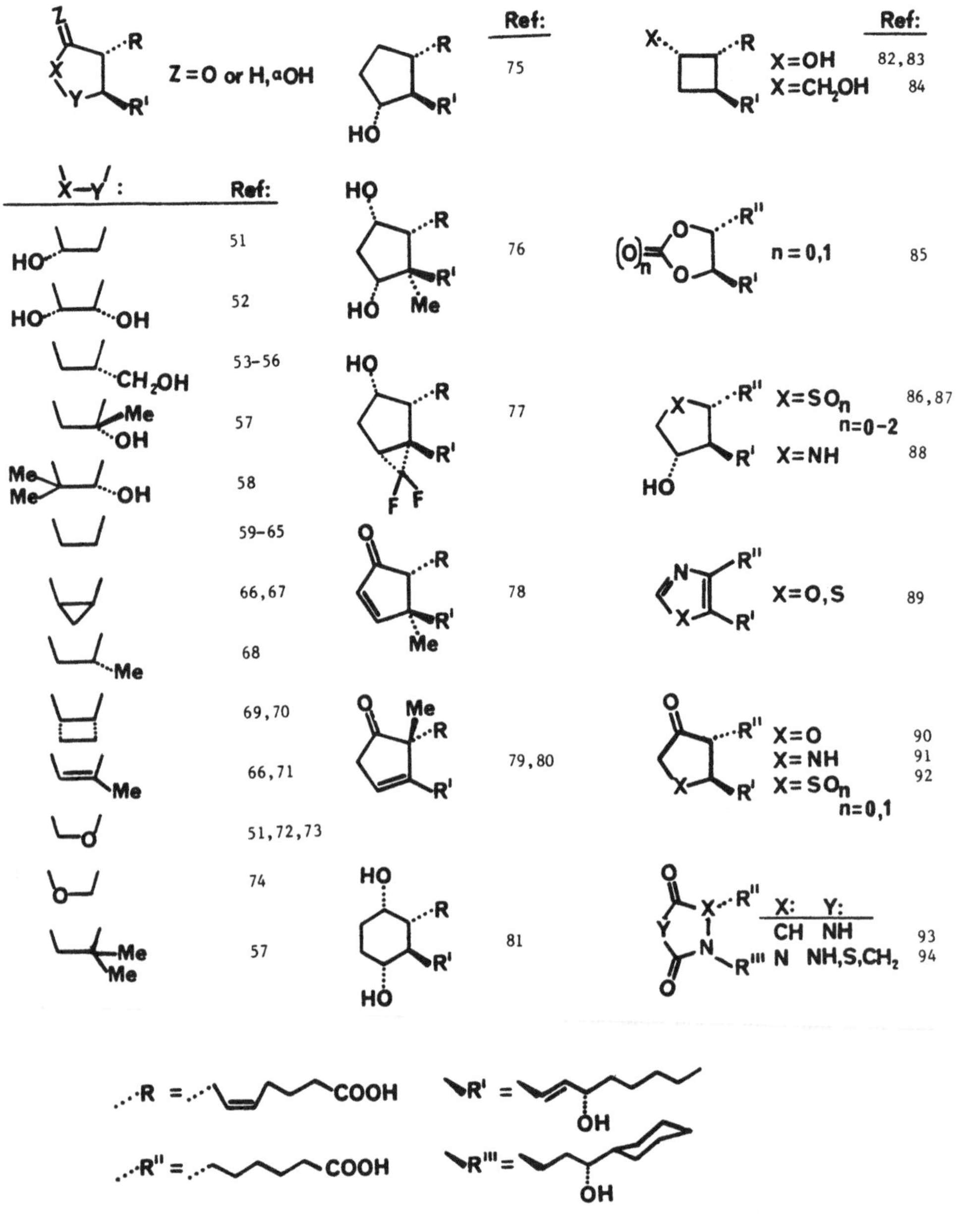

FIGURE 3. Ring modifications of prostaglandins.

PROSTACYCLINS

The molecular structure of prostacyclin (PGI$_2$) was established in 1976[31] and subsequently many synthetic analogs were described. The chemical synthesis of the naturally occurring substance was achieved very early[32-38] by using as a starting material the readily available PGF$_{2\alpha}$. The great potential of PGI$_2$-like compounds as antihypertensive, anti-asthmatic, or antithrombotic agents prompted extensive synthetic effort in this area, resulting in a wide range of synthetic materials (e.g., Figure 10).

The carbocyclic series of PGI$_2$ compounds (represented by carboprostacyclin:6,9-methanoprostacyclin) is most promising because of high chemical stability and ready access.

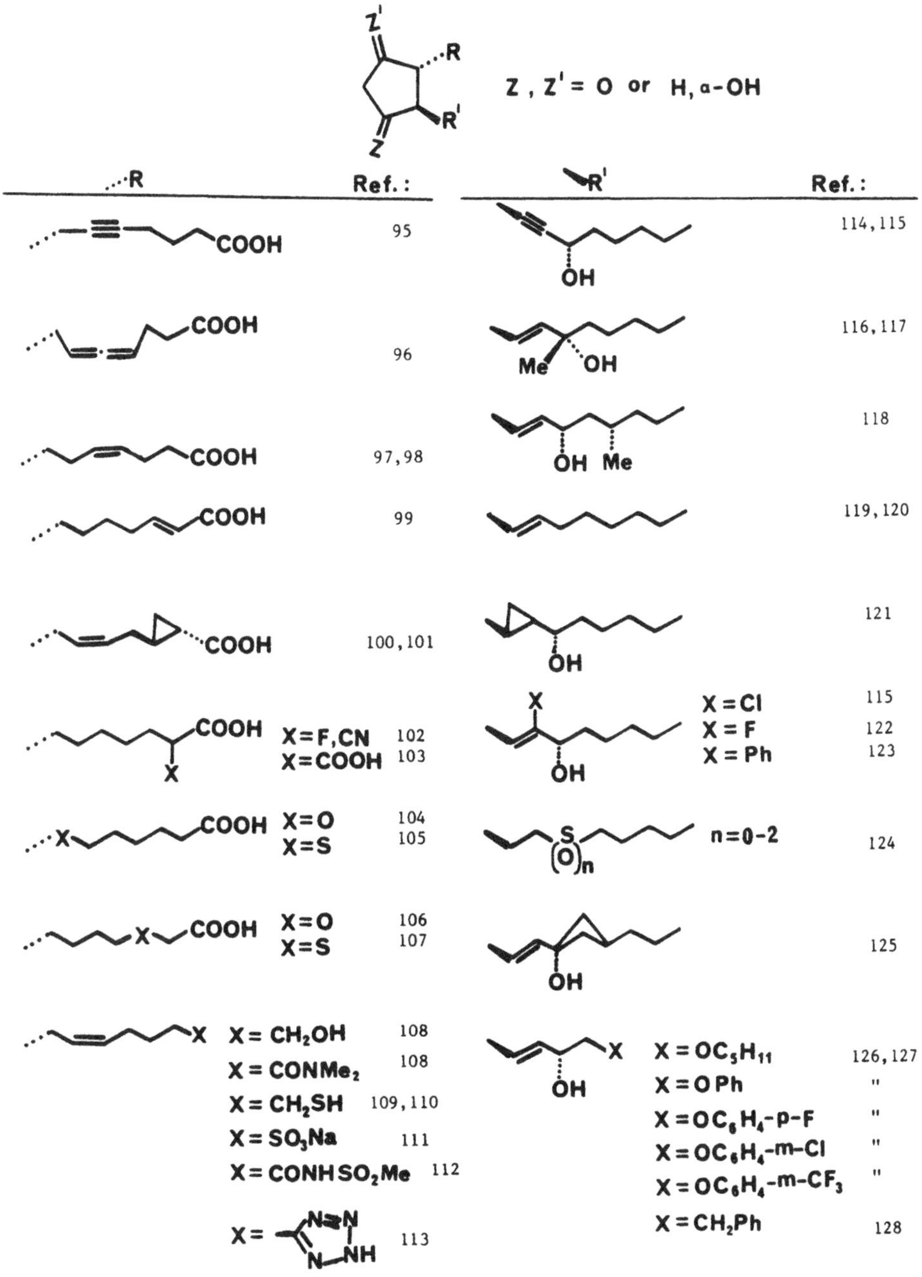

FIGURE 4. Chain modifications of prostaglandins.

Several chemical approaches to this important PGI_2 analog have been reported and two of these are shown in Figure 11. In the first approach by Nicolaou et al.,[39] the readily available *cis*-bicyclo-3.3.0-octane-3,7-dione is used as the starting material. This procedure, which builds the two chains following Corey's approach, produces racemic carbaprostacyclin and is easily adaptable to analog preparation. In the second approach by Aristoff[40] (Figure 10), advantage is taken of the Corey lactone, which is readily available in optically active form (see Figure 5) to secure all five asymmetric centers of the molecule. The oxygen atom in the second 5-membered ring was interchanged with a methylene group by an ingenious maneuver involving a ketophosphonate reaction as the key step.

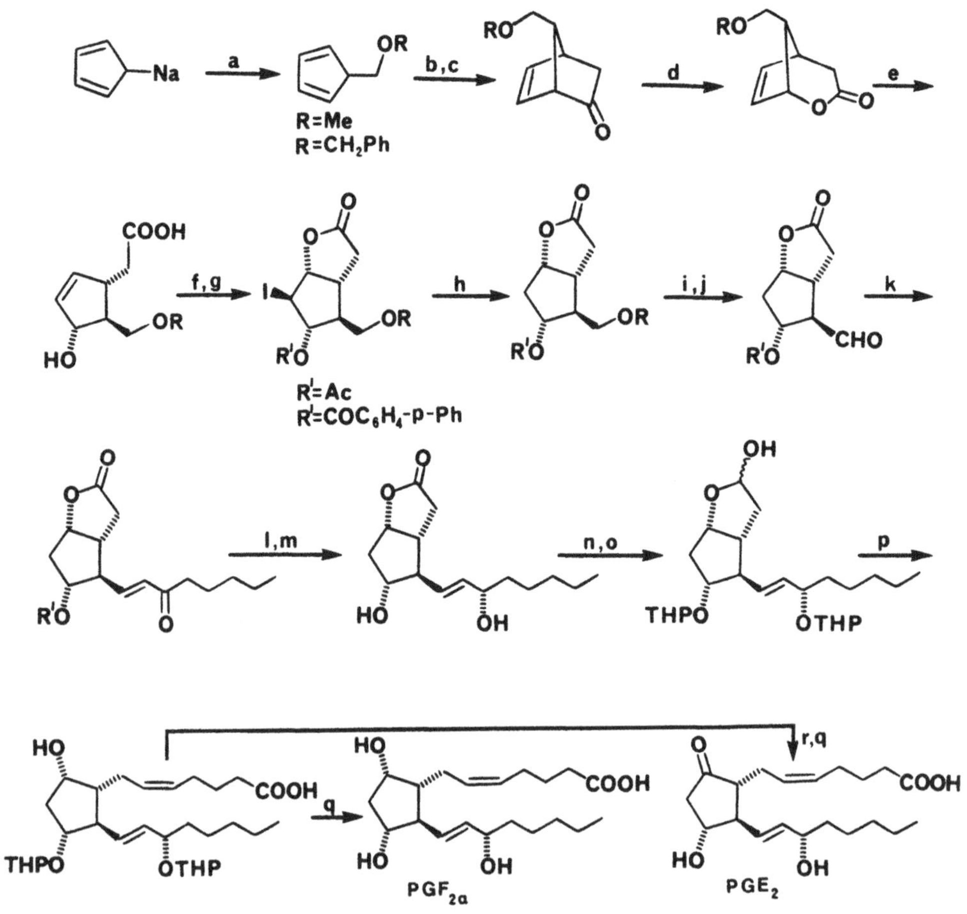

FIGURE 5.   Corey's bicycloheptane approach to prostaglandins PGF$_{2\alpha}$ and PGE$_2$.[11,12] Reagents: (a) ClCH$_2$OMe or ClCH$_2$OCH$_2$Ph, THF, $-55°$; (b) CH$_2$=C(Cl)CN, Cu(BF$_4$)$_2$, 0°; (c) KOH, DMSO, 25°; (d) m-Cl C$_6$H$_4$CO$_3$H, CH$_2$Cl$_2$, NaHCO$_3$; (e) aqNaOH, 0° and then CO$_2$; (f) aqKI-I$_2$, 0°; (g) Ac$_2$O,Py (R'=Ac) or p-Ph-C$_6$H$_4$-COCl,Py (R'=COC$_6$H$_4$-p-Ph); (h) Bu$_3$SnH, C$_6$H$_6$, AIBN, 55°; (i) BBr$_3$, CH$_2$Cl$_2$ (R=Me) or H$_2$, 5%Pd-C, EtOAc-EtOH, HCl(R=CH$_2$Ph); (j) CrO$_3$·2Py, CH$_2$Cl$_2$, 0°; (k) (MeO)$_2$P(O)CH$_2$COC$_5$H$_{11}$, NaH, DME, 25°; (l) Zn(BH$_4$)$_2$, DME, 20° or LiBR$_3$H; (m) K$_2$CO$_3$, MeOH, 25°; (n) DHP, CH$_2$Cl$_2$, p-TsOH, 25°; (o) DIBAL, PhMe, $-60°$; (p) Ph$_3$P=CH(CH$_2$)$_3$COOH, NaH-DMSO; (q) AcOH-THF, 37°; (r) Jones reagent, $-10°$.

LEUKOTRIENES

These are the more recent additions to the list of oxygenated metabolites of AA. They were first described by Borgeat and colleagues[41-44] and found to be components of the slow reacting substances of anaphylaxis (SRS-A).[45] The complete structural elucidation of leukotrienes was realized by comparison with various synthetic substances which were unambiguously synthesized by Corey and others.[46-50] Intensive work in this area by the groups of Corey, Rokach, and others led to the total synthesis of almost all of the naturally occurring leukotrienes as well as of many stereochemical and structural analogs (see Figure 12).

The first total synthesis of leukotriene C$_4$ (LTC$_4$) by Corey et al.[46] is shown in Figure 13. This route allowed the preparation of the correct antipode of LTC$_4$ also having the correct stereochemistry of all four double bonds. The fact that the natural product was identical with the synthetic material, but was distinctly different from other synthetic isomers, provided the unambiguous proof of the leukotriene structure. Using the methyl ester of leukotriene

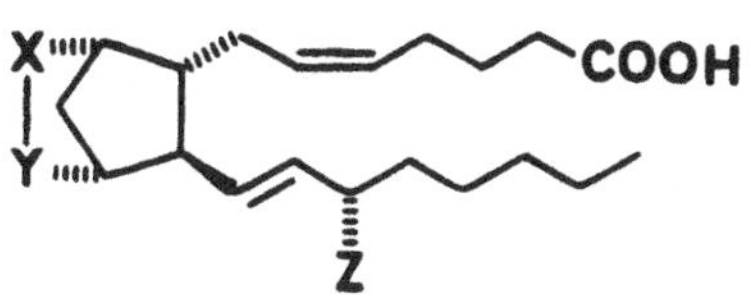

X–Y	Z	Ref:
O–O	OOH (PGG₂)	16
O–O	OH (PGH₂)	13
N=N	"	17, 129
CH=CH	"	130–132
CH₂-CH₂	"	131
O–CH₂	"	53
CH₂–O	"	53
O–CHMe–O	"	133
NH–CH₂	"	134
S–S	"	135
S₂–S₂	"	136
N=N	H	137
NH–O	H	138
O–NH	H	138

FIGURE 6.   Prostaglandin endoperoxides (PGH₂, PGG₂) and analogues.

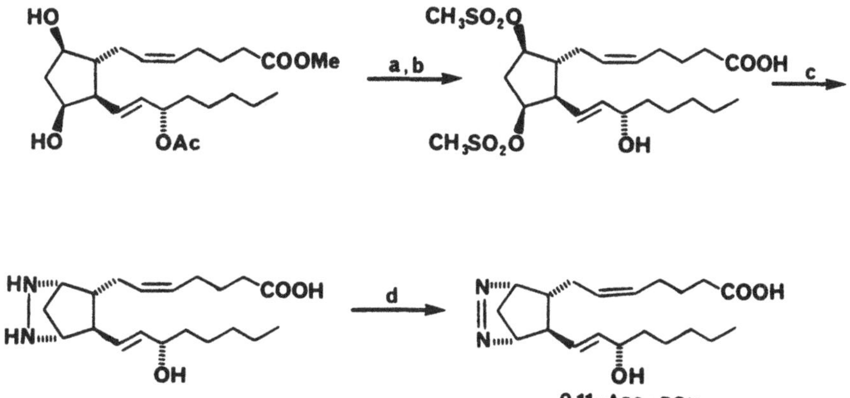

FIGURE 7.   Synthesis of 9,11-azo-PGH₂.[17] Reagents: (a) CH₃SO₂Cl, Et₃N, CH₂Cl₂, $-20°$; (b) LiOH, H₂O, MeOH, 25°C; (c) NH₂NH₂, EtOH-BuOH, Δ; (d) O₂-Cu(OAc)₂.

A₄ as a precursor, the amino acid segment of LTC₄ was introduced in a biomimetic fashion by regiospecific nucleophilic opening of the epoxide unit with the corresponding glutathione derivative. Finally, the polyene skeleton of the LTA₄ methyl ester was constructed stereospecifically by the use of a Wittig reaction, while the correct epoxide unit was efficiently prepared by chirality transfer from a precursor derived from D-(−)-ribose.

EPILOGUE

During the last 20 years or so, extensive studies on the biological properties of PGs and

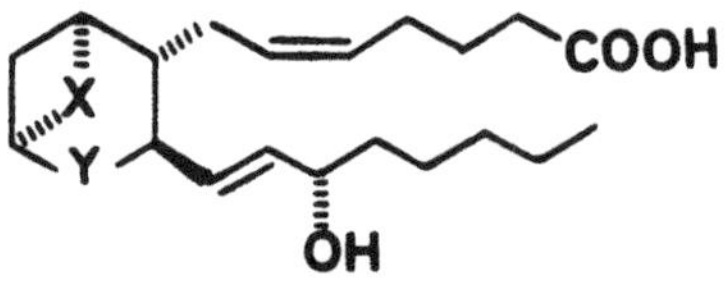

X	Y	Ref:
O	O(TA$_2$)	*
CMe$_2$	CH$_2$	139,140
CH$_2$	CH$_2$	29,141
CH$_2$	O	142
O	CH$_2$	143
CH$_2$	S	144
S	CH$_2$	145
S	S	146

FIGURE 8.   Thromboxane A$_2$ (TXA$_2$) and analogs. *See Reference 206.

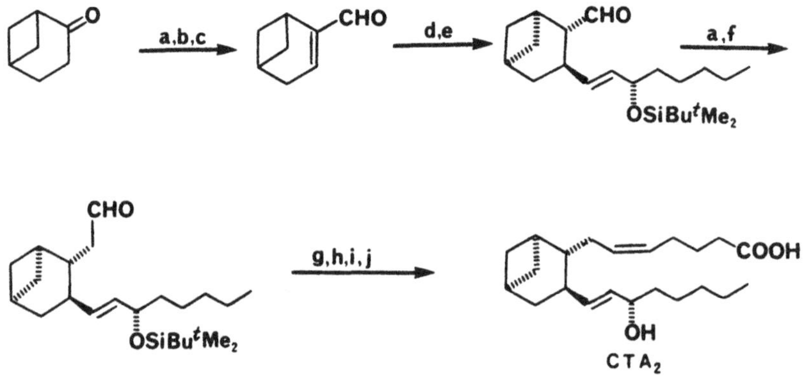

FIGURE 9.   Total synthesis of carbocyclic thromboxane A$_2$ (TXA$_2$).[27] Reagents: (a) Ph$_3$P=CHOMe, THF, PhMe; (b) PhSeCl, K$_2$CO$_3$; (c) m-Cl-C$_6$H$_4$CO$_3$H, CH$_2$Cl$_2$, $-78°\rightarrow0°$C; (d) LiCu(CH=^{t}CHCH(OS^{t}BuMe$_2$)(C≡OC$_3$H$_7$), Et$_2$O, $-78°$C; (e) K$_2$CO$_3$, MeOH; (f) Hg(OAc)$_2$, KI, THF, H$_2$O; (g) Ph$_3$P = CH(CH$_2$)$_3$COONa, DMSO; (h) CH$_2$N$_2$, Et$_2$O; (i) AcOH, THF, H$_2$O, 45°C; (j) LiOH, THF, H$_2$O.

other eicosanoids have been made possible by the contributions of chemical synthesis. Thus, the natural eicosanoids became available for study and numerous structural analogs were produced as potential therapeutic agents. So far, only a few synthetic eicosanoids have been introduced as drugs (e.g., PGF$_{2\alpha}$, PGE$_2$, PGE$_1$, PGI$_2$, sulprostone, fluprostenol). However, it is probable that the pharmaceutical industry will discover eicosanoid analogs with a pharmacological profile more selective than for natural compounds and with good bioavailability. Such compounds would undoubtedly be of great importance to medicine.

ACKNOWLEDGMENTS

We wish to express our many thanks and appreciation to our able collaborators, W. E. Barnette, D. A., Claremon, R. L., Magolda, S. P., Seitz, and W. J. Sipio for their contributions to the work carried out in these laboratories and described in this article. Our many thanks are also due to Professors J. B. Smith and A. M. Lefer of Thomas Jefferson

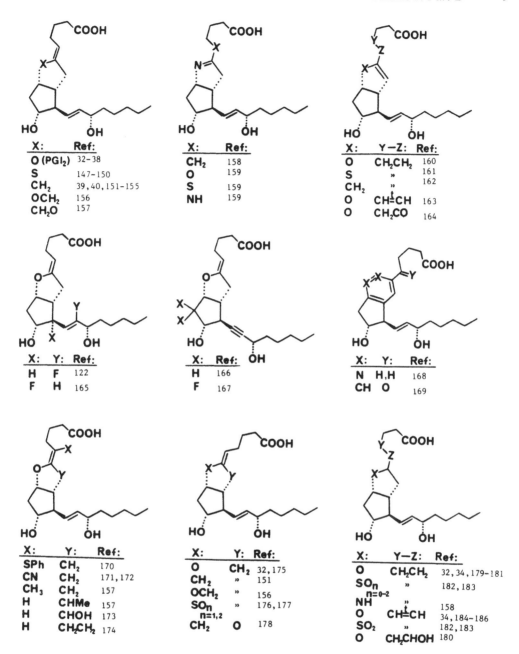

FIGURE 10.   Prostacyclin (PGI₂) and analogs.

University for their contributions to the biological aspects of this work and encouraging discussions. The research in our laboratories pertaining to this area was assisted by the National Institutes of Health (U.S.), Merck Sharp and Dohme (U.S.), Teijin Company (Japan), Ono Pharmaceutical Company (Japan), Grunenthal Chemie (West Germany), and the University of Pennsylvania (U.S.).

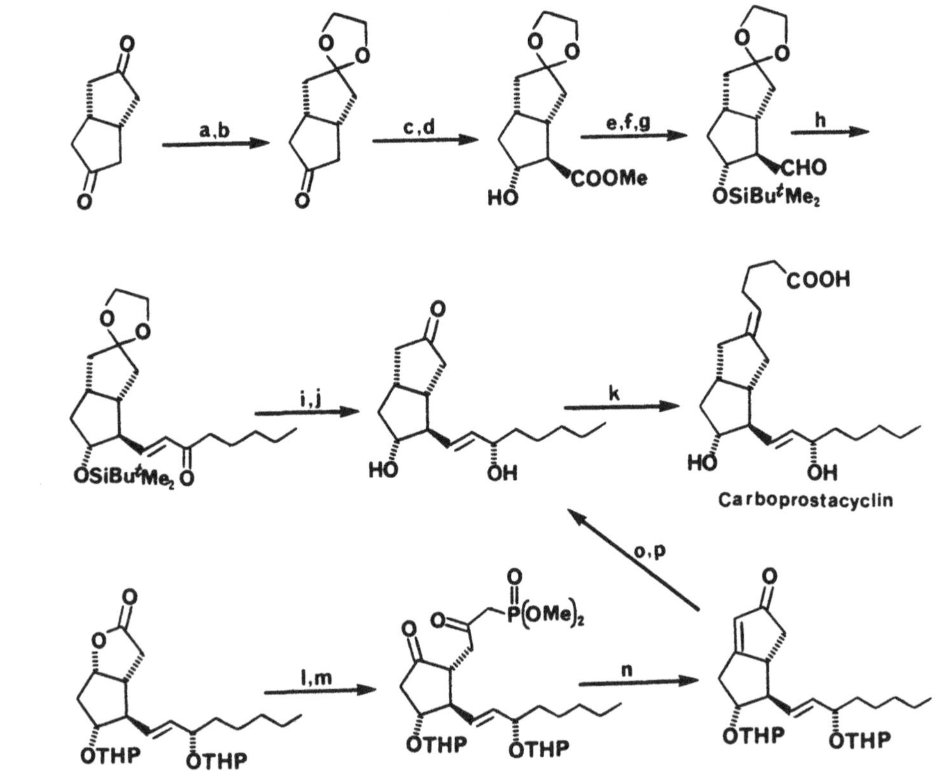

FIGURE 11. Total synthesis of carboprostacyclin.[39,40] Reagents: (a) $(CH_2OH)_2$, C_6H_6, reflux; (b) AcOH-THF-H_2O, 45°C; (c) NaH and then $(MeO)_2CO$; (d) $NaBH_4$, EtOH, −45°C; (e) $^tBuMe_2SiCl$, imidazole, DMF; (f) DIBAL, CH_2Cl_2, −78°C; (g) $CrO_3 \cdot Py \cdot HCl$, CH_2Cl_2; (h) $(MeO)_2P(O)$-$CH_2COC_5H_{11}$, NaH, DME; (f) $Zn(BH_4)_2$, MeOH, −40°C; (j) AcOH-THF-H_2O, 45°C; (k) $Ph_3P{=}CH(CH_2)_3COONa$, DMSO; (l) $(MeO)_2P(O)CH_2Li$, THF, −78°→25°C; (m) $CrO_3 \cdot 2Py$, CH_2Cl_2; (n) K_2CO_3, 18-Crown-6, PhMe, 78°C; (o) HCOOH-Et_3N 5% Pd/C, PhMe, 85°C; (p) AcOH-THF-H_2O, 45°C.

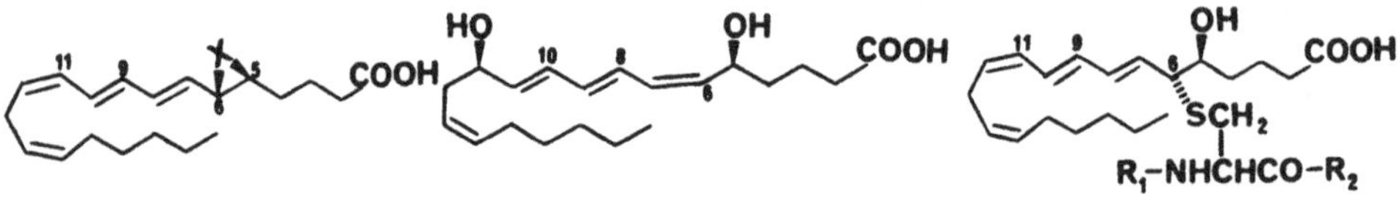

X:	Ref:		Ref:	R_1:	R_2:	Ref:
O (LTA₄)	46, 187–195	LTB₄	47, 198	Glu	Gly (LTC₄)	46, 201, 189–190
O , 9Z	46, 193, 194	6E	199	"	" , 11E	202
O , 9Z,11E	194	6E,8Z	200	"	" , 9Z,11E	203
O , 5-epi	191	6E,10Z	200	"	" , 9Z	203
O , 5-epi,6epi	191			"	" , 6epi	204
O , 6-epi	195			H	Gly (LTD₄)	46, 190
S	196			H	" , 6epi	204
CH₂	197			H	Ala et al	205
				H	OH (LTE₄)	46, 192

FIGURE 12. Leukotrienes and analogs.

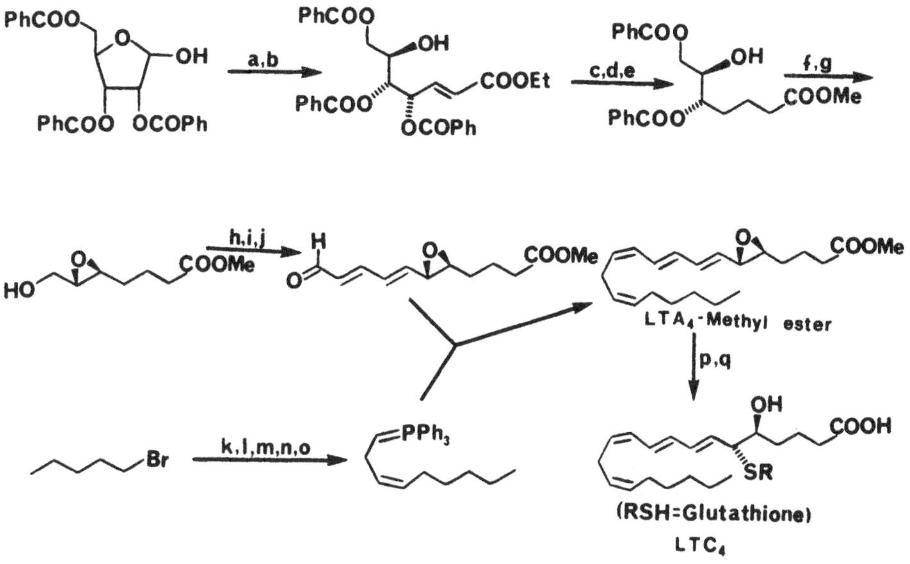

FIGURE 13. Total synthesis of leukotrienes LTA$_4$ and LTC$_4$.[46] Reagents: (a) ϕ_3P=CHCOOEt, DME, 84°C; (b) Ac$_2$O, H$_2$SO$_4$; (c) Zn-Hg, Et$_2$O, HCl, 15°C; (d) H$_2$-Pd/C, MeOH; (e) HCl, MeOH; (f) TsCl, Py; (g) K$_2$CO$_3$, MeOH, (h) CrO$_3$·2Py, CH$_2$Cl$_2$; (i) LiCH=CHCH=CHOEt, THF, −78°C and ag. NaHCO$_3$; (j) MsCl, Et$_3$N, CH$_2$Cl$_2$, −45°C; (k) Mg, CuBr·Me$_2$S and HC≡CH, LiC≡CC$_3$H$_7$, HMPA, −70°C and then (CH$_2$)$_2$O, −78°; (l) TsCl, py, 0°C; (m) NaI, acetone; (n) Ph$_3$P, C$_6$H$_6$, 80°C; (o) nBuLi, THF, −78°C, HMPA; (p) n-trifluoroacetylglutathione dimethyl ester, Et$_3$N, MeOH; (q) K$_2$CO$_3$, H$_2$O-MeOH.

REFERENCES

1. **Corey, E. J., Niwa, H., Falck, J. R., Mioskowski, C., Arai, Y., and Marfat, A.**, Recent studies on the chemical synthesis of eicosanoids, in *Advances in Prostaglandin and Thromboxane Research*, Vol. 6, Samuelsson, B., Ramwell, P. W., and Paoletti, R., Eds., Raven Press, New York, 1980, 19—25.
2. **Bently, P. H.**, Total syntheses of prostanoids, *Chem. Soc. Rev.*, 2, 29—48, 1973.
3. **Axen, U., Pike, J. E., and Schneider, W. P.**, The total synthesis of prostaglandins, in *The Total Synthesis of Natural Products*, Vol. 1, ApSimon, J., Ed., Wiley-Interscience, New York, 1973, 81—142.
4. **Bartmann, W.**, Prostaglandins, *Angew. Chem.*, 14, 337—344, 1975.
5. **Bindra, J. S. and Bindra, R.**, *Prostaglandin Synthesis*, Academic Press, New York, 1977.
6. **Mitra, A.**, *The Synthesis of Prostaglandins*, Wiley-Interscience, New York, 1977.
7. **Garcia, G. A., Maldonado, L. A., and Crabbé, P.**, Total syntheses of prostaglandins and modified prostaglandins, in *Prostaglandin Research*, Crabbé, P., Ed., Academic Press, New York, 1977, 121—313.
8. **Nicolaou, K. C., Gasic, G. P., and Barnette, W. E.**, Synthesis and biological properties of prostaglandin endoperoxides, thromboxanes, and prostacyclins, *Angew. Chem.*, 17, 293—312, 1978.
9. **Bindra, J. S.**, The synthesis of prostaglandins, in *The Total Synthesis of Natural Products*, Vol. 4, ApSimon, J., Ed., Wiley-Interscience, New York, 1981, 353—449.
10. **Caton, M. P. L.**, A survey of novel and useful reactions discovered through research on prostaglandins, *Tetrahedron*, 35, 2705—2742, 1979.
11. **Corey, E. J., Weinshenker, N. M., Schaaf, T. K., and Huber, W.**, Stereocontrolled synthesis of prostaglandins F$_{2\alpha}$ and E$_2$ (dl), *J. Am. Chem. Soc.*, 91, 5675—5677, 1969.
12. **Corey, E. J., Albonico, S. M., Koelliker, U., Schaaf, T. K., and Varma, R. K.**, New reagents for stereoselective carbonyl reduction. An improved synthetic route to the primary prostaglandins, *J. Am. Chem. Soc.*, 93, 1491—1493, 1971.
13. **Johnson, R. A., Nidy, E. G., Baczynskyj, L., and Gorman, R. R.**, Synthesis of prostaglandin H$_2$ methyl ester, *J. Am. Chem. Soc.*, 99, 7738—7740, 1977.

14. **Porter, N. A., Byers, J. D., Holden, K. M., and Menzel, D. B.,** Synthesis of prostaglandin H$_2$, *J. Am. Chem. Soc.,* 101, 4319—4322, 1979.
15. **Porter, N. A., Byers, J. D., Membane, R. C., Gilmore, D. W., and Nixon, J. R.,** Prostaglandin H$_2$ methyl ester, *J. Org. Chem.,* 43, 2088—2090, 1978.
16. **Porter, N. A., Byers, J. D., Ali, A. E., and Eling, T. E.,** Prostaglandin G$_2$, *J. Am. Chem. Soc.,* 102, 1183—1184, 1980.
17. **Corey, E. J., Nicolaou, K. C., Machida, Y., Malmsten, C. L., and Samuelsson, B.,** Synthesis and biological properties of 9,11-azo-prostanoid: highly active biochemical mimic of prostaglandin endoperoxides, *Proc. Natl. Acad. Sci., U.S.A.,* 72, 3355—3358, 1975.
18. **Hamberg, M., Svensson, J., and Samuelsson, B.,** Thromboxanes: a new group of biologically active compounds derived from prostaglandin endoperoxides, *Proc. Natl. Acad. Sci. U.S.A.,* 72, 2994—2998, 1975.
19. **Nelson, N. A. and Jackson, R. W.,** Total synthesis of thromboxane B$_2$, *Tetrahedron Lett.,* 3275—3278, 1976.
20. **Kelly, R. C., Schletter, I., and Stein, S. J.,** Synthesis of thromboxane B$_2$, *Tetrahedron Lett.,* 3279—3282, 1976.
21. **Schneider, W. P. and Morge, R. A.,** A synthesis of crystalline thromboxane B$_2$ from a derivative of prostaglandin F$_{2\alpha}$, *Tetrahedron Lett.,* 3283—3286, 1976.
22. **Hanessian, S. and Lavalle, P.,** A stereospecific, total synthesis of thromboxane B$_2$, *Can. J. Chem.,* 55, 562—565, 1977.
23. **Hanessian, S. and Lavalle, P.,** Total synthesis of (±) thromboxane B$_2$ and D-glucose. A detailed account, *Can. J. Chem.,* 59, 870—877, 1981.
24. **Corey, E. J., Shibasaki, M., Knolle, J., and Sugahara, T.,** A direct total synthesis of thromboxane B$_2$ (±), *Tetrahedron Lett.,* 785—788, 1977.
25. **Corey, E. J., Shibasaki, M., and Knolle, J.,** Simple, stereocontrolled synthesis of thromboxane B$_2$ from D-glucose, *Tetrahedron Lett.,* 1625—1626, 1977.
26. **Ohrui, H. and Emoto, S.,** A synthesis of optically active key intermediate for the synthesis of optically active thromboxanes, *Agric. Biol. Chem.,* 41, 1773—1778, 1977.
27. **Hernandez, O.,** Chiral synthesis of thromboxane B$_2$ intermediates, *Tetrahedron Lett.,* 219—222, 1978.
28. **Kelly, A. G.,** Roberts 35. A simple stereocontrolled synthesis of a thromboxane B$_2$ sython, *J. Chem. Soc. Chem. Commun.,* 228—229, 1980.
29. **Nicolaou, K. C., Magolda, R. L., and Claremon, D. A.,** Carbocyclic thromboxane A$_2$, *J. Am. Chem. Soc.,* 102, 1404—1409, 1980.
30. **Lefer, A. M., Smith, E. F., III, Arai, H., Smith, J. B., Aharony, D., Claremon, D. A., Magolda, R. L., and Nicolaou, K. C.,** Dissociation of vasoconstrictor and platelet aggregatory activities of thromboxane by carbocyclic thromboxane A$_2$, a stable analog of thromboxane A$_2$, *Proc. Natl. Acad. Sci. U.S.A.,* 77, 1706—1710, 1980.
31. **Johnson, R. A., Morton, D. R., Kinner, J. H., Gorman, R. R., McGuire, J. C., Sun, F. F., Whittaker, N., Bunting, S., Salmon, J., Moncada, S., and Vane, J. R.,** The chemical structure of prostacyclin, *Prostaglandins,* 12, 915—928, 1976.
32. **Johnson, R. A., Lincoln, F. H., Thompson, J. L., Nidy, E. G., Mizsak, S. A., and Axen, U.,** Synthesis and stereochemistry of prostacyclin and synthesis of 6-ketoprostaglandin F$_{1\alpha}$, *J. Am. Chem. Soc.,* 99, 4182—4184, 1977.
33. **Johnson, R. A., et al.,** Synthesis and characterization of prostacyclin, 6-ketoprostaglandin F$_{1\alpha}$, prostaglandin I$_1$, and prostaglandin I$_3$, *J. Am. Chem. Soc.,* 100, 7690—7705, 1978.
34. **Corey, E. J., Keck, G. E., and Székely, I.,** Synthesis of Vane's prostaglandin X, 6,9α-oxido-9α-15α-dihydroxy-prosta(2)5,(E)13-dienoic acid, *J. Am. Chem. Soc.,* 99, 2006—2008, 1977.
35. **Nicolaou, K. C., Barnette, W. E., Gasic, G. P., Magolda, R. L., and Sipio, W. J.,** Simple efficient synthesis of prostacyclin (PGI$_2$), *J. Chem. Soc. Chem. Commun.,* 630—631, 1977.
36. **Nicolaou, K. C., Barnette, W. E., and Magolda, R. L.,** Synthesis and chemistry of prostacyclin, *J. Chem. Res.,* (S) 202, (M) 2437—2445, 1979.
37. **Tömösközi, I., Galambos, G., Simonidesz, V., and Kovàcs, G.,** A simple synthesis of PGI$_2$, *Tetrahedron Lett.,* 2627—2628, 1977.
38. **Whittaker, N.,** A synthesis of prostacyclin sodium salt, *Tetrahedron Lett.,* 2805—2808, 1977.
39. **Nicolaou, K. C., Sipio, W. J., Magolda, R. L., Seitz, S., and Barnette, W. E.,** Total synthesis of carboprostacyclin, a stable and biologically active analogue of prostacyclin (PGI$_2$), *J. Chem. Soc. Chem. Commun.,* 1067—1068, 1978.
40. **Aristoff, P. A.,** Practical synthesis of 6a-carbaprostacyclin I$_2$, *J. Org. Chem.,* 46, 1954—1957, 1981.
41. **Borgeat, P. and Samuelsson, B.,** Transformation of arachidonic acid by rabbit polymorphonuclear leukocytes: formation of a novel dihydroxyeicosatetraenoic acid, *J. Biol. Chem.,* 254, 2643—2646, 1979.
42. **Murphy, R. C., Hammarström, S., and Samuelsson, B.,** Leukotriene C: a slow-reacting substance from murine mastocytoma cells, *Proc. Natl. Acad. Sci. U.S.A.,* 76, 4275—4279, 1979.

43. **Samuelsson, B. and Hammarström, S.**, Nomeclature of leukotrienes, *Prostaglandins,* 19, 645—648, 1980.

44. **Samuelsson, B.**, The leukotrienes: a new group of biologically active compounds, *Pure Appl. Chem.,* 53, 1203—1213, 1981.

45. **Brocklehurst, W. E.**, The forty year quest of "slow reacting substance of anaphylaxis," *Prog. Lipid Res.,* 20, 709—712, 1982.

46. **Corey, E. J., Clark, D. A., Goto, G., Marfat, A., Mioskowski, C., Samuelsson, B., and Hammarström, S.**, Stereospecific total synthesis of a "slow reacting substance" of anaphylaxis, leukotriene C-1, *J. Am. Chem. Soc.,* 102, 1436—1439, 1980.

47. **Corey, E. J., Marfat, A., Goto, G., and Brion, F.**, Leukotriene B. Total synthesis and assignment of stereochemistry, *J. Am. Chem. Soc.,* 102, 7984—7985, 1980.

48. **Hammarström, S., Samuelsson, B., Clark, D. A., Goto, G., Marfat, A., Mioskowski, C., and Corey, E. J.**, Stereochemistry of leukotriene C-1, *Biochem. Biophys. Res. Commun.,* 92, 946—953, 1980.

49. **Radmark, O., Malmsten, C., Samuelsson, B., Clark, D. A., Goto, G., Marfat, A., and Corey, E. J.**, Leukotriene A: stereochemistry and enzymatic conversion to leukotriene C, *Biochem. Biophys. Res. Commun.,* 92, 954—961, 1980.

50. **Morris, H. R., Taylor, G. W., Rokach, J., Girard, Y., Piper, P. J., Tippins, J. R., and Samhoun, M. N.**, Slow reacting substance of anaphylaxis, SRS-A: assignment of the stereochemistry, *Prostaglandins,* 20, 601—607, 1980.

51. **Crabbé, P., Cervantes, A., and Meana, M. C.**, Synthesis of 11-deoxy-10α-hydroxyprostaglandins, *J. Chem. Soc. Chem. Commun.,* 119—120, 1973.

52. **Crabbé, P., Guzmàn, A., and Velande, E.**, Synthesis of 10α-hydroxyprostaglandins, *J. Chem. Soc. Chem. Commun.,* 1126—1127, 1972.

53. **Bundy, G. L.**, The synthesis of prostaglandin endoperoxide analogs, *Tetrahedron Lett.,* 1957—1960, 1975.

54. **Guzmàn, A. and Muchowski, J. M.**, Synthesis of 11-hydroxy-methyl prostaglandins, *Tetrahedron Lett.,* 2053—2056, 1975.

55. **Sakai, K., Ide, J., and Oda, O.**, Total synthesis of 11-deoxy-11α-hydroxymethyl prostaglandin E$_2$, *Tetrahedron Lett.,* 3021—3024, 1975.

56. **Naruto, M., Ohno, K., Naruse, N., and Takeuchi, H.**, Synthesis of prostaglandins and their congeners. I. (±)-11-Deoxy-11α-hydroxymethyl prostaglandin F$_{2α}$ from aucubin, *Tetrahedron Lett.,* 251—254, 1979.

57. **Guzmàn, A. and Crabbé, P.**, Synthesis of methylated prostaglandins, *Chem. Ind.,* 635—636, 1973.

58. **Pernet, A. G.**, Prostaglandin analogs modified at the 10 and 11 positions, *Tetrahedron Lett.,* 3933—3936, 1979.

59. **Corey, E. J. and Ravindranathan, T.**, A simple route to a key intermediate for the synthesis of 11-desoxyprostaglandins, *Tetrahedron Lett.,* 4753—4755, 1971.

60. **Bagli, J. and Bogri, T.**, Prostaglandins. V. Utility of the nef reaction in the synthesis of prostanoic acids. A total synthesis of (±)-11-deoxy-PGE$_1$, -PGE$_2$, and their C-15 epimers, *Tetrahedron Lett.,* 3815—3817, 1972.

61. **Crabbé, P. and Guzmàn, A.**, Synthesis of 11-desoxyprostaglandins, *Tetrahedron Lett.,* 115—117, 1972.

62. **Abraham, N. A.**, Prostaglandins. VI. An efficient synthesis of 11-desoxyprostaglandins, *Tetrahedron Lett.,* 451—452, 1973.

63. **Abraham, N. A.**, Prostaglandins. IX. A simple synthesis of optically active 11-deoxyprostaglandins, *Tetrahedron Lett.,* 1393—1394, 1974.

64. **Grieco, P. A. and Reap, J. J.**, Prostaglandins. A total synthesis of (±)-11,15-dideoxy-PGE$_2$ and (±)-11-deoxy-PGE$_2$ methyl ester, *J. Org. Chem.,* 38, 3413—3415, 1973.

65. **Patterson, J. W., Jr. and Fried, J. H.**, Synthesis of prostaglandins by conjugate addition and alkylation of a directed enolate ion. 11-Deoxyprostaglandins, *J. Org. Chem.,* 39, 2506—2509, 1974.

66. **Vogel, P. and Crabbé, P.**, Methylène-10,11-et méthyl-11-prostaglandine, *Helv. Chim. Acta,* 56, 557—560, 1973.

67. **Guzmàn, A. and Muchowski, J. M.**, 10,11-Methylene-prostaglandins of unambiguous stereochemistry, *Chem. Ind.,* 790, 1975.

68. **Grudzinskas, C. V. and Weis, M. J.**, Prostaglandins and cogeners. IV. The synthesis of certain 11-substituted derivatives of 11-deoxyprostaglandin E$_2$ and F$_{2α}$ from 15-O-acetylprostaglandin A$_2$ methyl ester, *Tetrahedron Lett.,* 141—144, 1973.

69. **Crabbé, P., Garcia, G. A., and Rius, C.**, Photochemical cycloadditions in the prostaglandin series, *Tetrahedron Lett.,* 2951—2954, 1972.

70. **Crabbé, P., Garcia, G. A., and Rius, C.**, Synthesis of novel bicyclic prostaglandins by photochemical cycloaddition reactions, *J. Chem. Soc. Perkin Trans. I,* 810—816, 1973.

71. **Guzmàn, A., Vera, M., and Crabbé, P.**, Prostaglandins. XXXIX. Synthesis of isomeric 11-hydroxy-11 methylprostaglandins, *Prostaglandins,* 8, 85—91, 1974.

72. **Vlattas, I. and Lee, A. O.**, Synthesis of 11-oxaprostaglandins, *Tetrahedron Lett.,* 4451—4454, 1974.

73. **Lourens, G. J. and Koekemoer, J. M.**, The novel stereospecific synthesis of 11-oxaprostaglandin $F_{2\alpha}$, *Tetrahedron Lett.*, 3719—3722, 1975.

74. **Ishida, A., Saijo, S., and Himizu, J.**, Heterocyclic prostaglandins. III. Synthesis of 10-oxa-11-deoxy-prostaglandins E_2, *Chem. Pharm. Bull.*, 28, 783—788, 1980.

75. **Guzmàn, A. and Crabbé, P.**, Total synthesis of 9-deoxyprostaglandin, *Chem. Lett.*, 1073—1075, 1973.

76. **Grieco, P. A., Pogonowski, C. S., Nishizawa, M., and Wang, C. L. J.**, 12-Methylprostaglandins. Total synthesis of 12-methyl $PGF_{2\alpha}$ and 12-methyl PGE_2, *Tetrahedron Lett.*, 2541—2544, 1975.

77. **Crabbé, P. and Cervantes, A.**, Synthesis of difluoromethylene prostaglandins, *Tetrahedron Lett.*, 1319—1321, 1973.

78. **Corey, E. J., Shiner, C. S., Volante, R. P., and Cyr, C. R.**, Total synthesis of 12-methylprostaglandins A_2, *Tetrahedron Lett.*, 1161—1164, 1975.

79. **Corey, E. J. and Sachder, H. S.**, A simple synthesis of 8-methylprostaglandin C_2, *J. Am. Chem. Soc.*, 95, 8483—8484, 1973.

80. **Schick, H., Welzel, H. P., Schwarz, H., Schwarz, S., Truckenbrodt, G., Weber, G., and Meyer, M.**, Studies of the total synthesis of 8-methylprostaglandin C_2, *Pharmazie*, 34, 359—360, 1979.

81. **Crossley, N. S.**, Cyclohexane analogues of the prostaglandins, *Tetrahedron Lett.*, 3327—3330, 1971.

82. **Greene, A. E., Deprés, J. P., Meana, M. C., and Crabbé, P.**, Total synthesis of 11-nor prostaglandins, *Tetrahedron Lett.*, 3755—3758, 1976.

83. **Reuschling, D., Kühlein, K., and Linkies, A.**, Synthese von cyclobutanprostaglandinen, *Tetrahedron Lett.*, 17—18, 1977.

84. **Guzmàn, A., Muchowski, J. M., and Vera, M. A.**, Synthesis of cyclobutano prostaglandins, *Chem. Ind.*, 884—885, 1975.

85. **Harrison, I. T. and Fletcher, V. R.**, Synthesis of bis-oxaprostaglandins, *Tetrahedron Lett.*, 2729—2732, 1974.

86. **Vlattas, I. and Della Vecchia, L.**, 9-Thiaprostaglandins. The synthesis of optically active 9,9-dioxide analogs, *Tetrahedron Lett.*, 4267—4270, 1974.

87. **Vlattas, I. and Della Vecchia, L.**, Synthesis of 9-thioprostaglandins, *Tetrahedron Lett.*, 4459—4462, 1974.

88. **Rozing, G. P., DeKoning, H., and Huisman, H. O.**, Synthesis of 9-azoprostaglandin analogs, *Hetero-cycles*, 5, 325—330,

89. **Ambrus, G. and Barta, I.**, Synthesis of heteroaramatic analogs of prostaglandins, *Prostaglandins*, 10, 661—666, 1975.

90. **Harrison, I. T., Fletcher, V. R., and Fried, J. H.**, Synthesis of a tetrahydrofuranone prostaglandin analog, *Tetrahedron Lett.*, 2733—2736, 1974.

91. **Rozing, G. P., DeKoning, H., and Huisman, H. O.**, Synthesis of 11-azaprostaglandin analogs, *Heter-ocycles*, 7, 123—129, 1977.

92. **Harrison, I. T., Taylor, R. J. K., and Fried, J. H.**, Synthesis of 11-thiaprostaglandins, *Tetrahedron Lett.*, 1165—1168, 1975.

93. **Caldwell, A. G., Harris, C. J., Stepney, R., and Whittaker, N.**, Hydantoin prostaglandin analogues, potent and selective inhibitors of platelet aggregation, *J. Chem. Soc. Chem. Commun.*, 561—562, 1979.

94. **Barraclough, P., Caldwell, A. G., Harris, C. J., and Whittaker, N.**, Heterocyclic prostaglandin ana-logues. IV. Piperazine-2,5-diones, pyrazolidine-3,5-diones, 1,2,4-triazolidinediones, 1,3,4-oxadiazolidi-nediones and 1,3,4-thiazolidinediones, *J. Chem. Soc. Perkin Trans. I*, 2096—2105, 1981.

95. **Lin, C. H., Stein, S. J., and Pike, J. E.**, The synthesis of 5,6-acetylenic prostaglandins, *Prostaglandins*, 11, 377—380, 1976.

96. **Crabbé, P. and Carpio, H.**, Synthesis of allenic prostaglandins, *J. Chem. Soc. Chem. Commun.*, 904—905, 1972.

97. **Johnson, R. A. and Nidy, E. G.**, Synthesis of cis-Δ^4-prostaglandin $F_{1\alpha}$ and E_1 analogs, in *Advances in Prostaglandin and Thromboxane Research*, Vol. 2, Samuelsson, B., Ramwell, P. W., and Paoletti, R., Eds., Raven Press, New York, 1976, 873.

98. **Yankee, E. W., Ayer, D. E., Bundy, G. L., Lincoln, F. H., Miller, W. L., Youngdale,** Synthesis and biological activities of new prostanoids, in *Advances in Prostaglandin and Thromboxane Research*, Vol. 1, Samuelsson, B., Ramwell, P. W., and Paoletti, R., Raven Press, New York, 1976, 195—203.

99. **Wakatsuka, H., Kori, S., and Hayashi, M.**, Synthesis of Δ^2-prostaglandins, *Prostaglandins*, 8, 341—344, 1974.

100. **Schaub, R. G. and Weiss, M. J.**, 16,16-Spirocycloalkylprostaglandins, *Ger. Offen.*, 2,629,644 250, 1977.

101. *Chem. Abstr.*, 86, 155256n.

102. **Van Dorp, D. A. and Christ, E. J.**, Specificity in the enzyme conversion of substituted cis-8, cis-11,cis-14-eicosatrienoic acids into prostaglandins, *Rec. Trav. Chim. Pays-Bas*, 94, 247—253, 1975.

103. **Miyake, H. and Hayashi, M.**, Synthesis of 2-carboxy prostaglandins, *Prostaglandins*, 4, 577—580, 1973.

104. **Fried, J., Mehra, M. M., and Kao, W. L.**, Synthesis of ($\pm$)- and ($-$)-7-oxaprostaglandin $F_{1\alpha}$ and their 15-epimers, *J. Am. Chem. Soc.*, 93, 5594—5595, 1971.

105. **Fried, J., Hehra, M. M., and Chan, Y. Y.,** Stereospecific synthesis of 7-thiaprostaglandins, *J. Am. Chem. Soc.,* 96, 6759—6761, 1974.

106. **Bernady, K. F., Poletto, J. F., and Weiss, M. J.,** Prostaglandins and congeners. VIII. An improved procedure for the conjugate addition of 3-oxy-E-1-alkenyl ligands via lithium alanate reagents. 11-Deoxyprostaglandin E₁ analogues, *Tetrahedron Lett.,* 765—768, 1985.

107. **Floyd, M. B., Jr., Weiss, M. J., and Grudzinskas, C. V.,** Prostaglandins *Ger. Offen.,* 2,813,305 301 pp *Chem. Abstr.,* 90, 86849b, 1978.

108. **Crabbé, P.,** Recent advances in new synthetic prostaglandins, *Arch. Invest. Med. Suppl.,* 151—172, 1972.

109. **Hayashi, M., Kori, S., and Iguchi, S.,** Prostaglandin-like compounds, *Jpn. Kokai,* 78(34), 747, 7 1978.

110. *Chem. Abstr.,* 89, 108346U.

111. **Iguchi, Y., Kori, S., and Hayashi, M.,** Synthesis of prostaglandins containing the sulfo group, *J. Org. Chem.,* 40, 521—523, 1975.

112. **Schaaf, T. K. and Hess, H. J.,** Synthesis and biological activity of carboxyl-terminus modified prostaglandin analogues, *J. Med. Chem.,* 22, 1340—1346, 1979.

113. **Nelson, N. A., Jackson, R. W., An, A. T.,** Synthesis and biological activity of 2-decarboxy-2-(tetrazol-5-yl) prostaglandins, *Prostaglandins,* 10, 303—306, 1975.

114. **Fried, J. and Lin, C. H.,** Synthesis and biological effects of 13-dehydro derivatives of natural prostaglandin F₂α and E₂ and their 15-epi enantiomers, *J. Med. Chem.,* 16, 429—430, 1973.

115. **Gandolfi, C., Doria, G., and Gaio, P.,** Prostaglandins. IV. Cis,trans-5,13,14-chloroprostadienoic acid and cis-5-prosten-13-ynoic acids, *Farm. Ed. Sci.,* 27, 1125—1129, 1972.

116. **Bundy, G. L., Lincoln, F., Nelson, N., Pike, J., and Schneider, W.,** Novel prostaglandin syntheses, *Ann. N.Y. Acad. Sci.,* 180, 76—90, 1971.

117. **Yankee, E. W., Axen, U., and Bundy, G. L.,** Total synthesis of 15-methylprostaglandins, *J. Am. Chem. Soc.,* 96, 5865—5876,

118. **Hayashi, M., Miyake, H., Tanouchi, T., Iguchi, S., Iguchi, Y., and Tanouchi, F.,** The synthesis of 16(R)- or 16(S)-methylprostaglandins, *J. Org. Chem.,* 38, 1250—1251, 1973.

119. **Sih, C. J., Salomon, R. G., Price, P., Peruzzotti, G., and Sood, R.,** Total synthesis of (±)-15-deoxyprostaglandin E₁, *J. Chem. Soc. Chem. Commun.,* 240—241, 1972.

120. **Rappo, R.,and Collins, P. W.,** Synthesis of prostaglandin analog by novel 1,4-additinal reactions, *Tetrahedron Lett.,* 2627—2630, 1972.

121. **Radüchel, B., Mende, V., Cleve, G., Hoyer, G. A., and Vorbrüggen, H.,** The synthesis of 13,14-dihydro-13,14-methylene-PGF₂α and PGE₂, *Tetrahedron Lett.,* 633—636, 1975.

122. **Grieco, P. A., Yokoyama, Y., Nicolaou, K. C., Barnette, W. E., Smith, J. B., Ogletree, M., and Lefer, A. M.,** Total synthesis of 14-fluoroprostaglandin F₂α and 14-fluoroprostacyclin, *Chem. Lett.,* 1001—1004, 1978.

123. **Buckler, R. T. and Garling, D. L.,** Synthesis of 14-phenylprostaglandins E₁, A₁, and F₂α, *Tetrahedron Lett.,* 2257—2260, 1978.

124. **Plattner, J. J. and Gager, A. H.,** Synthesis of optically active 15-thiaprostaglandins, *Tetrahedron Lett.,* 1629—1632, 1977.

125. **Niwa, H. and Kurono, M.,** Synthesis of 15,17-methylene prostaglandins, *Chem. Lett.,* 23—26, 1979.

126. **Binder, D., Bowler, J., Brown, E. D., Crossley, N. S., Hutton, J., Senior, M., Slatter, L., Wilkinson, P., and Wright, N. C. A.,** 16-Aryloxyprostaglandins. New class of potent luteolytic agents, *Prostaglandins,* 6, 87—90, 1974.

127. **Crossley, N. S.,** Synthesis and biolgoical activity of potent selective luteolytic prostaglandins, *Prostaglandins,* 10, 5—18, 1975.

128. **Miller, W. L., Weeks, J. R., Lauderdale, J. W., and Kirton, K. T.,** Biological activities of 17-phenyl-18,19,20-trinorprostaglandins, *Prostaglandins,* 9, 9—18, 1975.

129. **Corey, E. J., Narasaka, K., and Shibasaki, M.,** A direct, stereocontrolled total synthesis of the 9,11-azo analogue of the prostaglandin endoperoxide PGH₂, *J. Am. Chem. Soc.,* 98, 6417—6418, 1976.

130. **Corey, E. J., Shibasaki, M., Nicolaou, K. C., Malmsten, C. L., and Samuelsson, B.,** Simple, stereocontrolled total synthesis of a biologically active analog of the prostaglandin endoperoxides (PGH₂, PGG₂), *Tetrahedron Lett.,* 737—740, 1976.

131. **Leeney, T. J., Marsham, P. R., Ritchie, G. A. F., and Senior, M. W.,** Inhibitors of prostaglandin biosynthesis. A bicyclo [2.2.1]heptene analog of 2′ series prostaglandins and related derivatives, *Prostaglandins,* 11, 953—960, 1976.

132. **Shimomura, H., Sugie, A., Katsube, J., and Yamamoto, H.,** Synthesis of 9,11-desoxy-9,11-vinyleno-PGF₂α and its diastereomer, analogs of the PG endoperoxide (PGH₂), *Tetrahedron Lett.,* 4099—4102, 1976.

133. **Portoghese, P. S., Carson, D. L., Abatjoglou, Dunham, E. W., Gerrard, J. M., and White, J. G.,** A novel prostaglandin endoperoxide mimic, prostaglandin F₂α acetal, *J. Med. Chem.,* 20, 320—321, 1977.

134. **Corey, E. J., Niwa, H., Bloom, M., and Ramwell, P. W.,** Synthesis of a new prostaglandin endoperoxide (PGH₂) analog and its function as an inhibitor of the biosynthesis of thromboxane A₂ (TBXA₂), *Tetrahedron Lett.,* 671—674, 1979.

135. **Miyake, H., Iguchi, S., Hoh, H., and Hayashi, M.,** Simple synthesis of methyl (4Z,9α-11α,13E,15S)-9,11-epidithio-15-hydroxyprosta-5,13-dienoate, endodisulfide analog of PGH$_2$, *J. Am. Chem. Soc.,* 99, 3536—3537, 1977.

136. **Nicolaou, K. C., Barnette, W. E., and Magolda, R. L.,** Synthesis of prostaglandin H$_2$ (PGH$_2$) and prostacyclin (PGI$_2$) analogs: tetrathia-PGH$_2$ and PGI$_2$-ketal methyl ester, *Prostaglandins Med.,* 1, 96—97, 1978.

137. **Gorman, R., Bundy, G. L., Peterson, D. C., Sun, F. F., Miller, O. V., and Fitzpatrick, F. A.,** Inhibition of human platelet thromboxane synthetase by 9,11-azoprosta-5,13-dienoic acid, *Proc. Natl. Acad. Sci. U.S.A.,* 74, 4007—4011, 1977.

138. **Bundy, G. L., and Peterson, D. C.,** The synthesis of 15-deoxy-9,11-(epoxyimino)-prostaglandins: potent thromboxane synthetase inhibitors, *Tetrahedron Lett.,* 41—44, 1978.

139. **Nicolaou, K. C., Magolda, R. L., Smith, J. B., Aharony, D., Smith, E. F., and Lefer, A. M.,** Synthesis and biological properties of pinane-thromboxane A$_2$, a selective inhibitor of coronary artery constriction, platelet aggregation, and thromboxane formation, *Proc. Natl. Acad. Sci. U.S.A.,* 76, 2566—2570, 1979.

140. **Ansel, M. F., Caton, M. P. L., Palfreyman, M. N., and Stuttle, K. A. J.,** Synthesis of structural analogues of thromboxane A$_2$, *Tetrahedron Lett.,* 4497—4498, 1979.

141. **Ohuchida, S., Hamanaka, N., and Hayashi, M.,** Synthesis of thromboxane A$_2$ analog DL-(9,11), (11,12)-dideoxa-(9,11), (11,12)-dimethylene thromboxane A$_2$, *Tetrahedron Lett.,* 3661—3664, 1979.

142. **Corey, E. J., Ponder, J. W., and Ulrich, P.,** Synthesis of a stable analog of thromboxane A$_2$ with methylene replacing the 9,11-bridging oxygen, *Tetrahedron Lett.,* 21, 137—140, 1980.

143. **Maxey, K. M. and Bundy, G. L.,** The synthesis of 11α- carbothromboxane A$_2$, *Tetrahedron Lett.,* 21, 445—448, 1980.

144. **Kosuge, S., Hamanaka, N., and Hayashi, M.,** Synthesis of thromboxane A$_2$ analog DL-(9,11), (11,12)-dideoxa-(9,11)-methylene-(11,12)-epithio-thromboxane A$_2$ methyl ester, *Tetrahedron Lett.,* 22, 1345—1348, 1981.

145. **Ohuchida, S., Hamanaka, N., and Hayashi, M.,** Synthesis of thromboxane A$_2$ analog D,L-(9,11), (11,12)-dideoxa-(9,11)-epithio-(11,12)-methylene-thromboxane A$_2$, *Tetrahedron Lett.,* 22, 1349—1352, 1981.

146. **Ohuchida, S., Hamanaka, N., and Hayashi, M.,** Synthesis of thromboxanes A$_2$ analogues: DL,9,11:11,12-dideoxa-9,11:11,12-diepithiothromboxane A$_2$, *J. Am. Chem. Soc.,* 103, 4597—4599, 1981.

147. **Nicolaou, K. C., Barnette, W. E., Gasic, G. P., and Magolda, R. L.,** 6,9-Thiaprostacyclin. A stable and biologically potent analogue of prostacyclin (PGI$_2$), *J. Am. Chem. Soc.,* 99, 7736—7738, 1977.

148. **Nicolaou, K. C., Barnette, W. E., and Magolda, R. L.,** Synthesis of (5Z)- and (5E)-6,9-thiaprostacyclins, *J. Am. Chem. Soc.,* 103, 3472—3480, 1981.

149. **Shibasaki, M. and Ikegami, S.,** Synthesis of 9(O)-thioprostacyclin, *Tetrahedron Lett.,* 559—562, 1978.

150. **Shimoji, K., Arai, Y., and Hayashi, M.,** A new synthesis of 6,9α-thiaprostacyclin, *Chem. Lett.,* 1375—1376, 1978.

151. **Kojima, K. and Sakai, K.,** Total synthesis of 9(O)-methanoprostacyclin and its isomers, *Tetrahedron Lett.,* 3743—3746, 1978.

152. **Shibasaki, M., Euda, J., and Ikegami, S.,** New synthetic routes to 9(O)-methanoprostacyclin. A highly stable and biologically potent analog of prostacyclin, *Tetrahedron Lett.,* 433—436, 1979.

153. **Sugie, A., Shimomura, H., Katsube, J., and Yamamoto, H.,** Stereocontrolled approaches to 9(O)-methanoprostacyclin, *Tetrahedron Lett.,* 2607—2610, 1979.

154. **Morton, D. R., Jr. and Brokaw, F. C.,** Total synthesis of 6α-carbaprostaglandins I$_2$ and related isomers, *J. Org. Chem.,* 44, 2880—2887, 1979.

155. **Barco, A., Benetti, S., Pollini, G. P., Baraldi, P. G., and Gandolfi, C.,** A new, elegant route to a key intermediate for the synthesis of 9(O)-methanoprostacyclin, *J. Org. Chem.,* 45, 4776—4778, 1980.

156. **Skuballa, W.,** Synthesis of 6,9-homo-prostacyclin. A stable prostaglandin analog, *Tetrahedron Lett.,* 21, 3261—3264, 1980.

157. **Ohuchida, S., Hashimoto, S., Wakatsuka, H., Arai, Y., and Hayashi, M.,** Synthesis and biological properties of some prostacyclin analogs, in *Advances in Prostaglandin and Thromboxane Research,* Vol. 6, Samuelsson, B., Ramwell, R. W., and Paolletti, R., Raven Press, New York, 1980, 337—340.

158. **Bundy, G. L. and Baldwin, J. M.,** The synthesis of nitrogen-containing prostacyclin analogs, *Tetrahedron Lett.,* 1371—1374, 1978.

159. **Bartmann, W., Beck, G., Knolle, J., and Rupp, R. H.,** New prostacyclin analogs, *Angew. Chem.,* 19, 819—820, 1980.

160. **Shimoji, K., Konishi, Y., Arai, Y., Hayashi, M., and Yamamoto, H.,** 6,9α-Oxido-11α,15α-dihydroxyprosta-6,(E)-13-dienoic acid methyl ester and 6,9α:6,11α-dioxido-15α-hydroxy-prost-(E)-13-enoic acid methyl ester. Two isomeric forms of prostacyclin (PGI$_2$), *J. Am. Chem. Soc.,* 100, 2547—2548, 1978.

161. **Shibasaki, M., Torisawa, Y., and Ikegami, S.,** Synthesis of the sulfur analog of Δ^6-PGI$_1$, *Chem. Lett.,* 1247—1250, 1980.

162. **Shibasaki, M., Iseki, K., and Ikegami, S.,** The synthesis of the carbon analog of Δ^6-PGE$_1$, *Tetrahedron Lett.,* 21, 169—172, 1980.

163. **Ohno, K. and Nishiyama, H.**, 4,5,6,7-Tetrahydro-PGI$_1$, a stable and potent inhibitor of blood platelet aggregation, *Tetrahedron Lett.*, 3003—3004, 1979.

164. **Nishiyama, H. and Ohno, K.**, Synthesis of 6,7-Dehydro-5-oxoprostaglandin I$_1$: a stable analog of prostacyclin, *Tetrahedron Lett.*, 3481—3484, 1979.

165. **Nicolaou, K. C., Barnette, W. E., Magold, R. L., Grieco, P. A., Owens, W., Wang, C. L. J., Smith, J. B., Ogletree, M., and Lefer, A. M.**, Synthesis and biological properties of 12-fluoroprostacyclins, *Prostaglandins*, 16, 789—794, 1978.

166. **Fried, J. and Barton, J.**, Synthesis of 13,14-dehydroprostacyclin methyl ester: a potent inhibitor of platelet aggregation, *Proc. Natl. Acad. Sci. U.S.A.*, 74, 2199—2203, 1977.

167. **Fried, J., Mitra, D. K., Nagarajan, M., and Mehrotra, N. M.**, 10,10-Difluoro-13-dehydroprostacyclin: a chemically and metabolically stabilized potent prostacyclin, *J. Med. Chem.*, 23, 234—237, 1980.

168. **Nicolaou, K. C., Barnette, W. E., and Magolda, R. L.**, 6,9-Pyridazaprostacyclin and derivatives: the first "aromatic" prostacyclins, *J. Am. Chem. Soc.*, 101, 766—768, 1979.

169. **Sipio, W. J.**, Ph.D. thesis, University of Pennsylvania, Philadelphia, 1981.

170. **Toru, T., Watanabe, K., Oba, T., Tanaka, T., Okamura, N., Bannai, K., and Kurozumi, S.**, Reaction of prostacyclin methyl ester with benzenesulfenyl chloride. Preparation of stable prostacyclin analogs, *Tetrahedron Lett.*, 21, 2539—2542, 1980.

171. **Vorbrueggen, H., Skuballa, W., Raduechel, B., Losert, W., Lose, O., Mueller, B., and Mannesmann, G.**, Prostacyclin derivatives, *Ger. Offen.*, 2,753,244 45 (1979).

172. *Chem. Abstr.*, 91, 912502.

173. **Bannai, K., Toru, T., Oba, T., Tanaka, T., Okamura, N., Watanabe, K., and Kurozumi, S.**, Stereocontrolled synthesis of 7-hydroxy- and 7-acetoxy-PGI$_2$: new stable PGI$_2$ analogs, *Tetrahedron Lett.*, 22, 1417—1420, 1981.

174. **Johnson, R. A. and Nidy, E. G.**, Synthesis and stereochemistry of 9-deoxy-5,9α-epoxyprostaglandins: a series of stable prostacyclin analogues, *J. Org. Chem.*, 45, 3802—3810, 1980.

175. **Corey, E. J., Székely, I., and Shiner, C. S.**, Synthesis of 6,9α-oxido-11α,15α-dihydroxyprosta-(E)5,(E)13-dienoic acid, *Tetrahedron Lett.*, 3529—3532, 1977.

176. **Nicolaou, K. C., Barnette, W. E., and Magolda, R. L.**, Organoselenium-induced ring closures. Sulfur-containing prostacyclins. Stereoselective synthesis of cyclic α,β-unsaturated sulfoxides, sulfones, and sulfides and synthesis of 6,9-sulfoxa-5(E)- and 5(Z)-prostacyclin, 6,9-sulfo-5(E) and -5(Z)-prostacyclin, 6,9-sulfo-6α- and 6β-4(E)-isoprostacyclin, and 6,9-thiaprostacyclin, *J. Am. Chem. Soc.*, 100, 2567—2570, 1978.

177. **Nicolaou, K. C., Barnette, W. E., and Magolda, R. L.**, Organoselenium-based synthesis of sulfur-containing prostacyclins, *J. Am. Chem. Soc.*, 103, 3486—3497,

178. **Shimoji, K., Arai, Y., Wakatsuka, H., and Hayashi, M.**, Synthesis of new prostacyclin analogs, in: *Advances in Prostaglandin and Thromboxane Research*, Vol. 6, Samuelsson, B., Ramwell, P. W., and Paoletti, R., Eds., Raven Press, New York 1980, 327—330.

179. **Nelson, N. A.**, Stereoconfiguration of 5,6-dihydroprostaglandins, *J. Am. Chem. Soc.*, 99, 7362—7363, 1977.

180. **Sih, J. C., Johnson, R. A., Nidy, E. G., and Graber, D. R.**, Synthesis of four isomers of 5-hydroxy-PGE$_1$, *Prostaglandins*, 15, 409—421, 1978.

181. **Finch, M. A. W., Roberts, S. M., and Newton, R. F.**, Total synthesis of ($\pm$)-6-prostaglandin I$_1$, *J. Chem. Soc. Chem. Commun.*, 589—590, 1980.

182. **Nicolaou, K. C., Magolda, R. L., and Barnette, W. E.**, Synthesis of sulfur-containing prostacyclin PGI$_1$ analogues 6,9-epithio-PGI$_1$, its S-oxide stereoisomers, and its SS-dioxide, *J. Chem. Soc. Chem. Commun.*, 375—377, 1978.

183. **Nicolaou, K. C., Barnette, W. E., and Magolda, R. L.**, Organoselenium-based synthesis of sulfur-containing prostacyclins, *J. Am. Chem. Soc.*, 103, 3486—3497, 1981.

184. **Corey, E. J., Pearce, H. L., Székely, I., and Ishiguro, M.**, Configuration at C-6 of 6,9α-oxido-bridged prostaglandins, *Tetrahedron Lett.*, 1023—1026, 1978.

185. **Nicolaou, K. C. and Barnette, W. E.**, Synthesis of (4E)-9-deoxy-6,9α-epoxy-Δ^4-PGF$_{1\alpha}$, a prostacyclin (PGX) isomer, *J. Chem. Soc. Chem. Commun.*, 331—332, 1977.

186. **Nicolaou, K. C., Barnette, W. E., and Magolda, R. L.**, Organoselenium-based synthesis of oxygen-containing prostacyclins, *J. Am. Chem. Soc.*, 103, 3480—3485, 1981.

187. **Corey, E. J., Arai, Y., and Mioskowski, C.**, Total synthesis of ($\pm$)-5,6-oxido-7,9-trans, 11,14-cis-eicosapentaenoic acid, a possible precursor of SRSA, *J. Am. Chem. Soc.*, 101, 6748—6749, 1979.

188. **Corey, E. J., Hashimoto, S., and Barton, A. E.**, Chirally directed synthesis of ($-$)-methyl 5(S),6(S)-oxido-7-hydroxyheptanoate (1), key intermediate for the total synthesis of leukotrienes A, C, D, and E, *J. Am. Chem. Soc.*, 103, 721—722, 1981.

189. **Rokach, J., Girard, Y., Guindon, Y., Atkinson, J. G., Larue, M., Young, R. N., Masson, P., and Holme, G.**, The synthesis of a leukotriene with SRS-like activity, *Tetrahedron Lett.*, 21, 1485—1488, 1980.

190. **Rokach, J., Young, R. N., Kukushima, M., Lau, C. K., Seguin, R. S., Frenette, R., and Guindon, Y.**, Synthesis of leukotrienes. New synthesis of natural leukotriene A₄, *Tetrahedron Lett.*, 22, 979—982, 1981.

191. **Rokach, J., Lau, C. K., Zamboni, R., and Guindon, Y.**, A C-glycoside route to leukotrienes, *Tetrahedron Lett.*, 22, 2763—2766, 1981.

192. **Rosenberger, M. and Neukom, C.**, Total synthesis of (5S,6R,7E,9E,11Z,14Z)-5-hydroxy-6-[(2R)-2-amino-2-(carboxy-ethyl)thio]-7,9,11,14-eicosatetraenoic acid, a potent SRS-A, *J. Am. Chem. Soc.*, 102, 5425—5426, 1980.

193. **Gleason, J. G., Bryan, D. B., and Kinzig, C. M.**, Convergent synthesis of leukotriene A methyl ester, *Tetrahedron Lett.*, 21, 1129—1132, 1980.

194. **Baker, S. R., Jamieson, W. B., McKay, S. W., Morgan, S. E., Rackham, D. M., Ross, W. J., and Shrubsall, P. R.**, Synthesis, separation and NMR spectra of the three double bond isomers of leukotriene A methyl ester, *Tetrahedron Lett.*, 4123—4126, 1980.

195. **Cohen, N., Banner, B. L., and Lopresti, R. J.**, Synthesis of optically active leukotriene (SRS-A) intermediates, *Tetrahedron Lett.*, 21, 4163—4166, 1980.

196. **Corey, E. J., Park, H., Barton, A., and Nii, Y.**, Synthesis of three potential inhibitors of the biosynthesis of leukotrienes of leukotrienes A-E, *Tetrahedron Lett.*, 21, 4243—4246, 1980.

197. **Nicolaou, K. C., Petasis, N. A., and Seitz, S. P.**, 5,6-Methanoleukotriene A₄. A stable and biologically active analogue of leukotriene A₄, *J. Chem. Soc. Chem. Commun.*, 1195—196, 1981.

198. **Corey, E. J., Marfat, A., Munroe, J., Kim, K. S., Hopkins, P. B., and Brion, F.**, A stereocontrolled and effective synthesis of leukotriene B, *Tetrahedron Lett.*, 22, 1077—1080, 1981.

199. **Corey, E. J., Marfat, A., and Hoover, D. J.**, Stereospecific total synthesis of 12-(R)- and 12-(S)-forms of 6-trans leukotriene B, *Tetrahedron Lett.*, 22, 1587—1590, 1981.

200. **Corey, E. J., Hopkins, P. B., Munroe, J. E., Marfat, A., and Hashimoto, S.**, Total synthesis of 6-trans,10-cis and (±)-6-trans, 8-cis isomers of leukotriene B, *J. Am. Chem. Soc.*, 102, 7986—7987, 1980.

201. **Corey, E. J., Barton, A. G., and Clark, D. A.**, Synthesis of the slow reacting substance of anaphylaxis leukotriene C-1 from arachidonic acid, *J. Am. Chem. Soc.*, 102, 4278—4279, 1980.

202. **Clark, D. A., Goto, G., Marfat, A., Corey, E. J., Hammastrӧm, S., and Samuelsson, B.**, 11-Trans-leukotriene C: a naturally occurring slow reacting substance, *Biochem. Biophys. Res. Commun.*, 94, 1133—1139, 1980.

203. **Baker, S. R., Jamieson, W. B., Osborne, D. J., and Ross, W. J.**, Synthesis of 92 and 92, 11E isomers of leukotriene C₄, *Tetrahedon Lett.*, 22, 2505—2508, 1981.

204. **Corey, E. J. and Goto, G.**, Total synthesis of slow reacting substances (SRS's): 6-epi leukotriene C and 6-epi-leukotriene D, *Tetrahedron Lett.*, 21, 3463—3466,

205. **Lewis, R. A., Austen, K. F., Drazen, J. M., Clark, D. A., Marfat, A., and Corey, E. J.**, Slow reacting substances of anaphylaxis: identification of leukotriene C-1 and D from human and rat sources, *Proc. Natl. Acad. Sci. U.S.A.*, 77, 3710—3714, 1981.

206. **Bhagwat, S. S., Hamann, P. R., and Still, W. C.**, Synthesis of thromboxane A₂, *J. Am. Chem. Soc.*, 107, 6372—6376, 1985.

SYNTHETIC PROSTANOIDS
(FROM CIRCA 1976)

Joseph M. Muchowski

The prostaglandins (PGs) are a class of 20-carbon, oxygenated, unsaturated acidic substances derived from dihomo-γ-linolenic acid (eicosa-8,11,14-trienoic acid), arachidonic acid (AA) (eicosa-5,8,11,14-tetraenoic acid), and all *cis*-eicosa-5,8,11,14,17-pentaenoic acid. The profound and varied biological actions of these compounds, which are formed widely in mammalian tissues and fluids, have fired the imagination of numerous investigators in academic and industrial circles for almost 2 decades. The resultant outpouring of biochemical, chemical (mainly synthetic in nature), pharmacological, and clinical pharmacology publications dealing with this class of compounds (prostanoids) has been, with the exception of the field of cyclic nucleotides, unparalleled. A computerized search of *Chemical Abstracts* (January 1, 1967 to January 31, 1983) and *Index Medicus* (January 1, 1966 to January 31, 1983) has revealed that approximately 20,600 and 23,000 papers, respectively, dealing with prostanoids have been abstracted by these reference sources during the time periods indicated.[1,22] Furthermore, prior to 1970, no journals specifically devoted to PGs existed; now there are 15. In addition, more than 50 books on various facets of PG research are currently available.[1,2,22] Whereas the appearance of publications on most aspects of prostanoids continues unabated, the number of purely synthetic papers is on the wane, the reason being that most of the synthetic problems associated with this field had been solved by the end of the 1970s.[3] The majority of the papers which are currently being published in this area are devoted to the synthesis of non-natural prostaglandins by well-established synthetic procedures, as well as to the biological actions associated with these compounds. It is these subjects which this article surveys, in tabular form, for the period January 1, 1976 to December 31, 1983.[4,5,23] A similar series of tables for data up to 1976 is found in the chapter by Willis and Stone.

The natural PGs are intimately associated with a formidable array of physiological processes[6,7] and the administration thereof, not unexpectedly, can elicit a complex spectrum of pharmacological responses.[6-8] Nevertheless, several of the natural PGs have found a modicum of clinical utility.[9] Thus, $PGF_{2\alpha}$, PGE_2 and the β-cyclodextrin clathrates thereof, are used to induce parturition and to terminate undesired pregnancies. PGE_1 and PGE_2 are utilized in delaying the closure of the ductus arteriosis in infants born with certain cardiac abnormalities until these defects have been surgically rectified. In addition, the potent platelet aggregating inhibitory activity of PGI_2 has found some application in cardiopulmonary bypass operations.[10]

The factors which limit the routine use of the natural PGs as therapeutic agents have long been recognized and it is with these in mind that research on the development of modified prostaglandins has been conducted. Thus, ideally for therapeutic purposes, a prostanoid should have oral activity, an appropriately long in vivo half life, be tissue selective, have no or minimal side effects, and have good chemical stability.

In mammals, including man, the natural PGs have both poor oral activity and a short half life. The latter observation is explicable almost entirely in terms of a very rapid metabolic inactivation. For example, in humans, PGs of the E and F types are subject to a series of metabolic processes which probably occur in the following sequence: oxidation of the allylic alcohol moiety of C-15 to the ketone by 15-hydroxyprostaglandin dehydrogenase, saturation of the 13,14-*trans* double bond by PG reductase, twofold degradation of the α-chain[11] by β-oxidation, and ω-oxidation of the terminal (C-20) methyl group to the corresponding alcohol and/or carboxylic acid. In addition, in some species, the ketone carbonyl group of

E PGs can be reduced to a 9α- or a 9β-hydroxy group.[12] Metabolism of PGs is further discussed by Jackson Roberts II in this volume. The recognition of these metabolic processes played a major role in the design of the modified PGs which are now in use or under development.

Inasmuch as 15-hydroxyprostaglandin dehydrogenase both usually initiates and is the fastest of the metabolic degradation processes, it was this site of the PG skeleton which was among the first to be subjected to chemical modification. Indeed, 15-methyl $PGF_{2\alpha}$ (and other 15-methyl PGs) is totally resistant to the dehydrogenase in vitro, has a much longer *in vivo* half life, and the metabolism is diverted mainly to products derived from degradation of the α-chain by β-oxidation.[13] A wide variety of other modifications of the ω-chain, in the vicinity of C-15, have been found which confer resistance to the dehydrogenase and such compounds usually also show an extended half life in vivo. These modifications include replacement of the 13-*trans* double bond by an acetylenic unit[14] (see also compound *14*, Table 3), monosubstitution at C-16 with a methyl,[15] fluoro,[16] or a phenyl group (compound *61*, Table 5), and disubstitution at C-16 with methyl[13] or fluoro groups.[16] In short, it would seem that any alteration, either electronic or steric, which reduces the ease of removal of H-15, will result in a reduced susceptibility to the dehydrogenase. Even a casual perusal of the data in Tables 3 through 7 will reveal numerous PG analogs with extended in vivo half lives. Most of these were designed, at least in part, to be resistant to inactivation at C-15 (e.g., 16-aryloxy compounds such as numbers *39* to *45*, Table 4) although rarely is the prolonged duration of action supported by published evidence of a derangement of metabolism.

Many modified PGs have been constructed which incorporate features expected to impart resistance to degradation of the α- and ω-chains by β- and terminal (as well as at C-19) oxidation, respectively. Thus, in Table 5, compounds *6*, *12* to *14*, and *52*, which contain (E)2,(E) 4-diene; 3-oxa-3,7-*m*-interphenylene; 3-thia; 4,5-diene; and 1-(tetrazol-5-yl) moieties, respectively, all have a prolonged duration of action and ought to be β-oxidation resistant. In addition, compounds *14* and *52* are ω-tetranor 16-phenoxy and 16-phenyl substituted and must be inert to ω-chain degradation as well. In the above context, it is of not inconsiderable significance that evidence of a diversion of the metabolism of the α- and ω-chains, from the normal, has been published for prostanoids with Δ^{-4} unsaturation[17,18] and 17-phenyl-ω-trinor[13] substitution.[24] Also, in passing it is worth emphasizing that inertness of a compound to the usual PG metabolism does not guarantee longevity in vivo. The stable 10,10-difluoro-13-acetylenic prostacyclin (PGI_2) derivative (compound *14*, Table 3) is illustrative of this point. This substance is resistant to the 15-dehydrogenase and shows prolonged survival in in vitro assays, yet the duration of action thereof is equivalent to PGI_2 itself because of a short residence time at the site of action[19] and an extremely rapid hepatic inactivation.

As mentioned earlier, chemical stability is also an important consideration in the development of a therapeutically useful PG. Nowhere is this more important than for PGI_2 (compound *1*, Table 3). This substance is extemely acid sensitive (because of the bicyclic enol ether moiety; it thus is ineffective orally) and has a very short half life in vivo. The stability problem has been solved by such stratagems as replacement of the enol ether oxygen by other atoms such as carbon, nitrogen, or sulfur (e.g., compounds *52*, *75*, and *79*, Table 3), electronic stabilization of the enol ether moiety (e.g., compounds *12* and *14*, Table 3), replacement of the cyclic enol ether unit with other cyclic systems with similar liphophilic and steric properties (e.g., compounds *77* and *86*), and the like.

Finally, one factor which is scarcely if ever mentioned with regard to the design of new therapeutic entities, but which is important to the pharmaceutical industry and self-evident to industrial medicinal chemists, is the necessity for structural novelty in the patent sense, if not the dictionary one. It would be difficult to overestimate the role that this requirement has played in the PG field.

FOREWORD TO THE TABLES

The purpose of this review was not to compile a list of every PG analog synthesized during the 1976 to 1983 period. The criteria for the selection of the prostanoids,[20] the structures of which are found in Tables 1 to 8, were the existence of published pharmacological or biochemical data, structural novelty, and in some instances, the whims of the reviewer. The patent literature, in general, was not used as a source from which new structural entities were chosen for inclusion in this article, because the nature of such documents frequently makes it difficult to ascertain which of the compounds disclosed actually were synthesized and which of these really were the important ones. Nevertheless, it is believed that representatives of most of the structural classes of the prostanoids synthesized during the review period, have been incorporated herein. Certainly, the majority of those non-natural PGs which have been subjected to detailed pharmacological evaluation, even if this data was reported prior to 1976, are to be found in the tables.

Encyclopedic coverage of the published biochemical and pharmacological activities of the compounds in the tables was not intended. Usually only a thumbnail sketch of the major actions is documented and monographs on individual compounds are cited whenever possible.

The nomenclature used throughout the tables is based on the prostanoic acid-prostane system.[21] The principle merits thereof are its great flexibility and that one system, not many, need be learned. There are, however, several structurally exotic members of the PGI family (e.g., compounds *66, 68* to *70*, Table 3) which are not trivial to name, even with this system, and the task of naming these compounds is left to more venturesome individuals.

The tables are very loosely organized according to the site at which the prostane skeleton was chemically modified, i.e., those modified at C-1 are listed first, those at C-2 second, and so on. For heteroatom containing PGs, in addition to this system, monoheteroatomic PGs precede those with two heteroatoms, etc. Nomenclature is discussed in Volume I in the chapter by Willis.

It will doubtless be noted that each table is provided with a separate set of references. This has necessitated some repetition, but it does have the distinct advantage of propinquity.

Finally, it is anticipated that the major use of these tables will be to provide chemists, biochemists, and pharmacologists, in industry and academia, rapid access to specific aspects of the voluminous literature on prostaglandins.

ACKNOWLEDGMENTS

The author is most grateful to the spirited efforts of Karlyn Jaime, Gina Costelli, and Jill Corsiglia to put this article in typewritten form in the face of an eleventh hour publisher's deadline (as it turned out, an ephemeral one). Updated material was typed by Judith Galyardt and Nicole Grinder whose efforts in this regard are greatly appreciated.

ABBREVIATIONS USED IN THE TABLES

AA	Arachidonic acid
ADP	Adenosine diphosphate
BP	Blood pressure
cAMP	Cyclic adenosine 3',5'-monophosphate
DDQ	2,3-Dichloro-5,6-dicyano-1,4-benzoquinone
DMF	Dimethylformamide
ECG	Electrocardiogram
HPRP	Human platelet-rich plasma
HR	Heart rate
i.a.	Intraarterial

i.g.   Intragastric
i.m.   Intramuscular
i.p.   Intraperitoneal
i.v.   Intravenous
LAH   Lithium aluminum hydride
MABP   Mean arterial blood pressure
MED   Minimum effective dose
MMM   Mastitis, metritis, agalactia
p.o.   per os (orally)
PRP   Platelet-rich plasma
q.i.d.   *quater in die* (four times a day)
s.c.   Subcutaneous
SH   Spontaneously hypertensive
STA_2 Sulfur thromboxane A_2
ST   S and T segment of an electrocardiogram
THP   Tetrahydropyranyl
t.i.d.   *ter in die* (three times a day)

GENERAL REFERENCES

1. These data were provided by Amal Moulik (Syntex Corporate Library) whose efforts in this regard are gratefully acknowledged.
2. *Subject Guide to Books in Print 1982—1983*, Vol. 3, R. R. Bowker Co., New York, 1983, 4539—4540.
3. Various aspects of the synthesis of prostanoids are dealt with in three excellent books: (1) **Bindra, J. S. and Bindra, R.**, *Prostaglandin Synthesis*, Academic Press, New York, 1977. (2) *Prostaglandin Research*, Crabbé, P., Ed., Academic Press, New York, 1977. (3) *New Synthetic Routes to Prostaglandins and Thromboxanes*, Roberts, S. M. and Scheinmann, F., Eds., Academic Press, London, 1982.
4. For a similar survey of this and other aspects of the prostaglandin literature up to the end of 1975 see **Willis, A. L. and Stone, K. J.**, *Handbook of Biochemistry and Molecular Biology*, 3rd ed., and *Physical and Chemical Data*, Vol. 2, Fasman, G. D., Ed., CRC Press, Cleveland, 1976, 312—423.
5. For a review directed mainly to prostaglandins with bronchial dilator activity which covers the literature through 1978, see **Grudzinskas, C. V., Skotnicki, J. S., Chen, S.-M. L., Floyd, M. B., Hallett, W. A., Schaub, R. E., Siuta, G. J., Wissner, A., and Weiss, M. J.**, in *Drugs Affecting the Respiratory System*, Temple, D.L., Ed., ACS Symposium Ser. 118, American Chemical Society, Washington, D.C., 1980, 301—377.
6. **Horton, E. W.**, in *Chemistry, Biochemistry and Pharmacological Activity of Prostanoids*, Roberts, S. M. and Scheinmann, F., Eds., Pergamon Press, Oxford, 1979, 1—16.
7. **Crabbé, P.**, Reference 36, pp. 6—16.
8. **Jones, R. L.**, Reference 36, pp. 65—87.
9. For a recent resumé of many of the prostanoids in current use or under development see **Nelson, N. A., Kelly, R. C., and Johnson, R. A.**, Prostaglandins and the arachidonic acid cascade, *Chem. Eng. News*, August 16, 1982, pp. 30—44.
10. For other applications of PGI_2, see Table 3, compound *1*.
11. The α-chain refers to carbons 1 to 7, the ω-chain to C-13 to 20.
12. **Samuelsson, B., Granström, E., Gréen, K., Hamberg, M., and Hammerström, S.**, Prostaglandins, *Ann. Rev. Biochem.*, 44, 674—676, 1975.
13. **Granström, E. and Hansson, G.**, Effect of chemical modification on the metabolic transformation of prostaglandins, in *Advances in Prostaglandin and Thromboxane Research*, Vol. 1, Samuelsson, B. and Paoletti, R., Eds., Raven Press, New York, 1976, 215—219, and references therein.
14. **Fried, J. and Lin, C. H.**, Synthesis and biological effects of 13-dehydro derivatives of natural prostaglandin $F_2\alpha$ and E_2 and their 15-epimers, *J. Med. Chem.*, 16, 429—430, 1973.
15. **Horton, E. W.**, Prostaglandins: advances by analogy, *New Sci.*, 69, 9—12, 1976.
16. **Magerlein, B. J. and Miller, W. L.**, 16-Fluoroprostaglandins, *Prostaglandins*, 9, 527—544, 1975.
17. **Gréen, K., Samuelsson, B., and Magerlein, B. J.**, Decreased rate of metabolism induced by a shift of the double bond in prostaglandin $F_2\alpha$ from the Δ^5 to the Δ^4 position, *Eur. J. Biochem.*, 62, 527—537, 1976.
18. **Hansson, G.**, Metabolism of two $PGF_2\alpha$ analogues in primates: 15(S)-15-methyl-Δ^4-cis-$PGF_1\alpha$ and 16,16-dimethyl-Δ^4-cis-$PGF_1\alpha$, *Prostaglandins*, 18, 745—771, 1979.

19. A similar suggestion has been made to account for the relatively short in vivo half life of compounds *31*, *32*, and *62* (Table 3). There are, however, two compounds (*61* and *76*) which appear to have a prolonged duration of action.

20. Except for PGI$_2$ and 6-keto-PGE$_1$, the natural prostaglandins were not included in these tables.

21. **Nelson, N. A.**, Prostaglandin nomenclature, *J. Med. Chem.*, 17, 911—918, 1974.

22. Updated to Jan. 1, 1985. *Chemical Abstracts* —25,200 references; *Index Medicus*—31,000 references. Journals devoted to prostaglandins and eicosanoids—19. Books in print—80. The computerized update was carried out by Julie Pieprzyk (Syntex Corporate Library) whose assistance is gratefully acknowledged. See Reference 23.

23. Because of the long time interval between the submission of this manuscript and the receipt of the galley proofs (ca. 2 years), it was decided to update this document at the time of the galley proof revision. As a consequence, an effort was made to include important new information which appeared during the period Jan. 1, 1984 to early 1986. The new data is found at the end of the Tables.

24. It has recently been demonstrated that the metabolism of iloprost (cpd. *61*, Table 3), a carbacyclin derivative with an extensively modified ω-chain, is completely shunted to β-oxidation products of the α-chain. Krause, W., Hümpel, M., Hoyer, G. A., Biotransformation of the stable prostacyclin analogue, iloprost, in the rat, *Drug Metabolism and Dispositon*, 12, 645—651, 1984.

Table 1
PGH ANALOGS

No.	Systematic name	Structure	Mol. formula (mol. wt)	Synthesis	Biological actions
1	9α,11α-Dioxyethylene-15α-hydroxy-prosta-5-*cis*-13-*trans*-dienoic acid		$C_{22}H_{36}O_5$ (380)	Acid catalyzed condensation of $PGF_{2\alpha}$ with acetaldehyde; a 3:2 mixture of epimers is obtained[1]	Stable endoperoxide mimic; induced aggregation of HPRP; $0.02 \times PGG_2$; Constricted isolated rabbit aorta (0.25 × PGG_2), gerbil colon (0.12 × PGG_2), and dog saphenous vein (32 × $PGF_{2\alpha}$)[1]
2	(−)9α,11α-Methano-epoxy-15α-hydroxy-prosta-5-*cis*-13-*trans*-dienoic acid (U-46,619)		$C_{21}H_{34}O_4$ (350)	From 11α-hydroxymethyl PGF PGF methyl ester *bis-t*-butyldimethyl-silyl ether derivative;[2] for another synthesis see Reference 3.	Stable endoperoxide mimic with TXA_2 agonistic properties; U-46619 was a proaggregatory agent almost equal in potency to PGH_2;[4,5] potent vasoconstrictor on dog saphenous vein ($ED_{50} = 1.5 \times 10^{-14}\ M$), cat coronary artery (threshold conc. $= 1.5 \times 10^{-10}\ M$), and was a potent vasoconstrictor in the dog (i.v., 0.25 µg/kg);[4,5] potent bronchial constrictor in the dog (i.v.);[4] substrate for human placental 15-hydroxyprostaglandin dehydrogenase[4]
3	9α,11α-Methano-epoxy-15α-hydroxy-prosta-13-*trans*-enoic acid		$C_{21}H_{36}O_4$ (352)	From 9α-hydroxymethyl PGF_2 methyl ester derivative via a process like that used to prepare U-46,619[7]	Powerful endoperoxide mimic; aggregated rabbit PRP (6.3 × PGH_2); contracted rabbit aorta (6.2 × PGH_2)[7]
4	9α,11α-Carbonyl-epoxy-15α-hydroxy-prosta-13-*trans*-dienoic acid		$C_{21}H_{34}O_5$ (366)	From *cis*-2-(6-methoxy-carbonylhexyl)-*trans*-3-methoxycarbonyl-4-hydroxycyclopentane carboxylic acid[7]	Potent endoperoxide mimic; aggregation of rabbit PRP (4.0 × PGH_2); contraction of rabbit aorta (5.6 × PGH_2)[7]

No.	Name	Structure	Formula (mass)	Synthesis	Properties
5	(−) 9α,11α-Epoxy-methano-15α-hydroxyprosta-5-*cis*-13-*trans*-dienoic acid (U-44,069)		$C_{21}H_{34}O_4$ (350)	From tosylate of 11α-hydroxymethyl PGF₂ 15-acetate methyl ester derivative on reaction with methanolic potassium hydroxide[2]	Stable endoperoxide analog with many properties in common with U-46,619;[5,24] inhibited human platelet thromboxane synthetase ($I_{50} = 32$ µM), whereas U-46,619 was much weaker ($I_{50} = 600$ µM)[6,11]
6	9α,11α-Epoxy-methano-15α-hydroxyprosta-13-*trans*-enoic acid		$C_{21}H_{36}O_4$ (352)	From 11α-hydroxy-methyl PGF₂ methyl ester derivative via a process similar to that used for the synthesis of U-46,619[2]	Inhibited AA-induced platelet aggregation of rabbit PRP (like PGH₁); contracted rabbit aorta (0.93 × PGH₂)[7]
7	9α,11α-Epoxy-carbonyl-15α-and 15β-hydroxyprosta-13-*trans*-enoic acid		$C_{21}H_{34}O_5$ (366)	From 11α-hydroxy-methyl PGF methyl ester derivative[7]	Inhibited AA-induced aggregation of rabbit PRP (like PGH₁); most potent known constrictor of rabbit aorta (31 × PGH₂)[7]
8	9α,11α-Imino-epoxy-prosta-5-*cis*-13-*trans*-dienoic acid		$C_{20}H_{33}NO_3$ (335)	From tosylate of 9β-hydroxy-11-THP-15-deoxy PGF methyl ester derivative[8]	Potent inhibitor of human platelet thromboxane synthetase[8,9]
9	9α,11α-Epoxy-imino-prosta-5-*cis*-13-*trans*-dienoic acid		$C_{20}H_{33}NO_3$ (335)	From mesylate of 11β-hydroxy-9-THP-15-deoxy PGF methyl ester derivative[8]	Antagonist of TXA₂ receptor in human platelets, i.e., antagonizes TXA₂ mediated aggregation of HPRP (IC₅₀ <200 ng/mℓ) without inhibiting TXA₂ synthetase or cyclooxygenase[8,9]
10	9α,11α-Imino-methano-15α-hydroxyprosta-5-*cis*-13-*trans*-dienoic acid		$C_{21}H_{35}NO_3$ (349)	From mesylate of 9β-hydroxy-11α-hydroxy-methyl PGF₂ 15-acetate methyl ester derivative and ammonia[10]	Inhibited TXA₂ synthetase in human platelets [1.6 × U-51605 (9,11-azo-prostadienoic acid)][10,11]

Table 1 (continued)
PGH ANALOGS

No.	Systematic name	Structure	Mol. formula (mol. wt)	Synthesis	Biological actions
11	9α,11α-Azo-15α-hydroxyprosta-5-cis-13-trans-dienoic acid (U-51,093)		$C_{20}H_{32}N_2O_3$ (348)	Reaction of the 9β,11β-dimesylate of the 9β,11β-dihydroxy PGF₂ derivative with hydrazine followed by aerial oxidation[12] or from the major Diels-Alder adduct of methyl cyclopentadiene carboxylate and diethyl azodicarboxylate[13]	Potent platelet aggregatory agent;[12] inhibited TXA₂ synthetase of human platelets (ID₅₀ = 2 μM);[11] potent constrictor of rabbit aorta, dog saphenous vein, and guinea pig trachea with potencies 1.4, 1.2, and 1.1 × 9,11-methanoepoxy PGH₂ (U-46,619)[24]
12	9α,11α-Azo-13-oxa-15α- and 15β-hydroxyprostanoic acid		$C_{19}H_{34}N_2O_4$ (354)	From Diels-Alder adduct obtained from diethyl azodicarboxylate and 3-(6-ethoxycarbonylhexyl) cyclopentadiene[14]	Potent inhibitor of thromboxane synthetase (washed human platelets); IC₅₀ ~10⁻⁶ M; PGH₂/TXA₂ receptor blocker; thus, blocks agonistic action of 9,11-epoxymethano PGH₂ (IC₅₀ = 9 × 10⁻⁷ M) and TXA₂ (IC₅₀ = 2.4 × 10⁻⁶ M) of HPRP[14,15]
13	9α,11α-Azoprosta-5-cis-13-trans-dienoic acid (U-51,605)		$C_{20}H_{32}N_2O_2$ (332)	Reaction of the dimesylate of 9β-11β-dihydroxy-15-deoxy PGF₂ with hydrazine followed by aerial oxidation[16]	Selective inhibitor of TXA₂ synthesis of human platelets by competitively inhibiting the binding of PGH₂ to TXA₂ synthetase[17]
14	9α,11α-Azo-13-oxa-prostanoic acid		$C_{19}H_{34}N_2O_3$ (338)	From one of the Diels-Alder adducts of dimethyl azodicarboxylate and the products of alkylation of lithium cyclopentadienide and methyl 7-bromoheptanoate[18]	

No.	Name	Structure	Molecular formula (MW)	Synthesis	Biological activity
15	(+)Methyl 9α,11α-epidithia-15α-hydroxyprosta-5-*cis*-13-*trans*-dienoate		$C_{21}H_{34}O_3S_2$ (398)	Manganese dioxide[19] or aerial[20] oxidatin of the corresponding 9,11-dimercapto compound	Potent endoperoxide mimic; caused aggregation of HPRP at a threshold dose of 0.6 μM; the aggregation was associated with serotonin release from platelets and was inhibited by PGI_2 methyl ester[19]
16	Methyl 9β,11β-epidithia-15α-hydroxyprosta-5-*cis*-13-*trans*-dienoate		$C_{21}H_{34}O_3S_2$ (398)	Manganese dioxide oxidation of the 9β,11β-dimercapto compound[19]	No effect on HPRP[19]
17	9α,11α-Epitrithiacarbona-15α-hydroxyprosta-5-*cis*-13-*trans*-dienoic acid		$C_{21}H_{32}O_3S_3$ (428)	Reaction of the mesylate of the 9β-hydroxy-11α-mercapto-11,15-diacetate PGF_2 methyl ester derivative and then saponification[20]	
18	9α,11α-Epitetrathia-15α-hydroxyprosta-5-*cis*-13-*trans*-dienoic acid		$C_{20}H_{32}O_3S_4$ (448)	Reaction of dimesylate of 9β,11β-dihydroxy PGF_2 derivative with disodium disulfide followed by aerial oxidation[21]	Potent endoperoxide mimic[21]
19	9α,11α-Ethano-15α-hydroxyprosta-5-*cis*-13-*trans*-dienoic acid		$C_{22}H_{36}O_3$ (348)	From the Diels-Alder adduct of maleic anhydride and cyclopentadiene;[22] also obtained as a mixture of epimers at C-15 from 3-*endo*-cyanomethylbicyclo [2.2.1]hept-5-ene-2-*exo*-carboxaldehyde[23]	Mixture of 15-epimers. Abolished synthesis of PGE_2 from arachidonic acid (ram seminal vesicles) at 0.5 mM;[23] partial TXA_2 agonist on rabbit aorta, dog saphenous vein, and guinea pig trachea[24]

Table 1 (continued)
PGH ANALOGS

No.	Systematic name	Structure	Mol. formula (mol. wt)	Synthesis	Biological actions
20	9α,11α-Ethano-15α-hydroxy-16-(4-fluorophenoxy)-ω-tetranorprosta-5-cis-13-trans-dienoic acid		$C_{24}H_{31}FO_4$ (402)	From Diels-Alder adduct of cyclopentadiene and maleic anhydride[25]	Potent endoperoxide analog; irreversible aggregatin of HPRP at 0.31 μM;[25] potent constrictor of rabbit aorta and dog saphenous vein with potencies 0.26 and 0.13 × 9,11-methanoepoxy PGH$_2$ (U-46619)[24]
21	3-endo-(6-Carboxyhex-(Z)2-enyl)bicyclo[3.1.1]heptan-2-exo-carboxaldehyde phenyl semicarbazone (EP 045)		$C_{22}H_{29}N_3O_3$ (383)		Thromboxane receptor antagonist; blocked constriction of rabbit aorta, dog saphenous vein, and guinea pig trachea induced by 9,11-methanoepoxy PGH$_2$, and other endoperoxide analogs;[26] inhibited aggregation of HPRP induced by TXA$_2$, PGH$_2$, and synthetic endoperoxide analogs, AA, and collagen but not that induced by ADP;[26,30,42] in the anesthetized guinea pig, it blocked the fall in platelet count induced by collagen (51% at 2.5 mg/kg i.v.) and 9,11-methanoepoxy PGH$_2$ (89.5% at 2.5 mg/kg), but the duration of blockade is short lived[30,42]
22	8-epi-9α,11α-Ethano-15α- and 15β-hydroxyprosta-5-trans-en-13-ynoic acid		$C_{22}H_{34}O_3$ (346)	From norbornene[41]	Inhibited platelet aggregation; 0.5 × PGE$_1$; specific inhibitor of thromboxane synthetase[41]
23	8-epi-12-epi-9α,11α-Ethano-15α-hydroxyprosta-5-cis-13-trans-dienoic acid		$C_{22}H_{36}O_3$ (348)	From (±) norcamphor[27]	Pronounced hypotension in anesthetized rat; very weak inhibitor of ADP-induced platelet aggregation and of TXA$_2$ synthetase[27]

No. / Name	Structure	Formula (MW)	Synthesis	Activity
24 9α,11α-Etheno-15α-hydroxy-prosta-5-*cis*-13-*trans*-dienoic acid		$C_{22}H_{34}O_3$ (346)	From Diels-Alder adduct of cyclopentadiene with methyl propiolate;[28,29] also synthesized from Diels-Alder adduct of cyclopentadiene and maleic anhydride[23]	Stable endoperoxide mimic; induced aggregation of HPRP ($0.1 \times PGG_2$) which was not inhibited by indomethacin; ^{14}C serotonin was released from platelets at aggregatory concentrations;[28] inhibited conversion of AA to PGE_2; 100% at 0.5 mmol (ram seminal vesicles),[23] partial agonist with regard to contraction of rat aorta, dog saphenous vein, and guinea pig trachea[24]
25 9α,11α-Etheno-15α-hydroxy-16-(4-fluorophenoxy)-ω-tetranorprosta-5-*cis*-13-*trans*-dienoic acid (EP 011)		$C_{24}H_{29}FO_4$ (398)	From Diels-Alder adduct of cyclopentadiene and maleic anhydride[25]	Potent endoperoxide mimic; irreversible aggregation of HPRP at 0.19 μM;[25] potent constrictor of rabbit aorta, dog saphenous vein, and guinea pig trachea with potencies 0.12, 0.09, 0.06 $\times$ 9,11-methanoepoxy PGH_2 (U-46,619)[24]
26 (−)8-*epi*-12-*epi*-9β,11β-Ethano-10-oxa-15α-hydroxy-5-*cis*-13-*trans*-prostadienoic acid (SQ-26271)		$C_{21}H_{34}O_4$ (350)	From *endo* Diels-Alder adduct of furan and maleic anhydride[31,34]	Potent prostacyclin-like compound; inhibited aggregation of HPRP by AA ($IC_{50} = 0.30\ \mu M$) and ADP ($IC_{50} = 1.6\ \mu M$);[32,34] inhibited AA-induced bronchial constriction in the guinea pig[31]
27 (−)8-*epi*-12-*epi*-9β,11β-10-oxa-15α-hydroxy-15-cyclohexyl-ω-pentanorprosta-5-*cis*-13-*trans*-dienoic acid (SQ,27,986)		$C_{22}H_{34}O_4$ (362)	From *endo* Diels-Alder adduct of furan and maleic anhydride[32]	Potent PGI_2-like compound; inhibited aggregation of HPRP by ADP ($IC_{50} = 0.03\ \mu M$; $0.33 \times PGI_2$) and AA ($IC_{50} = 0.01\ \mu M$; $0.2 \times PGI_2$); stimulated platelet adenyl cyclase and elevated platelet cAMP; hypotensive in normotensive anesthetized rats (racemate $0.1 \times PGI_2$, *i.v.*); inhibited bronchial constriction in anesthetized guinea pigs induced by AA or histamine[32]

Table 1 (continued)
PGH ANALOGS

No.	Systematic name	Structure	Mol. formula (mol. wt)	Synthesis	Biological actions
28	(+)12-epi-9β,11β-Ethano-10-oxa-15α-hydroxyprosta-5-cis-13-trans-dienoic acid (SQ-26536)		$C_{21}H_{34}O_4$ (350)	From exo Diels-Alder adduct of furan and maleic anhydride[31]	Stable antagonist of the human blood platelet thromboxane receptor; inhibited AA-induced aggregation of HPRP; $IC_{50} = 0.6 \mu M$; did not inhibit PG synthetase activity (bovine seminal microsomes) or thromboxane synthetase activity (lysed human blood platelets); inhibited aggregation induced by epinephrine (secondary phase), 9,11-azo PGH_2, and collagen, but did not inhibit primary phase of epinephrine or ADP-induced aggregation[33,34]
29	(+)12-epi-9β,11β-Ethano-10-oxa-15β-hydroxyprosta-5-cis-13-trans-dienoic acid (SQ-26,238)		$C_{21}H_{34}O_4$ (350)	From exo Diels-Alder adduct of furan and maleic anhydride[31]	Stable agonist of the human blood platelet thromboxane receptor; induced aggregation of HPRP; $I_{50} = 2.0 \mu M$ which was not blocked by indomethacin nor a thromboxane synthesis inhibitor (SQ 80338, N-cinnamylimidazole); the aggregation was, however, blocked by the 15α-isomer (SQ-26,536)[34,38]
30	8-epi-12-epi-9α,11α-Ethano-10-oxa-15α-hydroxyprosta-5-cis-13-trans-dienoic acid		$C_{21}H_{34}O_4$ (350)	From exo Diels-Alder adduct of furan and maleic anhydride[31,34,36]	Stable agonist of human platelet thromboxane receptor. Induced aggregation of HPRP($I_{50} = 30mM$).[34] See also Reference 36.

No.	Name	Structure	Formula (MW)	Synthesis	Biological activity
31	9α,11α-Ethano-10-oxa-15α-hydroxyprosta-13-trans-enoic acid (GBR-30730)		$C_{21}H_{36}O_4$ (352)	From *exo* adduct of furan and maleic anhydride[35,37]	Contracted isolated guinea pig trachea and rat fundus, but relaxed rat uterus; vasopressor and bronchial constrictor in anesthetized guinea pig (i.v.); in the anesthetized rat (intra-arterial) the vasopressor response was very weak; no effect on ADP-induced platelet aggregation in the rat by the arterial route[35,38]
32	9α,11α-Ethano-10-oxa-15β-hydroxyprosta-13-trans-enoic acid (GBR-30731)		$C_{21}H_{36}O_4$ (352)	From *exo* Diels-Alder adduct of furan and maleic anhydride[35,37]	Contracted isolated guinea pig trachea and rat fundus, but much less so than the 15α-isomer (GBR-30730); relaxes rat uterus; vasopressor and bronchial constrictor in anesthetized guinea pig (i.v.), but was a mild hypotensive in anesthetized rats (intra-arterial); Intra-arterial injection (250 µg) in the rat caused a long-lasting (≥30 min) inhibition of ADP-induced platelet aggregation[35,38]
33	8-Aza-9α,11α-ethano-7-keto-15α-hydroxyprosta-13-trans-enoic acid		$C_{21}H_{35}NO_4$ (364)	From Diels-Alder adduct of cyclopentadiene and the oxime tosylate of ethyl oxalylcyanide[39,43]	Did not contract rabbit aortic strip; no activity in platelet aggregation tests[43]
34	8-Aza-9α,11α-ethano-15α- and 15β-hydroxyprosta-5-cis-13-trans-dienoic acid		$C_{21}H_{35}NO_3$ (349)	From Diels-Alder adduct of cyclopentadiene and *n*-butyl-*N*-tosyliminoglyoxalate[40]	
35	9α,11α-Ethano-10a-homo-15α-hydroxyprosta-5-cis-13-trans-dienoic acid		$C_{23}H_{38}O_3$ (362)	From *endo* Diels-Alder adduct of cyclohexa-1,3-diene and maleinaldehyde pseudo ethyl ester[25]	Potent irreversible inhibitor of HPRP[25]

Table 1 (continued)
PGH ANALOGS

No.	Systematic name	Structure	Mol. formula (mol. wt)	Synthesis	Biological actions
36	8-epi-9β,11β-Ethano-10-oxa-14,15,17-triaza-16-keto-17-phenyl-ω-trinor prosta-5-cis-enoic acid (SQ-29,548)		$C_{21}H_{29}N_3O_4(387)$		Thromboxane receptor antagonist in platelets and vasculature; inhibited in vitro aggregation of rabbit PRP induced by arachidonic acid (IC_{50} = 0.82 μ*M*), collagen (IC_{50} = 2.94 μ*M*) and U-46619 (IC_{50} = 0.31 μ*M*), but not by ADP; inhibited arachidonic acid induced sudden death in rabbits at doses from 0.2—2 mg/kg (i.v.) with 100% survival rate at doses of 1 or 2 mg/kg[44,45]

REFERENCES

1. **Portoghese, P. S., Larson, D. L., Abatjoglu, A. G., Dunham, E. W., Gerrard, J. M., and White, J. G.,** A novel prostaglandin endoperoxide mimic, prostaglandin $F_{2\alpha}$ acetal, *J. Med. Chem.*, 20, 320—321, 1977.

2. **Bundy, G. L.,** The synthesis of prostaglandin endoperoxide analogs, *Tetrahendron Lett.*, 1957—1960, 1975.

3. **Trost, B. M., Timko, J. M., Stanton, J. L.,** An enantioconvergent approach to prostanoids, *J. Chem. Soc. Commun.*, 436—438, 1978.

4. U-46,619, *Drugs of the Future*, 9, 453-458, 1980.

5. U-46,619, *Drugs of the Future*, 827—829, 1983.

6. **Sun, F. F.,** Biosynthesis of thromboxanes in human platelets. I. Characterization and assay of thromboxane synthetase, *Biochem. Biophys. Res., Commun.*, 74, 1432—1440, 1977.

7. **Sakai, K., Inoue, K., Amemiya, S., Morita, A., and Kojima, K.,** Synthetic studies on prostanoids. XX. Synthesis of stable ($\pm$)-prostaglandin H_1 analogs, *Chem. Pharm. Bull.*, 28, 1814—1819, 1980.

8. **Bundy, G. L. and Peterson, D. C.,** The synthesis of 15-deoxy-9,11-(epoxyimino)prostaglandins-potent thromboxane synthesis inhibitors, *Tetrahedron Lett.*, 41—44, 1978.

9. **Fitzpatrick, F. A., Bundy, G. L., Gorman, R. R., and Honohan, T.,** 9,11-Epoxyiminoprosta-5,13-dienoic acid is a thromboxane A_2 antagonist in human platelets, *Nature (London)*, 275, 764—766, 1978.

10. **Corey, E. J., Niwa, M., Bloom, M., and Ramwell, P. W.,** Synthesis of a new prostaglandin endoperoxide (PGH$_2$) analog and its function as an inhibitor of the biosynthesis of thromboxane A_2 (TBXA$_2$), *Tetrahedron Lett.*, 671—674, 1979.

11. **Diczfalusy, V. and Hammerström, S.,** Inhibitors of thromboxane synthetase in human platelets, *FEBS Lett.*, 82, 107—110, 1977.

12. **Corey, E. J., Nicolaou, K. C., Machida, Y., Malmsten, C. L., and Samuelsson, B.,** Synthesis and biological properties of a 9,11-azo-prostanoid; highly active biochemical mimic of prostaglandin endoperoxides, *Proc. Natl. Acad. Sci. U.S.A.*, 72, 3355—3358, 1975.

13. **Corey, E. J., Narasaka, K., and Shibasaki, M.,** A direct stereocontrolled total synthesis of the 9,11-azo analog of the prostaglandin endoperoxide PGH$_2$, *J. Am. Chem. Soc.*, 98, 6417—6418, 1976.

14. **Kam, S. S., Portoghese, P. S., Gerrard, J. M., and Dunham, E. W.,** Synthesis and biological evaluation of 9,11-azo-13-oxa-15-hydroxyprostanoic acid, a potent inhibitor of platelet aggregation, *J. Med. Chem.*, 22, 1402—1408, 1979.

15. **Kam, S. S., Portoghese, P. S., Gerrard, J. M., and Dunham, E. W.,** 9,11-Azo-13-oxa-15-hydroxy-prostanoic acid. A potent thromboxane synthetase inhibitor and a PGH$_2$/TXA$_2$ receptor antagonist, *Prostaglandins Med.*, 3, 279—290, 1979.

16. **Gorman, R. R., Bundy, G. L., Peterson, D. C., Sun, F. F., Miller, O. V., and Fitzpatrick, F. A.,** Inhibition of human platelet thromboxane synthetase by 9,11-azoprosta-5,13-dienoic acid, *Proc. Natl. Acad. Sci. U.S.A.*, 74, 4007—4011, 1977.

17. U-51605, *Drugs of the Future*, 4, 298—301, 1979.

18. **Ansell, M. F., Caton, M. P. L., and North, P. C.,** A stereospecific synthesis of 9,11-azo-PGH$_1$ derivatives-potential inhibitors of blood platelet aggregation, *Tetrahedron Lett.*, 23, 4113—4114, 1983.

19. **Miyake, H., Iguchi, S., Itoh, H., and Hiyashi, M.,** Simple synthesis of methyl (5Z, 9α,11α,13E,15S)-9,11-epidithio-15-hydroxyprosta-5,13-dienoate, endodisulfide analogue of PGH$_2$, *J. Am. Chem. Soc.*, 99, 3536—3537, 1977.

20. **Greene, A. E., Padilla, A., and Crabbé, P.,** Synthesis of thio analogoues of prostaglandin H$_2$ and prostaglandin F$_2$ from prostaglandin A$_2$, *J. Org. Chem.*, 43, 4377—4379, 1978.

21. **Nicolaou, K. C., Gasic, G. P., and Barnette, W. E.,** Synthesis and biological properties of prostaglandin endoperoxides, thromboxanes and prostacyclins, *Angew. Chem. Int. Ed. Engl.*, 17, 293—312, 1978; see especially pp. 299, 306, and 307.

22. **Kametani, T., Suzuki, T., Kamada, S., and Unno, K.,** Synthesis of the prostaglandin H$_2$ analogue DL-9,11-Ethano-9,11-dideoxaprostaglandin H$_2$, *J. Chem. Soc. Perkin Trans. I*, 3101—3105, 1981.

23. **Leeney, T. J., Marsham, P. R., Ritchie, G. A., and Senior, M. W.,** Inhibitors of prostaglandin biosynthesis. A bicyclo-[2.2.1]-heptene analogue of "2" series of prostaglandins and related derivatives, *Prostaglandins*, 11, 953—960, 1976.

24. **Jones, R. L., Peesapati, V., and Wilson, N. H.,** Antagonism of the thromboxane-sensitive contractile systems of the rabbit aorta, dog saphenous vein and guinea-pig trachea, *Br. J. Pharmacol.*, 76, 423—438, 1982.

25. **Wilson, N. H., Peesapati, V., Jones, R. L., and Hamilton, K.,** Synthesis of prostanoids with bicyclo[2.2.1]heptane, bicyclo[3.1.1]heptane, and bicyclo[2.2.2]octane ring systems. Activities of 15-hydroxy epimers on human platelets, *J. Med. Chem.*, 25, 495—500, 1982.

26. **Jones, R. L. and Wilson, N. H.,** Partial agonism of prostaglandin H$_2$ analogs and 11-deoxy-prostaglandin $F_{2\alpha}$ to thromboxane sensitive preparations, *Adv. Prostaglandin Thromboxane Res.*, 6, 467—475, 1975.

27. **Barraclough, P.,** A simple synthesis of a stable thromboxane A$_2$ analogue, *Tetrahedron Lett.*, 21, 1897—1900, 1980.

28. **Corey, E. J., Shibasaki, M., Nicolaou, K. C., Malmsten, C. L., and Samuelsson, B.,** Simple, stereocontrolled total synthesis of a biologically active analog of the prostaglandin endoperoxides (PGH$_2$, PGG$_2$), *Tetrahedron Lett.*, 737—740, 1976.

29. **Shimomura, H., Sugie, A., Katsube, J., and Yamamoto, H.,** Synthesis of 9,11-desoxy-9,11-vinyleno-PGF$_{2\alpha}$ and its diastereoisomer, analogs of the PG endoperoxide (PGH$_2$), *Tetrahedron Lett.*, 4099—4102,

30. **Jones, R. L., Wilson, N. W., Armstrong, R. A., Peesapati, V., and Smith, G. W.,** Effects of thromboxane antagonist EP 045 on platelet aggregation, *Adv. Prostaglandin Thromboxane Res.*, 11, 345—350, 1983.

31. **Sprague, P. W., Heikes, J. E., Harris, D. N., and Greeberg, R.,** Stereocontrolled synthesis of 7-oxabicyclo[2.2.1]heptane prostaglandin analogs as thromboxane A$_2$ antagonists, *Adv. Prostaglandin Thromboxane Res.*, 6, 493—496, 1980.

32. **Haslanger, M. F., Sprague, P. W., Snitman, D., Vu, T., Harris, D. N., Greenberg, R., and Powell, J.,** Novel 7-oxabicyclo[2.2.1]heptane prostacyclin agonists, *Adv. Prostaglandin Thromboxane Res.*, 11, 293—297, 1983.

33. **Harris, D. N., Phillips, M. B., Michel, I. M., Goldberg, H. J., Heikes, J. E., Sprague, P. M., and Antonaccio, M. J.,** 9α-Homo-9,11-epoxy-5,13-prostadienoic acid analogues: specific stable agonist (SQ 26,538) and antagonist (SQ 26,536) of the human platelet thromboxane receptor, *Prostaglandins*, 22, 295—307, 1981.

34. **Sprague, P. W., Heikes, J. E., Gougoutas, J. Z., Mally, M. F., Harris, D. N., and Greenberg, R.,** Synthesis and in vitro pharmacology of 7-oxabicyclo [2.2.1] heptane analogues of thromboxane A$_2$/PGH$_2$, *Prostaglandins*, 28, 1580—1590, 1985.

35. It is likely that the stereochemistry of these compounds has been assigned erroneously. See Reference 34, especially footnote 33, page 1586.

36. **Kametani, T., Suzuki, T., Tomino, A., Kamada, S., and Unno, K.,** Studies on syntheses of heterocyclic and natural compounds. CMXLVI. Stereocontrolled total synthesis of a thromboxane A$_2$ analog, ($\pm$)-9α-homo-(11,12)-deoxathromboxane A$_2$, *Chem. Pharm. Bull.*, 30, 796—801, 1982 These authors report that this compound precipitates aggregation of HPRP at 0.08 μM..

37. **Eggelte, T. A., de Köning, H., and Huisman, H. O.,** Synthesis of 9,11-epoxy-9α-homoprostaglandin analogues, *J. Chem. Soc. Perkin Trans. I*, 980—989, 1978.

38. **Hall, D. W. R., Funcke, A. B. H., and Jaitly, K. D.,** Some biological activities of a series of 9α-homo-9,11-epoxy prostanoic acid analogoues, *Prostaglandins*, 18, 317—330, 1979.

39. **Blondet, D. and Morin, C.,** Synthese totale d'un analogue azote d'endoperoxide de prostaglandine, *Tetrahedron Lett.*, 23, 3681—3682, 1982.

40. **Barco, A., Benetti, S., Baraldi, P. G., Moroder, F., Pollini, G. P., and Simoni, D.,** Stereoselective synthesis of 8-aza-9,11-ethenoprostaglandin H$_1$, *Liebigs Ann. Chem.*, 960—965, 1982.

41. **Larock, R. C., Burkhart, J. P., and Oertle, K.,** Organopalladium approaches to prostaglandins. II. Synthesis of prostaglandin endoperoxide analogs via II-allylpalladium additions to bicyclic olefins, *Tetrahedron Lett.*, 23, 1071—1074, 1982.

42. **Armstrong, R. A., Jones, R. L., Peesapati, V., Will, S. G., and Wilson, N. H.,** Competitive antagonism at thromboxane receptors in human platelets, *Br. J. Pharmac.*, 84, 595—607, 1985.

43. **Blondet, D. and Morin, C.,** Total synthesis of ($\pm$)-8-aza-9a,9b-dicarbaprostaglandin H$_1$, *J. Chem. Soc. Perk. Trans. I*, 1085—1090, 1984.

44. **Darius, H., Smith, J. B., and Lefer, A. M.,** Beneficial effects of a new potent and specific thromboxane receptor antagonist (SQ-29,548) in vitro and in vivo, *J. Pharmacol. Exp. Ther.*, 235, 274—278, 1985.

45. **Ogletree, M. L., Harris, D. N., Greenberg, R., Haslanger, M. F., and Nakane, M.,** Pharmacological actions of SQ-29,548, a novel selective thromboxane antagonist, *J. Pharmacol. Exp. Ther.*, 235, 435—441, 1985.

Table 2
TXA ANALOGS

No.	Systematic name	Structure	Mol. formula (mol. wt.)	Synthesis	Biological actions
1	9α,11α-Carba-15α-hydroxythromba-(Z)5,(E)13-dienoic acid		$C_{21}H_{34}O_4$ (350)	From 1-(3-hydroxy-*trans*-1-propenyl)cyclobutan-3-one[1]	Inhibitor of PGH_2-induced aggregation of HPRP[3]
2	9α,11α-Epoxy-15α-hydroxy-11a-carbathromba-(Z)5,(E)13-dienoic acid		$C_{21}H_{34}O_4$ (350)	Via a Demjanov-Tiffenau[2] ring expansion of PGA_2 methyl ester 15-*t*-butyldimethylsilyl ether to a 6-membered enone which was conjugated, epoxidized, reduced to a 9,11-diol and cyclized to the oxetane[3]	
3	Methyl 9α,11α-epithia-15α-hydroxy-thromba(Z)5,(E)13-dienoate		$C_{21}H_{34}O_4S$ (382)	From TXB_2 methyl ester 11-methyl ether 15-benzoate by inversion of 11α-OH, mesylation, formation of 11-methoxycarbonylethylthio derivative, and cyclization[4]	Contracted rat aortic strip; CD_{50} = 10^{-7} M[4]
4	9α,11α-Carba-15α-hydroxy-11α-carbathromba-(Z)5,(E)13-dienoic acid (carbocyclic TXA_2)		$C_{22}H_{36}O_3$ (348)	From 2-formylbicyclo[3.1.1]hept-2-ene;[5] for other syntheses see Refs. 6, 7	Potent TXA_2-like agonist on the vasculature but profound TXA antagonist on the aggregation of platelets; profound constriction of isolated perfused coronary arteries (e.g., cat; $10^5 \times$ TXB_2). Exacerbated ischemic damage in the heart (without induction of platelet aggregation) during coronary insufficiency (anesthetized cat, i.v.)

Table 2 (continued)
TXA ANALOGS

No.	Systematic name	Structure	Mol. formula (mol. wt.)	Synthesis	Biological actions
5	9α,11α-Dimethylcarba-15α-hydroxy-11α-carbathromba(Z)5,(E)13-dienoic acid (pinane TXA$_2$)		C$_{24}$H$_{40}$O$_3$ (376)	From (−) myrtenol[9] or nopol[20]	When given i.v. to anesthetized or conscious rabbits (100—125 μg/kg) produced sudden death in 8—12 min characterized by rapid reduction in MABP, dramatic increase in ST segment, indicative of acute myocardial ischemia. Respiration becomes rapid and shallow and eventually stops; these changes can be prevented by PTA$_2$ (pinane TXA$_2$; compound 5); completely prevented platelet aggregation induced by AA, ADP, 9,11-azo PGH$_2$, 9,11-methanoepoxy PGH$_2$, and 9,11-epoxymethano PGH$_2$ at 4—5 μ*M* Effective inhibitor of TX synthesis in washed platelets (rabbit) but has no effect on PGI$_2$ synthesis (ram seminal vesicles)[8] Potent TX synthesis inhibitor and TX receptor antagonist; at 0.5—1 μ*M* abolished contraction of isolated cat coronary artery induced by 9,11-azo PGH$_2$, 9,11-methanoepoxy-PGH$_2$ and carbocyclic TXA$_2$ (compound 4)

During acute myocardial ischemia in anesthetized cats, infusion (0.5 μM/kg/hr) 30 min postocclusion of left anterior descending coronary artery, prevented extension of ischemic damage, and was antiarrhythmic

At 1—15 μM, inhibited aggregation of HPRP induced by 9,11-azo-PGH_2, 9,11-epoxy-methano-PGH_2, and 9,11-methanoepoxy-PGH_2

At 10 μM, abolished aggregation induced by AA, abolished the second wave, but not the first, of aggregation induced by ADP or epinephrine

Inhibited TXB_2 formation (washed rabbit platelets) by 83% at 100 μM, but had no effect on PGI_2 synthesis (sheep vesicular glands)[8]

See Reference 20 for the synthesis and biological actions of isomeric 10,11α-dimethylcarba compounds

Inhibited vasoconstriction of isolated perfused cat coronary artery induced by carbocyclic TXA_2 by 76% at 1 μM; inhibited aggregation of HPRP induced by stable PG endoperoxide analogs[11]

From (−) myrtenol[11]

$C_{27}H_{38}O_3$ (410)

6 2,3,4-Trinor-1,5-inter-*m*-phenylene-9α,11α-dimethyl-carba-15α-hydroxy-11α-carbathromba-(Z)5,E(13)-dienoic acid

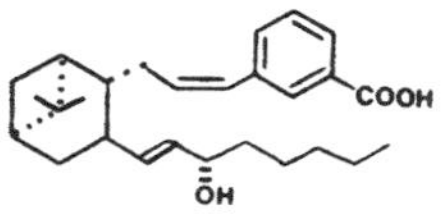

Table 2 (continued)
TXA ANALOGS

No.	Systematic name	Structure	Mol. formula (mol. wt.)	Synthesis	Biological actions
7	2,3,4-Trinor-1,5-inter-*m*-phenylene-9α,11α-di-methylcarba-15α-hy-droxy-11α-carbathromba-(E)5,(E)13-dienoic acid		$C_{27}H_{38}O_3$ (410)	From (−) myrtenol[11]	Inhibited vasoconstriction of isolated perfused cat coronary artery induced by carbocyclic TXA₂ by 72% at 1 μM[11]
8	9α,11α-Dimethylcarba-15α-hydroxy-16-(4-fluorophenoxy)-ω-tetra-nor-11α-carbathromba-(Z)5,(E)13-dienoic acid		$C_{26}H_{35}FO_4$ (430)	From nopol[12]	Precipitated irreversible aggre-gation of HPRP at 0.31 μM[12]; partial agonist in the rabbit aorta, dog saphenous vein, and guinea pig trachea assays[19] TXA₂ antagonist on platelets and vasculature; inhibited ag-gregation of HPRP induced by STA₂ (compound *10*), 9,11-methanoepoxy PGH₂, AA, col-lagen, and ADP (second phase but not first) with IC₅₀ values of 0.11—0.38 μM; also inhib-ited TXA₂ induced aggretation of HPRP but had no effect on cyclooxygenase, TX synthe-tase, or PGE₂ synthetase
9	9α,11α-Dimethylcarba-13-aza-13H-15α- and 15β-hydroxy-20-*n*-pro-pyl-11α-carbothroma-(Z)5-enoic acid		$C_{26}H_{47}NO_3$ (421)		I.V. infusion in guinea pigs or cats (1—10 μg/kg/min) caused ex vivo inhibition of AA or STA₂-induced aggreation Inhibited vasoconstrictior re-sponse in vitro (rat aorta) and in vivo (guinea pig, i.v.) in-duced by stable endoperoxide and TXA₂ analogs[21]

No.	Name	Structure	Formula (MW)	Synthesis	Activity
10	9α,11α-Epithia-15α-hydroxy-11α-carbathromba-(Z)5,(E)13-dienoic acid (STA$_2$)		$C_{21}H_{34}O_3S$ (366)	From *trans*-1-formyl-2-methoxy-carbonylcyclohex-4-ene;[13] also from PGA$_2$ methyl ester 15-acetate[14]	Potent, full TXA$_2$ agonist on platelets and vasculature Caused rapid irreversible aggregation of HPRP; $IC_{50} = 0.6$ μM Constricted isolated rat aorta, $CD_{50} = 0.4$ nM, and was a vasoconstrictor in vivo (i.v., guinea pig)[21]
11	Methyl 9α,11α-carba-15α-hydroxy-11α-thiathromba-(Z)5,(E)13-dienoate		$C_{22}H_{36}O_3S$ (380)	From 3-vinylcyclobutanone[15]	Contracted isolated rat aorta; $CD_{50} = 5 \times 20^{-7}$ g/mℓ; no effect on HPRP[15]
12	9α,11α-Carba-15α-hydroxy-11α-iminothromba-(Z)5,(E)13-dienoic acid		$C_{21}H_{35}NO_3$ (349)	From *cis*-3-formylcyclobutanol[17]	Contracted isolated rat aorta; $CD_{50} = 3 \times 10^{-8}$ g/mℓ; no effect on HPRP[17]
13	9α,11α-Epithia-15α-hydroxy-11α-thiathromba-(Z)5,(E)13-dienoic acid		$C_{20}H_{32}O_3S_2$ (384)	From 4,4-dimethoxy-2-(6-chloro-2-hexyn-1-yl)-acetoacetate[16]	Contracted isolated rat aorta, $CD_{50} = 7 \times 10^{-10}$ M; irreversible aggregation of HPRP; $IC_{50} = 4.3 \times 10^{-6}$ M[16]
14	9α,11α-Methanoepoxy-15α- and 15β-hydroxy-thromba (Z)5,(E)13-dienoic acid		$C_{21}H_{34}O_5$ (366)	From the product of methylenation of *trans-cis-cis-N,N*-dimethyl-2-benzyloxy-6-methoxy-tetrahydropyran-4-one-3-acetamide which was obtained from the Corey lactone[22,23]	Weak TXA$_2$ agonist; contraction of rabbit aortic strips; 0.04 × TXA$_2$; aggregation of rabbit PRP; 0.001 × TXA$_2$; did not antagonize aggregatory effect of TXA$_2$, nor was TXA$_2$ synthetase inhibited[22]
15	9α,11α-Methanoepoxy-10-oxa-11α-carba-15α- and 15β-hydroxy-thromba-5-*cis*-13-*trans*-dienoic acid		$C_{21}H_{34}O_5$ (366)	From *trans-cis-cis-N,N*-dimethyl-2,4-*bis*-hydroxymethyl-6-methoxytetrahydropyran-3-acetamide[22]	Contraction of rat aortic strips; 0.02 × TXA$_2$; did not aggregate rabbit PRP, antagonize the aggregatory effect of TXA$_2$, or inhibit TXA$_2$ synthetase[22]
16	9α,11α-Epoxy-10-homo-15α- and 15β-hydroxy-11α-carbathromba-(Z)5,(E)13-dienoic acid		$C_{22}H_{36}O_4$ (364)	From [4 + 3] adduct of furan and tetrabromoacetone[18]	No effect on blood platelets; weak vasoconstrictor activity[18]

Table 2 (continued)
TXA ANALOGS

No.	Systematic name	Structure	Mol. formula (mol. wt.)	Synthesis	Biological actions
17			$C_{22}H_{34}O_5(378)$	From 1-dimethoxymethyl-2-styryl-8-oxabicyclo [3.2.1] oct-6-en-3-ol[24]	Mixture of C-15 epimers has strong TXA_2-like activity[24]

REFERENCES

1. **Corey, E. J., Ponder, J. W., and Ulrich, P.,** Synthesis of a stable analog of thromboxane A$_2$ with methylene replacing the 9,11-bridging oxygen, *Tetrahedron Lett.,* 21, 137—140, 1980.
2. **Smith. P. A. S. and Baer, D. R.,** The Demjanov and Tiffeneau-Demjanov ring expansions, *Org. React.,* 11, 157—188, 1960.
3. **Maxey, K. M., and Bundy, G. L.,** The synthesis of 11α-carbathromboxane A$_2$, *Tetrahedron Lett.,* 21, 445—448, 1980.
4. **Ohuchida, S., Hamanaka, N., Hashimoto, S., and Hayashi, M.,** Synthesis of 9α-11α-thiathromboxane A$_2$ methyl ester, *Tetrahderon Lett.,* 23, 2883—2886, 1982.
5. **Nicolaou, K. C., Magolda, R. L., and Claremon, D. A.,** Carbocyclic thromboxane A$_2$, *J. Am. Chem. Soc.,* 102, 1404—1409, 1980.
6. **Ohuchida, S., Hamanaka, N., and Hayashi, M.,** Synthesis of thromboxane A$_2$ analog DL-(9,11), (11,12)-dideoxa-(9,11), (11,12)-dimethylene thromboxane A$_2$, *Tetrahedron Lett.,* 3661-3664, 1979; Synthesis of thromboxane A$_2$ Analogs. I., (±)dimethanothromboxane A$_2$, *Tetrahedron,* 39, 4257—4261, 1983.
7. **Ansell, M. F., Caton, M. P. L., and Stuttle, K. A. J.,** A simple approach to carbocyclic thromboxane A$_2$ from a cyclobutane, 23, 1955—1956, 1982.
8. **Nicolaou, K. C., Smith, J. B., and Lefer, A. M.,** Chemistry and pharmacology of a series of new thromboxane analogs, *Drugs of the Future,* 7, 331—340, 1982.
9. **Nicolaou, K. C., Magolda, R. L., Smith, J. B., Aharony, D., Smith, E. F., and Lefer, A. M.,** Synthesis and biological properties of pinane-thromboxane A$_2$, a selective inhibitor of coronary artery constriction, platelet aggregation, and thromboxane formation, *Proc. Natl. Acad. Sci. U.S.A.,* 76, 2566—2570, 1979.
10. **Ansell, M. F., Caton, M. P. L., Palfreyman, M. N., and Stuttle, K. A. J.,** Synthesis of structural analogues of thromboxane A$_2$, *Tetrahedron Lett.,* 4497—4498, 1979.
11. **Roth, D. M., Lefer, A. M., Smith, J. B., and Nicolaou, K. C.,** Anti-thromboxane A$_2$ actions of pinane thromboxane A$_2$ derivatives, *Prostaglandins, Leukotrienes Med.,* 9, 503—509, 1982.
12. **Wilson, N. H., Peesapati, V., Jones, R. L., and Hamilton, K.,** Synthesis of prostanoids with bicyclo[2.2.1]heptane, bicyclo[3.1.1]heptane, and bicyclo[2.2.2]octane ring systems. Activites of 15-hydroxy epimers on human platelets, *J. Med. Chem.,* 25, 495—500, 1982.
13. **Ohuchida, S., Hamanaka, N., and Hayashi, M.,** Synthesis of thromboxane A$_2$ analog DL-(9,11), (11,12)-dideoxa-(9,11)-epithio-(11,12)-methylene thromboxane A$_2$, *Tetrahedron Lett.,* 22, 1349—1352, 1981; Synthesis of thromboxane A$_2$ analogs. (±) Thiathromboxane A$_2$, *Tetrahedron,* 39, 4263—4268, 1983.
14. A convenient route to (+)-(9,11)-epithia-(11,12)-methano-thromboxane A$_2$ from prostaglandin E$_2$ methyl ester, *Tetrahedron Lett.,* 22, 5301—5302, 1981; Synthesis of thromboxane A$_2$ analogs. III. (±) Thiathromboxane A$_2$, *Tetrahedron,* 39, 4269—4272, 1983.
15. **Kosuge, S., Hamanaka, N., and Hayashi, M.,** Synthesis of thromboxane A$_2$ analog DL-(9,11), (11,12)-dideoxa-(9,11)-methylene-(11,12)epithio-thromboxane A$_2$ methyl ester, *Tetrahedron,* 1345—1348, 1983.
16. **Ohuchida, S., Hamanaka, N., and Hayashi, M.,** Synthesis of thromoboxane A$_2$ analogues: DL-9,11:11,12-dideoxa-9,11:11,12-diepithiothromboxane A$_2$, *J. Am. Chem. Soc.,* 103, 4597—4599, 1981; Synthesis of thromboxane A$_2$ analogs. I. (±) Dithiathromboxane A$_2$ sodium salt, *Tetrahedron,* 39, 4273—4280, 1983.
17. **Kosuge, S., Hayashi, M., and Hamanaka, N.,** Synthesis of thromboxane A$_2$ analog, DL-(9,11)-methano-(11,12)-amino thromboxane A$_2$, *Tetrahedron Lett.,* 23, 4027—4030, 1982.
18. **Ansell, M. F., Caton, M. P. L., and Mason, J. S.,** Synthesis of a stable analogue of thromboxane A$_1$ (±)-9α-Homo-(11,12)-deoxa-(11,12)-methylene thromboxane A$_1$, *Tetrahedron Lett.,* 22, 1141—1142, 1981.
19. **Jones, R. L., Peesapati, V., and Wilson, N. H.,** Antagonism of the thromboxane-sensitive contractile systems of the rabbit aorta, dog saphenous vein and guinea pig trachea, *Br. J. Pharmacol.,* 76, 423—438, 1982.
20. **Bounameaux, Y., Coffey, J. W., O'Donnell, M., Kling, K., Quinn, R. J., Schönholzer, P., Szente, A., Tobias, L. D., Tschopp, T., Welton, A. F., and Fischli, A.,** Synthese Stereisomerer Pinanthromboxane und Evaluation der Verbindungen als Plättchenaggregationsinhibitoren, *Helv. Chim. Acta,* 66, 989—1008, 1983.
21. **Katsura, M., Miyamoto, T., Hamanaka, N., Kondo, K., Terada, T., Ohgaki, Y., Kawasaki, A., and Tsuoboshima, M.,** *In vitro* and *in vivo* effects of new powerful thromboxane antagonists (3-alkylamino pinane derivatives), in *Advances in Prostaglandin, Thromboxane, and Leukotriene Research,* Samuelsson, B., Paoletti, R., and Ramwell, P., Eds., Raven Press, New York, 1983, 351—357.
22. **Schaaf, T. K., Bussolotti, D. L., Parry, M. J., and Corey, E. J.,** Synthesis of 11α,9α-epoxymethanothromboxane A$_2$; a stable optically active TXA$_2$ antagonist, *J. Am. Chem. Soc.,* 103, 6502—6505, 1981.
23. For a detailed discussion of the "Corey", "Stork" and other prostaglandin syntheses, see References 3a and 3b of the textual section of this review. See also, Chapter by Nicolaou and Petasis of this volume.
24. **Bowers, K. G. and Mann, J.,** Oxyallyls in synthesis: preparation of tricyclic thromboxane A$_2$ analogues, *Tetrahedron Lett.,* 26, 4412, 1985.

Table 3
PGI ANALOGS

No.	Systematic name	Structure	Mol. formula (mol. wt.)	Synthesis	Biological Actions
1	6,9α-Oxido-11α,15α-dihydroxyprosta-(Z)5,(E)13-dienoic acid (Epoprostenol)		$C_{20}H_{32}O_5$ (352)	Haloetherification of $PGF_{2\alpha}$ methyl ester or 11,15-protected derivatives by N-bromosuccinimide, or iodine followed by saponification;[1-4] the substance must be stored as a salt (sodium) or the methyl ester	Potential utility in cardiopulmonary bypass operations, hemoperfusion for kidney patients, prevention of persistent fetal circulation, treatment of peripheral vascular disease, preinfarctional angina, and stroke; i.v. infusion in man (2, 4, or 8 ng/kg/min) caused dose-dependent inhibition (ex vivo) of ADP-induced platelet aggregation which in some instances persisted for 2 hr after termination of infusion; these doses were accompanied by small transient decreases in BP and increases in HR[5] I.V. infusion (22 ± 11 ng/kg/min) to patients with severe congestive heart failure, resistant to digitalis and diuretics, caused decreases in pulmonary capillary wedge pressure, mean arterial pressure, systemic vascular resistance, pulmonary vascular resistance, and pulmonary arteriolar resistance; also caused increase in heart rate (slight), cardiac index, and stroke index[6]

In patients with chronic renal failure on dialysis, i.v. infusion (5 ng/kg/min) reduced heparin requirements, and prevented activation and consumption of platelets[7-8]

In patients with coronary artery disease i.v. infusion (8 ng/kg/min) caused increased HR and cardiac output and decreased systolic and diastolic BP, systemic vascular resistance, and coronary vascular resistance; at 6 ng/kg/min, pacing time to angina was prolonged in 60% of cases[7]

In patients undergoing coronary bypass surgery, infusion (2.5—20 ng/kg/min) produced potent arteriolar vasodilation with tachycardia[8]

In patients with pulmonary hypertension, infusion (2—16 ng/kg/min) decreased systemic and pulmonary vascular resistance and increased cardiac output.

Doses above 8 ng/kg/min caused nausea in all subjects[8]

In patients with Reynaud's disease infusion of 2.5—10 ng/kg/min over 72 hr caused subjective improvement lasting 4—28 weeks in most (21/24) cases and complete response for 6—22 weeks in some subjects (8/24)[8]

Table 3 (continued)
PGI ANALOGS

No.	Systematic name	Structure	Mol. formula (mol. wt.)	Synthesis	Biological Actions
2	6,9α-Oxido-11α,15α-dihydroxyprosta-(E)5,(E)13-dienoic acid		$C_{20}H_{32}O_5$ (352)	Dehydrobromination of major adduct from 5-*trans*-PGF$_{2\alpha}$ methyl ester and *N*-bromosuccinimide followed by saponification;[11] also obtainable from minor haloetherification product of PGF$_{2\alpha}$ methyl ester;[4,11,12] must be stored as salts or methyl ester	In patients undergoing coronary vein grafting, infusion of PGI$_2$ preserved platelet number and function, decreased blood loss by 50% and reduced heparin requirements[8] In vitro, inhibited ADP-induced aggregation of HPRP, IC$_{50}$ = 3×10^{-9} M Relaxed bovine coronary artery, ED$_{50}$ = 5.9 ng/mℓ[9] For reviews on PGI, see Refs. 9, 10
3	Methyl 3ξ-methyl-6,9α-oxido-11α,15α-dihydroxyprosta-(Z)5(E)13-dionoate		$C_{22}H_{36}O_5$ (380)	Bromo etherification, dehydrobromination of PGF$_{2\alpha}$ derivative[13]	Inhibited ADP-induced platelet aggregation (rat PRP); 0.14 × PGE$_1$ Very weak or no effect on BP (anesthetized dog), uterus (rat, i.v.), or coronary artery (rabbit)[13]
4	Methyl 4-keto-6,9α-oxido-11α,15α-dihydroxyprosta-(Z)5,(E)13-dienoate		$C_{21}H_{32}O_6$ (380)	Condensation of Corey lactol with intact α-chain with methyl 5-phenylsulfinyl-4-oxopentanoate[90]	Very weak inhibition of ADP-induced aggregation of rabbit PRP[90]

#	Compound	Formula (MW)	Structure	Synthesis	Activity
5	5ξ-Nitro-6ξ,9α-oxido-11α,15α-dihydroxy-prosta-(E)13-enoic acid	$C_{20}H_{33}NO_7$ (399)		Condensation of Corey lactol with methyl 5-nitropentanoate and subsequent saponification[100,129]	Extremely potent vasodilation ($1.2 \times PGI_2$) with no effect on blood platelets[100,129]
6	Methyl 5-methyl-6,9α-oxido-11α,15α-dihydroxyprosta-(Z)5 (E)13-dienoate	$C_{22}H_{36}O_5$ (380)		Synthesis analogous to that of 3-methyl isomer[13]	Inhibition of ADP-induced platelet aggregation (rat PRP); $1.5 \times PGE_1$. Vasodepression, anesthetized dog (i.v.); $0.4 \times PGE_1$ Uterine contraction (rat, i.v.); $5 \times PGE_1$ *In vitro* coronary vasodilation (rabbit); $0.15 \times PGE_1$[13]
7	(+) 5-Cyano-6,9α-oxido-11α,15α-dihydroxy-16ξ-methylprosta-(E)5,(E)13-dienoic acid (Nileprost)	$C_{22}H_{33}NO_5$ (391)		Reaction of 16-methyl PGE_2 methyl ester diacetate with chlorosulfonyl isocyanate followed by treatment with DMF and saponification[14] Stable oil	Gastric secretion inhibitor, bronchial dilation, and cytoprotective agent Inhibited histamine induced bronchial constriction in guinea pigs by aerosol ($18 \times PGI_2$) or i.v. ($60 \times PGI_2$) routes with a duration approx. 12 min Coughing was not induced in cats at doses several times therapeutic dose Inhibited development of indomethacin induced lesions in gastric mucosa of rat ($ED_{50} = 61$ µg/kg p.o.); gastric secretion was inhibited at much higher doses ($ED_{50} = 1.7$ mg/kg p.o.)[14] Inhibited ADP-induced aggregation of HPRP; $0.005 \times PGE_2$

Table 3 (continued)
PGI ANALOGS

No.	Systematic name	Structure	Mol. formula (mol. wt.)	Synthesis	Biological Actions
8	Methyl 5-phenylthio-6,9α-oxido-11α,15α-dihydroxyprosta-E(S)E(13)-dienoate		$C_{27}H_{38}O_5S$ (474)	Addition of phenylsulfenyl chloride (2 eq.) to PGE_2 methyl ester at $-78°C$. Stable in acid solution, even at pH 1.5 (1 hr. room temp.)[15]	Weak inhibitory activity on platelet aggregation; IC_{50} = 1.7 µg/ℓ (0.003 × PGI_2)[106]
9	Ethyl 5-aza-6,9α-oxido-11α,15α-dihydroxy-prosta-5,(E)13-dienoate		$C_{21}H_{35}NO_5$ (381)	From 5-aza-6-oxo-$PGF_{1α}$ ethyl ester 11,15-bis-THP; more stable than PGI_2, e.g., t½ at pH 4.5 = 11.5 hr[91]	Decreases systolic BP in rat by 17% at 1 mg/kg (p.o.) with a duration of 1 hr; no effect on rat or human PRP.[91]
10	Methyl 6,9α-oxido-7β-methyl-11α, 15α-dihydroxyprosta-(Z)5,(E)13-dienoate		$C_{22}H_{36}O_5$ (380)	From 7-methyl $PGF_{2α}$ methyl ester by the same process as used for 3-methyl compound (3)[13]	Inhibition of ADP-induced platelet aggregation (rat PRP); 2.1 × PGI_1[13] Vasodepression, anesthetized dog (i.v.); 1.7 × PGE_1[13]
11	Methyl 6,9α-oxido-7β-acetoxy-11α,15α-dihydroxyprosta-(Z)5,(E)13-dienoate		$C_{23}H_{36}O_7$ (424)	Addition of phenylsulfenyl chloride to protected PGI_2 methyl ester in the presence of triethylamine, followed by oxidation to the allylic sulfoxide; reaction of the sulfoxide with diethylamine-tris-dimethylaminophosphine gave the rearranged alcohol which was acetylated[16,106]	Weak inhibitor of platelet aggregation; more potent hypotensive agent; Free 7β-hydroxy compound not active.[16,106]

No.	Name	Structure	Synthesis	Activity
11a	6,9α-Oxido-7β-fluoro-11α,15α-dihydroxy-prosta-(Z)5,(E)13-dien-oic acid	$C_{20}H_{31}FO_5$ (369)	From 7β-hydroxy PGI$_2$ 11,15-diacetate methyl ester and diethylaminosulfurtrifluoride followed by alkaline hydrolysis; sodium salt stable at room temperature at pH 7.4 for more than 1 month[106,108]	Inhibited ADP-induced aggregation of rabbit PRP with IC$_{50}$ = 0.05 μg/mℓ (0.1 × PGI$_2$)[106,108]
12	(+) 6,9α-Oxido-7-keto-11α,15α-dihydroxy-prosta-(Z)5,(E)13-dien-oic acid	$C_{20}H_{30}O_6$ (366)	From Δ^6-isomer of PGI$_2$ methyl ester diacetate,[17] or from PGF$_{2α}$ methyl ester;[18] very stable at low and high pHs[18]	Inhibited ADP induced platelet aggregation (HPRP) with IC$_{50}$ = 15 ng/mℓ (PGI$_2$ = 1 ng/mℓ); strong activator of human platelet adenylate cyclase — 0.5 PGI$_2$; in cats, 6 μg/kg (i.v.) has same disaggregatory action as 2 μg/kg PGI$_2$. Relaxed isolated bovine coronary artery; 0.1 × PGI$_2$. I.V. infusion in anesthetized dogs elicited a fall in BP (0.1 × PGI$_2$), decreased cardiac ouput (0.05 × PGI$_2$), decreased total peripheral resistance (0.1 × PGI$_2$), decreased coronary resistance (at 1000 ng/kg/min), but had variable effects on heart rate[18]
13	6,9β-Oxido-8,12-*epi*-11α,15α-dihydroxy-prosta(Z)5,(E)13-dien-oic acid	$C_{20}H_{32}O_5$ (352)	Iodoetherification dehydroiodination of PGF$_2$ methyl ester derivative[12]	
14	6,9α-Oxido-10,10-difluoro-11α,15α-dihydroxyprosta-(Z)5-en-13-ynoic acid	$C_{20}H_{28}F_2O_5$ (386)	From *i*-butyl ether of cyclopentane-1,3-dione[19]	Inhibited platelet aggregation (HPRP) induced by collagen (0.1 × PGI$_2$) and 9,11-azo PGH$_2$; the inhibition was elicited via elevation of cAMP levels[20]

Table 3 (continued)
PGI ANALOGS

No.	Systematic name	Structure	Mol. formula (mol. wt.)	Synthesis	Biological Actions
15	Methyl 6,9α-oxido-15α-hydroxyprosta-(Z)5,10,(E)13-trienoate		$C_{22}H_{34}O_4$ (362)	Bromoetherification dehydrobromination of Δ^{10}-PGF methyl ester derivative[13]	In anesthetized dogs (i.v. bolus), equivalent to PGI_2 in lowering BP, decreasing peripheral resistance, and increasing renal blood flow; the in vivo effects are short lived ($\sim PGI_2$) even though the in vitro half-life is much greater ($150 \times PGI_2$) and the compound is not a substrate for 15-hydroxyprostaglandin dehydrogenase;[21] short duration of action in vivo apparently due to rapid hepatic inactivation and/or rapid diffusion from the sites of action[92]
16	6,9α-Oxido-11β,15α-dihydroxyprosta-(Z)5,(E)13-dienoic acid		$C_{20}H_{32}O_5$ (352)	Iodoetherification dehydroiodination of PGF methyl ester derivative[12]	Inhibition of ADP-induced platelet aggregation (rat PRP); $0.023 \times PGE_1$
17	6,9α-Oxido-11α-fluoro-15α-hydroxyprosta-(Z)5,(E)13-dienoic acid		$C_{20}H_{31}FO_4$ (354)	Reaction of 15-acetate methyl ester of PGI_2 with morpholinosulfur trifluoride followed by hydrolysis[22]	Increased BP (rat, i.v.); little effect on rat uterus or isolated rabbit coronary artery[13]

No.	Name	Structure	Formula (MW)	Synthesis	Activity
18	Methyl 6,9α-oxido-15α-hydroxyprosta-(Z)5,11,(E)13-prostatrienoate		$C_{21}H_{32}O_4$ (362)	Iodoetherification dehydroiodination of Δ_{11}-PGF derivative[13]	Inhibited ADP-induced aggregation of rat PRP; 0.003 × PGI$_2$; contracted isolated rat coronary artery[13]
19	6,9α-Oxido-11α,15α-dihydroxy-12α-fluoro-prosta-(Z)5,(E)13-dienoic acid		$C_{20}H_{31}FO_5$ (370)	Iodoetherification, dehydroiodination of corresponding PGF$_{2\alpha}$ methyl ester followed by saponification[23]	Equipotent with PGI$_2$ with regard to inhibition of ADP-induced aggregation of HPRP and dilation of isolated cat coronary artery[23]
20	6,9α-Oxido-11α,15α-dihydroxy-13-oxa-prosta-(Z)5-enoic acid		$C_{19}H_{32}O_6$ (356)	Iodoetherification, dehydroiodination of corresponding F derivative[24]	Inhibited ADP-induced platelet aggregation (rabbit); 1 × PGI$_2$ Hypotensive; 0.01 × PGI$_2$[24]
21	6,9α-Oxido-11α,15α-dihydroxy-13-thia-prosta-(Z)5-enoic acid		$C_{19}H_{32}O_5S$ (372)	Synthesis analogous to that of the 13-oxa derivative (20)[24]	Inhibited ADP-induced aggregation of rabbit platelets; 0.04 × PGI$_2$ Hypotensive; 0.01 × PGI$_2$[24]
22	(+) 6,9α-Oxido-11α,15α-dihydroxy-20-methylprosta-(Z)5-en-13-ynoic acid (K-13,415)		$C_{21}H_{32}O_5$ (362)	Iodocyclization dehydroiodination of corresponding F derivative as the carboxylic acid[25] Less stable in solution than PGI$_2$[36]	Inhibited ADP-induced aggregation of rabbit PRP (4 × PGI$_2$) in vitro, and deaggregation of cat heparinized blood in vivo (3.3 × PGI$_2$, i.v. infusion); relaxed isolated bovine coronary artery; 8.5 × PGI$_2$ Hypotensive in cats; 3 × PGI$_2$ (i.v. infusion);[36] more effective stimulant of cAMP than PGI$_2$[25]
23	6,9α-Oxido-11α,15α-dihydroxyprosta-(Z)5-enoic acid		$C_{20}H_{34}O_5$ (354)	Prepared from PGF derivative by iodoetherification dehydroiodination[12]	Inhibition of ADP-induced platelet aggregation (HPRP); 0.4 × PGI$_2$ Rat blood pressure (anesthetized i.v.); ~0.2 × PGI$_2$[28]

Table 3 (continued)
PGI ANALOGS

No.	Systematic name	Structure	Mol. formula (mol. wt.)	Synthesis	Biological Actions
24	6,9α-Oxido-11α,15α-dihydroxy-14-fluoro-prosta-(Z)5,(Z)13-dienoic acid		$C_{20}H_{31}FO_5$ (370)	Iodoether formation on methyl ester of F_2 derivative followed by elimination of hydrogen iodide and saponification[26]	Equivalent to PGI_2 with regard to inhibition of ADP-induced aggregation of HPRP and dilation of isolated cat coronary artery[26]
25	6,9α-Oxido-11α,15α-dihydroxy-15-cyclohexyl-ω-pentanorprosta-(Z)5,(E)13-dienoic acid		$C_{21}H_{32}O_5$ (364)		Inhibition of ADP-induced platelet (HPRP) aggregation; 0.15 × PGI_2. Anesthetized rat vasodepression (i.v.); 0.85 × PGI_2; for methyl ester, duration of vasodepression (18 min) was 3 × greater than that of PGI_2 methyl ester; relaxation of isolated bovine coronary artery; 0.28 × PGI_2[9,27]
26	6,9α-Oxido-11α,15α-dihydroxy-16-(thiophen-3-yloxy)-ω-tetranor-prosta-(Z)5,(E)13-dienoic acid		$C_{20}H_{26}O_6S$ (394)		Inhibition of ADP-induced platelet aggregation (HPRP); 0.6 × PGI_2. Anesthetized rat vasodepression (i.v.); 0.92 × PGI_2; Relaxation of isolated bovine coronary artery; 1.3 × PGI_2[9,27]
27	6,9α-Oxido-11α,15α-dihydroxy-16,16-dimethyl-prosta-(Z)5,(E)13-dienoic acid		$C_{22}H_{36}O_5$ (380)	Iodoetherification, dehydroiodination of $F_{2α}$ derivative followed by saponification[12]	Inhibition of ADP-induced aggregation of HPRP; 0.03 × PGI_2; hypotensive in anesthetized rat (i.v.); 0.01 × PGI_2[28]

No.	Structure	Name	Formula (MW)	Synthesis	Biological activity
28		6,9α-Oxido-11α,15α-di-hydroxyprosta-(Z)5,(E)13,Z(16)-tri-enoic acid	$C_{20}H_{30}O_5$ (350)	Iodoetherification, dehydroiodination of $PGE_{3\alpha}$ methyl ester followed by saponification Compound isolated as sodium salt[29]	Equivalent to PGI_2 with regard to inhibition of ADP-induced aggregation of HPRP[29]
29		6,9α-Oxido-11α,15α-dihydroxy-17,17-dimethylprosta-(Z)5,(E)13-dienoic acid	$C_{22}H_{36}O_5$ (380)	From $PGF_{2\alpha}$ derivative by same sequence as described above for 16,16-diemthyl derivative (27)[12]	Inhibition of ADP-induced platelet aggregation (HPRP); $0.33 \times PGI_2$ Rat blood pressure (anesthetized, i.v.); $\sim 0.4 \times PGI_2$[28]
30		6,9α-Oxido-11α,15α-dihydroxy-20-methyl-prosta-(Z)5,(E)13-dienoic acid	$C_{21}H_{34}O_5$ (366)	Iodoetherification, dehydroiodination of 20-methyl PGF analog;[30] more stable in solution than PGI_2[36]	Inhibited ADP-induced aggregation of rabbit ($1 \times PGI_2$)[36] and rat (2—$3 \times PGI_2$)[30] PRP; equal to PGI_2 in deaggregation of cat heparinized blood in vivo (i.v. infusion)[36] Relaxed isolated bovine coronary artery ($5 \times PGI_2$)[36] and increased coronary blood flow in isolated perfused rat heart ($3 \times PGI_2$)[30] Equal to PGI_2 as a hypotensive agent in anesthetized cats (i.v. infusion)[36]
31		6,9α-Oxido-2,3,4-trinor-1,5-inter-m-phenylene-11α,15α-dihydroxy-15β-methylprosta-(Z)5,(E)13-dienoic acid (SE 51)	$C_{24}H_{32}O_5$ (400)	Iodoetherification, dehydroiodination of $PGF_{2\alpha}$ methyl ester derivative and subsequent saponification; stable for days at neutral pH[31]	Inhibited arachidonic-induced aggregation of HPRP; IC_{50} = 0.21 μmol/ℓ Stimulated cAMP formation 3-fold at 1 μmol/ℓ (washed horse platelets)

Table 3 (continued)
PGI ANALOGS

No.	Systematic name	Structure	Mol. formula (mol. wt.)	Synthesis	Biological Actions
32	6,9α-Oxido-2,3,4-trinor-1,5-inter-*m*-phenylene-11α,15α-dihydroxy-15-cyclohexyl-ω-pentanor-prosta-(Z)5,(E)13-dien-oic acid (CG 4203)		$C_{24}H_{30}O_5$ (398)	Iodoetherification, dehydroiodination of $PGF_{2\alpha}$ methyl ester derivative and subsequent saponification[31]	Reduced diastolic BP by 20% in SH rats at 0.05 mg/kg (i.v.); maximum BP reduction in anesthetized rats observed at 1 mg/kg i.v. (80—90 min), which is maintained for about 10 min at this dose and then declines to nonsignificant levels in about 30 min[31] Inhibited arachidonic-induced aggregation of HPRP; IC_{50} = 0.017 μmol/ℓ Stimulated cAMP formation 2-fold at less than 0.1 μmol/ℓ Reduced diastolic BP in SH rats by 20% at 0.01 mg/kg i.v.; in anesthetized rats an i.v. bolus dose of 1 mg/kg gave maximum lowering of diastolic BP which declined very slowly and was still observable after 1 hr

33	Methyl 6,9α-oxido-11α,15α-dihydroxy-prosta-6,(E)13-dienoate		C$_{21}$H$_{34}$O$_5$ (366)	Acetic acid methanolysis of PGI$_2$ methyl ester gave a methoxy lactol which on pyrolysis at 180°C in HMPA lost the elements of methanol Also reaction of 6-oxo-PGF$_{1α}$ with excess trimethylsilyldiethylamine followed by removal of the silyloxy groups with methanolic potassium carbonate[32]	In anesthetized rats 1 mg/kg i.v. gave profound ex vivo inhibition of ADP induced aggregation with t$_{1/2}$ = 37 min;[31] in man, i.v. infusion of the max tolerated dose (160 ng/kg/min for 0.5 hr) resulted in ex vivo inhibition of ADP-induced aggregation by 56% for 0.5—1.5 hr and was accompanied by an increase in heart rate (23%), facial flushing and headache[107]
34	6,9α-Oxido-11α,15α-dihydroxyprosta-(E)4,6,E(13)-trienoic acid		C$_{20}$H$_{30}$O$_5$ (350)	Acid promoted cyclization of 6-oxo-Δ^4-PGF methyl ester 11,15-diacetate followed by methanolysis and saponification; more stable than PGI$_2$[35]	Inhibited ADP-induced rat platelet aggregation; 11.7 × PGE$_2$[32] Nearly equipotent with PGI$_2$ with regard to inhibition of AA-induced aggregation of rabbit PRP[35]
35	5-Keto-6,9α-oxido-11α,15α-dihydroxy-prosta-6,(E)13-dienoic acid		C$_{20}$H$_{30}$O$_6$ (366)	Bromoetherification of 5-hydroxy-6,7-didehydro PGF$_{1α}$-11,15-diacetate, followed by Collins oxidation,[103] elimination of HBr, and hydrolysis[33]	Weak inhibitor of rabbit platelet aggregation, 0.01 × PGE$_2$[33]

Table 3 (continued)
PGI ANALOGS

No.	Systematic name	Structure	Mol. formula (mol. wt.)	Synthesis	Biological Actions
36	Methyl 6,9α-oxido-7-phenylthio-11α,15α-dihydroxyprosta-6,(E)13-dienoate		$C_{27}H_{38}O_5S$ (474)	Addition of phenylsulfenyl chloride (1 eq) to PGI_2 methyl ester at $-78°C$, gave a mixture containing the 7-phenylthio compound Stable in acid solution even at pH 1.5 (1 hr, room temp.)[15]	Weak inhibitor of platelet aggregation[15]
37	6β,9α-Oxido-11α,15α-dihydropxyprosta-7,(E)13-dienoic acid		$C_{20}H_{32}O_5$ (352)	From Corey lactone containing protected ω-chain and phenylselenyl group at C-8[34]	
38	6α,9α-Oxido-11α,15α-dihydroxyprosta-7,(E)-13-dienoic acid		$C_{20}H_{32}O_5$ (352)	Synthesis analogous to that of the 6β-epimer (38)[44]	
39	6,9-Oxido-11α,15α-dihydroxyprosta-6,$\Delta^{8,9}$(E)13-trienoic acid		$C_{20}H_{30}O_4$ (334)	From the Corey lactone via elaboration of the furan ring and α-chain first[93]	
40	Methyl 5,9α-carba-7-oxa-11α,15ξ-dihydroxy-prosta-(Z)5,(E)-dienoate		$C_{21}H_{34}O_5$ (366)	Iodoetherification, dehydrohalogenation of the corresponding "PGF like" compound[87]	Much weaker inhibitor of ADP-induced aggregation of rabbit PRP than PGI_2[87]

41	6β,9α-Oxido-11α,15α-dihydroxyprosta-(E)13-enoic acid		C$_{20}$H$_{34}$O$_5$ (354)	Reaction of PGF$_{2\alpha}$ methyl ester with phenylselenyl chloride gave a mixture of epimeric phenylselenyl ethers, one of which was converted into title compound on reduction with tri-*n*-butyltin hydride[37] For other syntheses see Ref. 3,12,38—40	Inhibited histamine-induced gastric acid secretion in the gastric fistula dog (0.07 × PGI$_2$, i.v. infusion) Inhibited pentagastrin-stimulated gastric acid secretion in the anesthetized rat (0.06 × PGI$_2$, i.v. infusion). Protected gastric mucosa of rats against indomethacin-induced erosions (1.4 × PGI$_2$, s.c.); considerably less hypotensive than PGI$_2$ Inhibition of ADP-induced aggregation of HPRP; 0.004 × PGI$_2$[41,42]
42	6α,9α-Oxido-11α,15α-dihydroxyprosta-(E)13-enoic acid		C$_{20}$H$_{34}$O$_5$ (354)	Obtained as one of the products in the processes used to synthesize 6β-isomer (*41*)	Inhibited indomethacin-induced gastric erosion in rat (0.7 × PGI$_2$, s.c.) Inhibited ADP-induced aggregation of HPRP; 0.001 × PGI$_2$[41,42]
43	6β,9α-Oxido-11α,15α-dihydroxyprosta-(E)4,E(13)-dienoic acid		C$_{20}$H$_{32}$O$_5$ (352)	Reaction of PGE$_{2\alpha}$ methyl ester with phenylselenyl chloride gave a mixture of epimeric phenylselenyl compounds, one of which was converted into the title compound after hydrogen peroxide oxidation and saponification[37] For other syntheses see Ref. 3,43	Inhibited aggregation of ADP-induced HPRP (0.001 × PGI$_2$)[94]
44	6β,9α-Oxido-11α,15α-dihydroxyprosta-(E)-13-en-4-ynoic acid		C$_{20}$H$_{30}$O$_5$ (350)	From addition product of Corey lactone with intact ω-chain and lithium-ω-*t*-butyldimethylsilyloxy-1-pentyne[43]	

Table 3 (continued)
PGI ANALOGS

No.	Systematic name	Structure	Mol. formula (mol. wt.)	Synthesis	Biological Actions
45	Methyl 6β,9α-oxido-4-thia-11α,15α-dihydroxyprosta-(E)-13-enoate		$C_{20}H_{34}O_5$ (386)	One of the products from addition of methoxycarbonylethylsulfenyl chloride to PGF derivative with allyl group at C-8 instead of normal 7-carbon fragment[45]	Cytoprotective; ethanol-induced gastric lesions[45]
46	6β,9α-Oxido-5β,11α,15α-trihydroxy-prosta-(E)13-enoic acid		$C_{20}H_{32}O_6$ (368)	Acetoxymercuration of $PGF_{2\alpha}$ methyl ester and subsequent conversion to the chloromercuri derivatives gave two isomeric compounds, each of which, on oxygenation with molecular oxygen in DMF in the presence of sodium borohydride, gave two isomeric hydroxy methyl esters. Saponification of the esters gave the free acids, one of which is the title compound[44]	
47	6β,9α-Oxido-11α,15α-dihydroxy-16,16,20-tri-methyl-17-oxaprosta-(E) 13-enoic acid (K-11907)		$C_{22}H_{38}O_6$ (398)	Haloetherification dehydrohalogentation of PGF derivative with ω-chain absent, followed by attachment ω-chain[46]	Proaggregatory, rabbit PRP; 0.01 × 11,9-epoxymethano PGH_2. Contraction of isolated rat mesenteric artery; 0.002 × 11,9-epoxymethano PGH_2. Contraction of isolated rat colon; 0.007 × PGE_2[46]

48	5,9α-Oxido-11α,15α-dihydroxyprosta-(Z)4,E-(13)-dienoic acid		$C_{20}H_{32}O_5$ (352)	Iodoetherification, dehydroiodination of Δ⁴-PGF derivative followed by saponification[47]	Similar potency to PGI₂ for inhibition of HPRP aggregation and about 0.5 × PGI₂ in lowering of rat BP[48]
49	5β,9α-Oxido-11α-15α-dihydroxyprosta-(E)13-enoic acid		$C_{20}H_{34}O_5$ (354)	Iodoetherification and subsequent tri-*n*-butyltin hydride reduction of Δ⁴-PGF derivative; the 6α,6β mixture was separated and then each saponified; also can use mercuric acetate and soidum borohydride for the cyclization reduction sequence[47]	About 0.5 × PGI₂, with regard to inhibition of aggregation of HPRP and in lowering rat BP[48]
50	5α,9α-Oxido-11,15α-dihydroxy-16-phenoxy-ω-tetranorprosta-(E)13-enoic acid		$C_{22}H_{30}O_6$ (360)		Inhibited pentagastrin-stimulated gastric acid secretion in anesthetized rats on i.v. infusion; ID₅₀ = 0.1 µg/kg/min Secretion returned to normal levels 40 min after infusion ceased. Inhibited indomethacin-induced gastric erosions in rats (s.c.); ID₅₀ = 4.5 µg/kg (120 × PGI₂) Inhibited indomethacin-induced intenstinal lesions in rats orally; ID₅₀ = 250 µg/kg[41] Inhibited histamine-stimulated gastric acid secretion in consious gastric fistula dog (i.v. infusion); 2 × PGI₂ Gastric acid levels returned to normal 0.5 hr after termination of infusion[49]

Table 3 (continued)
PGI ANALOGS

No.	Systematic name	Structure	Mol. formula (mol. wt.)	Synthesis	Biological Actions
51	6a-Homo-6a,9α-oxido-11α,15α-dihydroxy-prosta-(Z)5,(E)13-dien-oic acid		$C_{21}H_{34}O_5$ (366)	Wittig[104] reaction on bicyclic ketone which in turn was obtained from the Corey lactone by a ring expansion process[50,66]	Inactive as inhibitor of ADP-induced platelet aggregation; weak hypotensive activity in the rat[50]
52	6,9α-Carba-11α,15α-dihydroxyprosta-(E)5,(E)13-dienoic acid (carbacyclin)		$C_{21}H_{34}O_4$ (350)	From bicyclo[3.3.0]octan-3,7-dione[51] For other syntheses see Ref. 52, 53, 66	I.V. infusion (20—80 mg/kg/min) for 1 hr to healthy volunteers caused ex vivo inhibition of ADP-induced platelet aggregation (12.5—100%); usually, aggregation returned to baseline within 1.5 hr of termination of infusion, but in two cases the aggregation (87.5%) was prolonged (>5 hr, but back to normal at 24 hr); in two patients given a single 25 mg oral dose, there was significant inhibition at 30 and 90 min (maximum of 59 and 77%, respectively) after administration. Both showed an increase in heart rate (34—44%), one had severe headache, flushing of face and extremities (5 min after dose) as well as an increase in systolic (100—142) and diastolic (72—85) blood pressure[54]

53   6,9α-Carba-12β-methyl-11α,15α-dihydroxy-prosta-(Z)5,(E)13-dien-oic acid (Ciprostene)

$C_{22}H_{36}O_4$ (364)

Addition of lithium dimethyl copper to the appropriate bicyclic enone containing protected ω-chain, followed by addition of α-chain[101]

Inhibited platelet aggregation in vitro, induced by ADP or collagen in human, dog, or rabbit plasma; 0.03 × PGI_2;

Inhibited ADP-induced aggregation (ex vivo) in anesthetized dogs (i.v. infusion), and conscious or anesthetized rabbits (i.v. infusion); in dogs the inhibition was accompanied by minimal changes in HR and BP; aggregation is not apparent 10 min after the infusion is stopped; thus, although chemically stable, this compound is metabolized at about the same rate as PGI_2;

This analog of PGI_2 is a substrate for 15-hydroxy PG dehydrogenase[52];

Inhibition of ADP-induced aggregation of HPRP; 0.02 × PGI_2[101]

I.v. infusion to the anesthetized monkey caused hypotension, tachycardia, and inhibition of ex vivo ADP-induced platelet aggregation; potency as a hypotensive and antiaggregatory agent was 72 times less than PGI_2 but the duration of the hypotensive response was very similar;[109] in humans, i.v. infusion caused a comparable degree of tachycardia, decrease in pre-ejection period, pre-ejection/left ventricular ejection

Table 3 (continued)
PGI ANALOGS

No.	Systematic name	Structure	Mol. formula (mol. wt.)	Synthesis	Biological Actions
					time ratio, and inhibition of ADP-induced platelet aggregation to PGI_2 but with 0.01 times the potency; headache, facial flushing, nasal stuffiness, abdominal discomfort and nausea were observed as side effects.[110]
54	6,9α-Carba-12β-ethynyl-11α,15α-dihydroxy-prosta-(Z)5,(E)13-dien-oic acid		$C_{23}H_{34}O_4$ (374)	Conjugate addition of trimethyl-silylethynyl lithium in the presence of diethylaluminum and a catalytic amount of nickel 2,4-pentanedionate and diisobutylaluminum hydride, to the appropriate bicyclic enone with a protected ω-chain, followed by desilylation and introduction of α-chain[101]	Inhibition of ADP-induced aggregation of HPRP; 2 × PGI_2[101]
55	6,9α-Carba-12β-cyano-11α,15α-dihydroxy-prosta-(Z)5,(E)13-dien-oic acid		$C_{22}H_{33}NO_4$ (375)	Conjugate addition of cyanide ion to the appropriate bicyclic enone with intact protected ω-chain, followed by introduction of α-chain[101]	Inhibition of ADP-induced aggregation of HPRP; 0.1 × PGI_2[101]
56	6,9α-Carbo-11-thia-15α-hydroxyprosta-(Z)5,(E)13-dienoic acid		$C_{20}H_{32}O_3S$ (352)	From 7-thiabicyclo [3.3.0]octan-3-one[95]	Very weak antiaggregatory activity[95]

	Name	Formula (MW)	Synthesis	Biological activity
57	4-Benzamidophenoxy 6,9α-carba-11α-15α-dihydroxy-13-thia-15-phenyl-ω-pentanor-prosta-(E)5-enoate (EMD 46335)	$C_{34}H_{37}NO_5S$ (571)	Wittig reaction on bicyclic ketone with attached ω-chain[102]	Highly active in lowering BP in renal hypertensive dogs by the oral route[102]
58	6,9α-Carba-11α, 15α-dihydroxyprosta-(E)5-en-13-ynoic acid (FCE 21258)	$C_{21}H_{32}O_4$ (348)		Inhibited ADP induced aggregation of rabbit PRP; 0.09 × PGI_2; in the guinea pig co-administration (i.v.) with ADP caused an inhibition of aggregation considerably less than PGI_2; the effect was no longer observable after 30 min. Relaxed isolated bovine coronary artery; 0.2 × PGI_2. Reduced BP in normotensive (0.02 × PGI_2) and SH (0.06 × PGI_2) rats (i.v.)[55]
59	6,9α-Carba-11α,15α-dihydroxy-20-methyl-prosta-(E)-5-en-13-ynoic acid (FCE 21292)	$C_{22}H_{34}O_4$ (362)		Inhibited aggregation of rabbit PRP induced by ADP; 0.11 × PGI_2; in the guinea pig co-administration (i.v.) with ADP caused inhibition of aggregation caused by the latter with a potency similar to PGI_2 with a duration of action longer than 30 min. Relaxed bovine coronary artery; 3.5 × PGI_2. Reduced BP in normotensive (0.08 × PGI_2) and SH (0.16 × PGI_2) rats (i.v.)[55]

Table 3 (continued)
PGI ANALOGS

No.	Systematic name	Structure	Mol. formula (mol. wt.)	Synthesis	Biological Actions
60	6,9α-Carba-11α,15α-dihydroxy-15-cyclopentyl-ω-pentanorprosta-(E)5,(E)13-dienoic acid (ONO 41483)		$C_{21}H_{32}O_4$ (348)		Inhibits ADP-induced platelet aggregation (HPRP; 5 × PGE_1; baboon PRP; 6.8 × PGE_1) in vitro; orally active in baboons (ED_{50} × 87 μg/kg) with duration up to 4 hr; no effect on HR or BP at least up to 100 μg/kg p.o., but is hypotensive i.v. (1.45 × PGE_1) Antagonized vasopressin-induced ECG changes in monkeys, indicative of reversal of coronary arterial spasm, within 2 min of an oral dose of 3.10 μg/kg; potentially useful in therapy of myocardial infarction[56] In man, administration of the maximum tolerated doses by i.v. infusion (2.5 ng/kg/min for 1 hr) and p.o. (200 μg) resulted in ex vivo inhibition of ADP-induced platelet aggregation of 27 and 56%, respectively; higher doses by i.v. infusion (5 and 10 ng/kg/min for 1 hr) and oral (400 μg) routes caused marked aggregation inhibition (39—100%), but was accompanied by flushing of face and extremities, headache, and phlebitis[57]

61   6,9α-Carba-11α,15α-dihydroxy-16ξ-methyl-prosta-(E)5,(E)13-dien-18-ynoic acid (Ciloprost, Iloprost, ZK 36374)

$C_{22}H_{32}O_4$ (360)

Reaction of Corey lactone with ethyl lithioacetate, and transformation of this product into a bicyclic ketone which was converted into the title compound by a "Corey like" sequence[58]

Inhibited ADP-induced aggregation of PRP of human (17 × PGI_2), rat (0.3 × PGI_2), cat, dog, and monkey;[59,60] I.V. infusion in normotensive and SH rats caused a prolongation of bleeding time (potency = 0.3 × PGI_2) and a prolonged drop in BP; similar results were obtained orally but at much higher doses (1—3 mg/kg);[60] inhibited pentagastrin stimulated gastric secretion in anesthetized rats (1 × PGI_2, i.v. infusion) and in pylorus ligated conscious rats (ED_{50} = 6.4 mg/kg, p.o.).

Prevented indomethacin-induced gastric lesions in the rat (ED_{50} = 60 μg/kg, p.o.; about 0.2 × PGE_2)[61]

I.v infusion in normal volunteers over 4 hr at doses up to 2 ng/kg/min caused decrease in rate of ex vivo ADP induced platelet aggregation of PRP at 2 and 4 hr without significant effect on BP and HR; side effects observed were facial flushing, headache, jaw trismus and nausea, especially at the highest dose[111]

Table 3 (continued)
PGI ANALOGS

No.	Systematic name	Structure	Mol. formula (mol. wt.)	Synthesis	Biological Actions
62	6,9α-Carba-2,3,4-trinor-1,5-inter-*m*-phenylene-11α,15α-dihydroxy-15-cyclohexyl-ω-pentanor-prosta-(E)5,(E)13-dien-oic prostadienoic acid (CG 4305)		$C_{25}H_{32}O_4$ (396)	Wittig reaction on a bicy-clo[3.3.0]octan-3-one deriva-tive, containing ω-chain at C-6 and OH at C-7, followed by saponification[31]	In 9 patients with advanced ob-literative arterial disease, i.v. infusion (0.5—8 ng/kg/min) for 72 hr caused a marked va-sodilation and compensatory increase in cardiac output; glo-merular filtration was increased by 45%, tubular reabsorption of sodium and water were re-duced by 80 and 107% respec-tively and urine excretion rate increased by 122%[112] I.v. infusion (2—6 ng/kg/min) in 6 patients with severe coro-nary artery disease caused chest pain and ST segment depression in 4, suggesting precipitation of myocardial is-chemia in these individuals[113] Inhibited arachidonic acid in-duced aggregation of HPRP; $IC_{50} = 0.07$ μg/mℓ; stimulated cAMP formation 3-fold at less than 0.1 μmol/ℓ (washed horse platelets).

No.	Name	Structure	Formula (MW)	Preparation	Activity
63	6,9α-Carba-11α-,15α-dihydroxyprosta-6,E(13)-dienoic acid		$C_{21}H_{34}O_4$ (350)	From a bicyclo[3.3.0]-3-one derivative containing the future ω-chain and 11-OH in protected forms[62]	Very weak inhibitor of collagen-induced platelet (HPRP) aggregation[62]
64	6,9α-Carba-11α,15α-dihydroxyprosta-Δ[6,6a],(E)13-dienoic acid		$C_{21}H_{34}O_4$ (350)	Deoxygenation of corresponding 6,6a-oxirane[96] or from 6a,β-trimethylsilyl-7α-hydroxy compounds[97]	Inhibited ADP-induced aggregation of rabbit PRP (0.1 × PGI₂); equipotent to PGE₁ as a cytoprotective agent[96]
65			$C_{23}H_{30}O_4$ (370)	From 1-hydroxymethyl-6-*anti*-benzymoxymethyl-bicyclo[3.3.0]octa-2,7-diene[64]	No PGI₂-like activity[64]

Reduced diastolic BP in SH rats by 20% at 0.09 mg/kg (i.v.); in anesthetized rats, i.v. bolus does of 1 mg/kg gave maximum lowering of diastolic BP which declined very slowly and was still observable after 1 hr; in anesthetized rats, i.v. bolus dose 1 mg/kg gave maximum lowering of diastolic BP which declined vey slowly and was still observable after 1 hr; in anesthetized rats, 1 mg/kg (i.v.) gave profound ex vivo inhibition of ADP-induced platelet aggregation with $t_{1/2} = 113$ min[31]

Table 3 (continued)
PGI ANALOGS

No.	Systematic name	Structure	Mol. formula (mol. wt.)	Synthesis	Biological Actions
66	6,9α-Carba-11α,15α-dihydroxyprosta(Z)4,(E)13-dienoic acid		$C_{21}H_{34}O_4$ (350)	Wittig olefination of corresponding bicyclic ketone[63]	Marginally active as inhibitor of ADP-induced platelet aggregation (HPRP) Hypotensive (rat); $1 \times PGE_1$[63]
67			$C_{23}H_{32}O_5$ (388)	From the alkylation product of the corresponding tricyclic phenol with methyl bromoacetate[65]	Potent inhibitor of platelet aggregation; $2 \times$ carbacyclin[65]
68			$C_{23}H_{30}O_5$ (386)	Methodology analogous to that described above for the saturated congener (68)[65]	Potent inhibitor of platelet aggregation[65]
69			$C_{23}H_{30}O_5$ (386)	Methodology analogous to that described above for the isomer with 1,2,3-trisubstitution of the aromatic ring[65]	No PGI_2-like activity[65]
70	11α-15α-Dihydroxy-1a-homo-6,9α-prosta-(Z)-and-(E)5,(E)13-diene-1-carboxylic acid		$C_{21}H_{34}O_4$ (350)	Wittig reaction on corresponding bicyclic ketone with appropriately functionalized ω-chain[66]	Inhibited collagen-induced platelet aggregation $ED_{50} = 0.8$ µg/mℓ[66]

71	Methyl 6ξ,9α-oxido-11-keto-12-aza-15ξ-hydroxyprostanoate		$C_{20}H_{35}NO_5$ (369)	Reaction of Δ^5-9-hydroxy-11-one with mercuric acetate and subsequent demercuration with sodium borohydride[78]	Potent inhibitor of ADP-induced platelet aggregation (HPRP)[78]
72	Methyl 6,9α-oxido-11α,15α-dihydroxy-13-aza-13-methylprosta-(Z)5,(E)-dienoate		$C_{21}H_{37}NO_5$ (383)	From 2-*endo*-methoxy-7-*syn*-(N-benzyl-N-methylamino) bicyclo[2.2.1]heptan-5-one[98]	Did not inhibit aggregation of HPRP induced by collagen or ADP at 10 μg/mℓ[98]
73	6-Aza-6,9α-carba-5-keto-11α,15α-dihydroxy-prosta-(E)13-enoic acid		$C_{20}H_{35}NO_5$ (369)	From the butadiene-maleic anhydride Diels-Alder adduct[67]	Weak inhibitor of platelet aggregation[67]
74	6-Aza-6,9α-carbonyl-7-keto-11α,15α-dihydroxyprosta-(E)13-enoic acid		$C_{20}H_{31}NO_6$ (381)	From the butadiene-maleic anhydride Diels-Alder adduct[67]	Weak inhibitor of platelet aggregation[67]
75	9α-Nitrilo-11α,15α-dihydroxyprosta(E)-enoic acid		$C_{20}H_{33}NO_4$ (351)	Thermolysis (70—80°) of the 9α-azido $F_{2\alpha}$ derivative[79]	Equal to PGI_2 with regard to inhibition of ADP or PGH_2 induced aggregation of HPRP;[79] in conscious newborn lambs injection (branch pulmonary artery) decreased vascular resistance (threshold dose; 1 μg/kg for normoxic or hypoxic animals) and caused pulmonar vasodilation in both normoxic (threshold dose; 10 μg/kg) and hypoxic (threshold dose 3 μg/kg) animals; like PGI_2, vascular resistance in injected lung was decreased more than systemic resistance; the title compound was less active than PGI_2 in this species[88]

**Table 3 (continued)
PGI ANALOGS**

No.	Systematic name	Structure	Mol. formula (mol. wt.)	Synthesis	Biological Actions
76	Methyl 5-thia-9α,6-ni-trilo-11α,15α-dihydrox-yprosta-(E)13-enoic acid (HOE 892)		$C_{20}H_{33}NO_4S$ (383)	Alkylation of "Corey like" bi-cylic thiolactam, containing protected ω-chain and 11-OH, with methyl 4-bromobutyrate; very stable in acid solution[68]	Inhibited aggregation of rabbit PRP induced by collagen or ADP (0.1 × PGI$_2$); in rabbits oral administration caused ex vivo inhibition of platelet ag-gregation induced by AA (ID$_{50}$ = 1.5 mg/kg) and collagen (ID$_{50}$ = 0.2 mg/kg) which lasted for more than 3 hr; there was no effect on BP at these doses; 1 month oral treatment of rabbits with a daily dose of 0.3 mg/kg results in marked anti-aggregatory effects Oral administration (0.25 mg/kg) to 2 kidney two wrapped hypertensive dogs caused a de-crease in diastolic (max., 41%) and systolic (max., 31%) BP which was still observable after 3 hr; oral administration of 0.5 mg/kg for 5 days caused BP decreases without appearance of tolerance; in the anesthe-tized rat i.v. administration caused a decrease in BP (ID$_{50}$ = 2.2 μg/kg) and stimulation of renin release[69]

77   6,9-*N*-Phenylimino-
11α,15α-dihydroxy-
prosta-6,Δ^{7,9}(E)13-tri-
enoic acid (U-60,257)

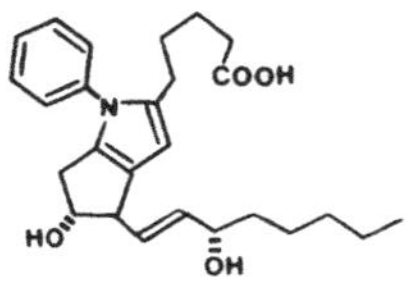

$C_{26}H_{35}NO_4$ (425)

From aniline and 6-keto-11,15-
bis-THP methyl ester of PGE[80]

Potent inhibitor of leukotriene
(LTC$_4$/LTD$_4$) synthesis in ion-
ophore stimulated rat peritoneal
mononuclear cells (ID$_{50}$ =
4.57 μM) and in human lung
fragments (ID$_{50}$ = 3.8 μM) induced by anaphylaxis; does not
affect cyclooxygenase pathway;[81] the principal site of inhibition of leukotriene
synthesis is arachidonate 5-li-
poxygenase; thus, in human
peripheral neutrophils LTB$_4$
synthesis is blocked with an
ID$_{50}$ = 1.8 μM[82]
In rhesus monkeys hypersensi-
tive to *Ascaris*, a 1% aerosol
inhibited the bronchial pulmonary effects of antigen chal-
lenge with a duration of 6 hr
Inhibition was also observed by
i.v. route[83]

78   Methyl 6,9α-carba-11-
keto-12-aza-15α-and-
15β-hydroxyprosta-5-
enoate

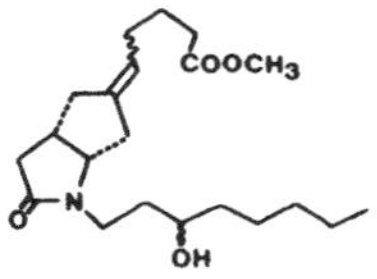

$C_{21}H_{35}NO_4$ (351)

From ethyl 1-ethoxycarbonyl-3-
pyrrolidon-2-acetate[99]

Very weak inhibitor of ADP-induced aggregation of rat PRP[99]

Reduced gastric acid secretion
in conscious, gastric fistula rats
on s.c. (ID$_{50}$ = 11.8 μg/kg; 4
× PGI$_2$) but not i.g. administration; reduced aspirin-induced
ulcer formation and protected
against ethanol induced gastric
mucosal necrosis (≈ 1 ×
PGI$_2$)[116]

Table 3 (continued)
PGI ANALOGS

No.	Systematic name	Structure	Mol. formula (mol. wt.)	Synthesis	Biological Actions
79	6,9α-Epithio-11α,15α-dihydroxyprosta-(Z)5,(E)13-dienoic acid		$C_{20}H_{32}O_4S$ (368)	Iodoetherification of 9-mercapto $PGF_{2\alpha}$ methyl ester followed by dehydroiodination and saponification[70,72] For other syntheses see Ref. 71, 75, 76 Much more stable than PGI_2	Inhibited platelet aggregation with a potency 0.3—1 × PGI_2; long-lasting decrease in perfusion pressure without change in systemic BP when administered close arterially in femoral circulation dog study (~0.1 × PGI_2)[73] Dilated isolated pig coronary artery, and coeliac and mesenteric vessels[73] In contrast, cat coronary artery was constricted;[72,73,76] in anesthetized cats, i.v. infusion (0.05 μmol/kg/min) decreased BP, increased HR, dilated mesenteric vasculature, and inhibited ADP-induced platelet aggregation (ex vivo 80%) These effects were very short-lived after termination of the infusion[74]
80	6,9α-Epithio-11α,15α-dihydroxyprosta-(Z)5,(Z)13-dienoic acid sulfone		$C_{20}H_{32}O_6S$ (400)	Oxidation of sulfide[72,76]	No significant effect on HPRP or cat coronary artery[76]

#	Name	Structure	Formula	Method	Activity
81	6β,9α-Epithio-11α,15α-dihydroxyprosta-(E)13-enoic acid		$C_{20}H_{32}O_4S$ (368)	By heating 9-deoxy-9-mercapto-PGF$_{2\alpha}$-11,15-*bis*-*t*-butyldimethylsilyl ether methyl ester in aqueous acetic acid followed by saponification[70,76] For other syntheses see Ref. 71, 72, 75	Weak inhibitor of aggregation of HPRP Weak constrictor of isolated cat coronary artery[86]
82	6,9α-Epithio-11α,15α-dihydroxyprosta-6,(E)13-dienoic acid		$C_{20}H_{32}O_4S$ (368)	Acetic acid-induced isomerization of exocyclic isomer; much more stable than PGI$_2$[77]	Inhibited ADP-induced rabbit platelet (PRP) aggregation; $0.01 \times$ PGI$_2$[77]
83	9α,5-Nitrilo-11α,15α-dihydroxyprosta-(E)13-enoic acid		$C_{20}H_{33}NO_4$ (351)	Reaction of 9-deoxy-9α-thioacetyl-11α,15α-dihydroxy-5H-prosta-13-trans-6-ynoic acid methyl ester with excess potassium carbonate[105]	
84	(+) Methyl 6-aza-6,9-nitrilo-11α,15α-dihydroxyprosta-7,(E)13-dienoate		$C_{20}H_{32}N_2O_4$ (364)	Thermolysis (70—80°) of the 9α-azido Δ^4-PGF methyl ester *bis*-*t*-butyldimethylsilyl ether followed by desilylation and saponification[79] Alkylation of the potassium salt of the bicyclic fully elaborated and protected pyrazole with methyl 5-iodopentanoate[84]	Inhibited ADP-indued aggregation of HPRP (0.003 × PGI$_2$)[89]
85	9,6-Nitrilo-7-thia-11α,15ξ-dihydroxyprosta-Δ^{8,9}(E)13-dienoic acid		$C_{19}H_{29}NO_4S$ (367)	From cyclopentadiene monoepoxide[85]	Inhibited ADP-induced platelet (HPRP) aggregation, 0.07 × PGE$_1$[85]

Table 3 (continued)
PGI ANALOGS

No.	Systematic name	Structure	Mol. formula (mol. wt.)	Synthesis	Biological Actions
86	6,9-Azo-11α,15α dihydroxyprosta-7,Δ,^{8,9} (E)13-trienoic acid		$C_{20}H_{32}N_2O_4$ (364)	Reaction of 6-oxo-PGE_1 with hydrazine followed by oxidation with platinic oxide[86]	Inhibited aggregation of HPRP and dilated the isolated perfused rat coronary artery; in these assays, the agent was more potent than PGI_2[86]
87	5,6,7-Trinor-4,8-inter-*m*-phenylene-6,9α-oxido-11α,15α-dihydroxy-16ξ-methyl-(E) 13-en-18-ynoic acid (TRK-100)		$C_{24}H_{30}O_5$(398)	From 6β-hydroxymethyl-7α-hydroxy-2-oxa-3,4-(2,4-dibromo)benzo-) bicyclo[3.3.0] octane[117]	Inhibited ADP induced aggregation of HPRP in vitro (IC_{50} = 1.4 ng/mℓ; 0.5 × PGI_2); induced disaggregation of a preexisting thrombus in the microvasculature of the hamster cheek pouch, after a single oral dose (50—200 μg/kg), which was maximal at 0.5—1 hr and lasted at least 2 hr (potency ≃ 1 × PGI_2)[118]
88	3-Oxa-4,5,6-trinor-3,7-inter-m-phenylene-5,9α-oxido-11α,15α-dihydroxy-(E)13-enoic acid		$C_{22}H_{30}O_6$(390)	Intramolecular Mitsunobu reaction[119] on optically active 4-membered precursor with intact, protected ω-chain, 11-OH (PG numbering) protected, free 9β-OH and α-chain as 2-hydroxy-3-tetrahydropyranyloxybenzyl group. Synthesis completed by deprotection of tricyclic product, alkylation of the phenolate with bromoacetate and saponification[119]	Equiactive with PGI_2 with regard to platelet and hypotensive effects[120]

89	6,9α-Carba-11-oxa-15α-hydroxyprosta-(Z)5,(E)13-dienoic acid	$C_{20}H_{32}O_4$(336)	From *cis* and *trans*-4-oxo-cyclopentane-1,2-dicarboxylic acid[121]	Weak inhibitor of collagen induced platelet aggregation (0.01 × carbacyclin)[121]
90	3-Oxa-6,9α-carba-11α,15α-dihydroxy-16;,20-dimethylprosta-(Z)5-en-13,18-diynoic acid (ZK-96,480)	$C_{22}H_{30}O_5$(374)	Sequential attachment of optically active ω-chain and α-chain to ethylene ketal of optically active 6β-formyl-7α-benzyloxybicyclo [3.3.0] octan-3-one[122]	Chemically and metabolically stable prostacyclin analog with oral activity and prolonged duration of action; plasma of rats dosed orally (1 mg/kg) inhibited human platelet aggregation for up to 24 hr (4 × iloprost); relaxed bovine coronary artery (ED_{50} = 3.0 nm; 20 × PGI_2), inhibited ADP induced aggregation of HPRP (1 × PGI_2) and decreased BP in SH rats at 0.01—1.0 mg/kg p.o; maximum reduction of MABD reached at 0.1 mg/kg; HR increased 151% at 1 mg/kg[123]
91	(1R,2R,3S,5R,3'S)-3-(3'-Hydroxyoct-1'-ynyl) bicyclo [3.2.0] heptan-2 ol-6-oximinoacetic acid	$C_{17}H_{25}NO_5$(323)	Opening of corresponding (±) bicyclic epoxy ketal with (S)-dimethyl-[3-(t-butyldimethylsilyloxy) oct-1-ynyl] aluminum, separation of diastereomeric pairs by chromatography of dicobalthexacarbonyl complexes, deprotection and oximination[124]	Potent inhibitor of ADP induced aggregation of HPRP (4.5 PGE_1) in vitro[124]

Table 3 (continued)
PGI ANALOGS

No.	Systematic name	Structure	Mol. formula (mol. wt.)	Synthesis	Biological Actions
92	1a-*nor*-6,9α-9a-*homo*-11α,15α-dihydroxy-15-cyclohexyl-ω-pentanor-prosta-(Z)5-en-13-ynoic acid (RS-93427)		$C_{21}H_{30}O_4$(346)	Cleavage of (±)-2α,3α-epoxybi-cyclo [4.2.0] octan-7-one ethyl-ene ketal with optically pure 1-cyclohexyl-3-lithiopropargyl al-cohol-t-butyldimethyl-siyl ether in the presence BF_3 etherate, followed by complex formation with dicobalt octacarbonyl, chromatographic separation of diastereomers, cleavage of co-balt complex and ethylene ke-tal, Wittig reaction, desiylation and saponification[125]	Chemically stable, potent inhibi-tor of ADP induced aggrega-tion of HPRP (IC_{50} = 4 nM; 14 × PGE_1); long lasting inhi-bition of platelet aggregation by the oral route (0.2—1 mg/kg) in guinea pigs, baboons, and other primates; long-acting hypotensive agent in SH rats by i.v. (ED_{20} = 16 μg/kg) and oral (ED_{20} = 480 μg/kg) routes; at low nanogram con-centrations, suppresses macro-phage accumulation of cholesterol esters from hyperli-pidemic serum; suppresses re-lease of mitogens from vascular endothelial cells, Macrophages, and human platelets at <1 ng/mℓ[125,126]

C$_{24}$H$_{34}$O$_5$(412)

Ozonolysis of the bis-THP methyl ester of optically pure compound 67, followed by attachment of the ω-chain in the Corey manner, THP cleavage, catalytic reduction and saponification.[127] For a more efficient synthesis from 5-methoxy-2-tetralone, see Reference 128

Potent antisecretory and cyctoprotective agent; inhibited pentagastrin stimulated gastric acid secretion in Heidenhain pouch dogs (ED$_{50}$ = 50 µg/kg) with a duration of 8—10 hr; inhibited gastric acid secretion in pylorus ligated rats (ED$_{50}$ = 35 µg/kg, p.o.) is cytoprotective to the stomach (ED$_{50}$ = 0.8 µg/kg, p.o.) and intestine (ED$_{50}$ = 22 µg/kg, p.o.) and inhibited aspirin induced gastric ulcer formation (ED$_{50}$ = 5 µg/kg, p.o.); not diarrheagenic in mice, even at oral doses of 5 mg/kg and is not uterotonic in second trimester pregnant monkeys at 2 mg/kg p.o.; inhibited ex vivo ADP induced platelet aggregation in the rat (ED$_{50}$ = 300 µg/kg, p.o.) and decreased MABP in the dog by 44% at 50 µg/kg p.o.[127]

REFERENCES

1. **Whittaker, N.**, A synthesis of prostacyclin sodium salt, *Tetrahedron Lett.*, 2805—2808, 1977.
2. **Nicolaou, K. C., Barnette, W. E., Gasic, G. P., Magolda, R. L., Sipio, W. J., Silver, M. J., Smith, J. B., and Ingerman, C. M.**, Rapid and easy preparation of prostacyclin, *Lancet*, 1058—1059, 1977; **Nicolaou, K. C. Barnette, W. E., Gasic, G. P., Magolda, R. L., and Sipio, W. J.**, Simple efficient synthesis of prostacyclin (PGI₂), *J. Chem. Soc. Chem. Commun.*,630—631, 1977.
3. **Corey, E. J., Heck. G. E., and Szeleky, I.**, Synthesis of Vane's prostaglandin X, 6,9-oxido-9α-15α-dihydroxyprosta-(Z)5,(E)13-dienoic acid, *J. Am. Chem. Soc.*, 99, 2006—2008, 1977.
4. **Johnson, R. A., Lincoln, F. H., Thompson, J. L., Nidy, E. G., Mizak, S. A., and Axen, U.**, Synthesis and stereochemistry of prostacyclin and synthesis of 6-ketoprostaglandin F₁ₐ, *J. Am. Chem. Soc.*, 4182—4184, 1977.
5. Prostacyclin epoprostenol, *Drugs of the Future*, 5, 525—527, 1980.
6. **Yui, Y., Nakajima, H., Kawai, C., and Murakami, T.**, Prostacyclin therapy in patients with congestive heart failure, *Am. J. Cardiol.*, 50, 320—324, 1980.
7. Epoprostenol Prostacyclin, *Drugs of the Future*, 6, 648—650, 1981.
8. Epoprostenol, *Drugs of the Future*, 7, 770—774, 1982.
9. **Schölkens, B. A., Bartman, W. Beck, G., Lerch, U., Konz, E., and Weithmann, U.**, Vasodilation and inhibition of platelet aggregation by prostacyclins with modified ω-side chains, *Prostaglandins Med.*, 3, 7—22, 1979.
10. **Moncada, S. and Vane, J. R.**, Biological significance and therapeutic potential of prostacyclin, *J. Med. Chem.*, 23, 591—593, 1980; Prostacyclin and blood coagulation, *Drugs*, 21, 430—437, 1981.
11. **Corey, E. J., Szekely, I., and Shiner, C.**, Synthesis of 6,9α-oxido-11α,15α-dihydroxyprosta-(E)5,(E)13-dienoic acid, an isomer of PGI₂ (Vane's PGX), *Tetrahderon Lett.*, 3529—3532, 1977.
12. **De, B., Andersen, N. H., Ippolito, R. M., Wilson, C. H., and Johnson, W. D.**, Synthesis and chiroptical characterization of prostacyclin diastereomers, *Prostaglandins*, 19, 221—247, 1980.
13. **Ohuchida, S., Hashimoto, S., Wakatsuka, H., Arai, Y., and Hayashi, M.**, Synthesis and biological activities of some prostacyclin analogs, *Adv. Prostaglandin Thromboxane Res.*, 6, 337—340, 1980.
14. Nileprost, *Drugs Future*, 7, 643—645, 1982.
15. **Toru, T., Watanabe, K., Oba, T., Tanaka, T., Okamura, K., Bannai, K., and Kurozumi, S.**, Reaction of prostacyclin methyl ester with benzenesulfenyl chloride. Preparation of stable prostacyclin analogs, *Tetrahedron Lett.*, 21, 2539—2542, 1980.
16. **Bannai, K., Toru, T., Oba, T., Tanaka, T., Okamura, N., Watanabe, K., and Korozumi, S.**, Stereocontrolled synthesis of 7-hydroxy- and 7-acetoxy-PGI₂: new stable PGI₂ analogues, *Tetrahedron Lett.*, 22, 1417—1420, 1981.
17. **Tömösközi, I., Kanai, K., Györy, P. and Kovacs, G.**, A convenient short synthesis of 7-oxo-PGI₂, *Tetrahedron Lett.*, 23, 1091—1094, 1982.
18. **Kovacs, G., Simonidesz, V., Tömösközi, I., Körmoczy, P., Szelely, I. Papp-Behr, A., Stadler, I., Szekeres, L., and Papp, G.**, A new stable prostacyclin mimic, 7-oxoprostaglandin I₂, *J. Med. Chem.*, 25, 105—107, 1982.
19. **Fried, J., Mitra, D. K., Nagarajan, M., and Mehrotra, M. M.**, 10,10-Difluoro-13-dehydroprostacyclin: a chemically and metabolically stabilized potent prostacyclin, *J. Med. Chem.*, 23, 234—237, 1980.
20. **Harris, D. N., Phillips, M. B., Michel, I. M., Goldenberg, H. J., Hasänger, M. F., Antonaccio, M. J., and Fried, J.**, Some biochemical activities of 10,10-difluoro-13-dehydroprostacyclin, a chemically stable analog of prostacyclin, in human blood, *Thrombosis Res.*, 23, 387—399, 1981.
21. **Hatano, Y., Kohli, J. D., Goldberg, L. I., Fried, J., and Mehrotra, M. M.**, Vascular relaxing activity and stability studies of 10,10-difluoro-13,14-dehydroprostacyclin, *Proc. Natl Acad. Sci. U.S.A.*, 77, 6846—6850, 1980.
22. **Bezuglov, V. V. and Bergelson, L. D.**, Synthesis of fluoroprostacyclins, *Dokl. Akad. Nauk. S.S.S.R.*, 250, 468—469, 1980; *Chem. Abstr.*, 93, 25949z, 1980.
23. **Nicolaou, K. C., Barnette, W. E., Magolda, R. L., Grieco, P. A., Owens, W., Wong, C. L. J., Smith, J. B., Ogletree, M., and Lefer, A. M.**, Synthesis and biological properties of 12-fluoroprostacyclins, *Prostaglandins*, 16, 789—794, 1978.
24. **Novak, L., Aszodi, J., and Szantay, C.**, Synthesis of 13-oxa- and 13-Thia-PGI₂, metabolically stable and biologically potent PGI₂ analogues, *Tetrahedron Lett.*, 23, 2135—2138. 1982; **Novak, L., Aszodi, J., Rohaly, J., Stadler, I., Kormoczy, P., Siminidosz, V., and Szantay, C.**, Biologically potent analogues of prostacyclin. I. Synthesis of 13-thiaprostacyclins, *Acta Chim. Hung.*, 113, 111—128, 1983.
25. K-13,415, *Drugs of the Future*, 6, 783, 1981.
26. **Grieco, P. A., Yokoyama, Y., Nicolaou, K. C., Barnette, W. E., Smith, J. B., Ogletree, M., and Lefer, A. M.**, Total synthesis of 14-fluoroprostaglandin F₂ₐ and 14-fluoroprostacyclin, *Chem. Lett.*, 1001—1004, 1978.

27. **Schökkens, B. A., Bartman, W., Beck, G., Lerch, U., Konz, E., and Weithmann, U.,** Vasodilator and antiplatelet activities of prostacyclins with modified ω-side chain, *Adv. Prostaglandin Thromboxane Res.,* 6, 341—345, 1980.

28. **Anderson, N. H., Inamoto, S., Subramanian, N., Picker, D. H., Ladner, D. W., De, B., Tynan, S. S., Eggerman, T. L., Harker, L. A., Robertson, R. P., Dien, H. G., and Rao, C. V.,** Molecular basis for prostaglandin potency. III. Tests of the significance of the "hairpin conformation" in biorecognition phenomena, *Prostaglandins,* 22, 841—856, 1981.

29. **Nidy, E. G. and Johnson, R. A.,** Synthesis of prostaglandin I₃ (PGI₃), *Tetrahedron Lett.,* 2375—2378, 1978.

30. **van Dorp, D. A., van Evert, W. C., and van der Wolf, L.,** 20-Methylprostacyclin a powerful "unnatural" platelet aggregation inhibitor, *Prostaglandins,* 6, 953—955, 1978.

31. **Föhle, L., Böhlke, H., Frankas, E., Kim, S. M. A., Lintz, W., Loschen, G., Müller, B., Schneider, J., Seipp, U., Vollenberg, W., and Wilsman, K.,** Designing prostacyclin analogues, *Arzneim, Forsch. Drug. Res.,* 33, 1240—1248, 1983.

32. **Shimoji, K., Konishi, Y., Arai, Y., Hayashi, M., and Yamamoto, H.,** 6,9α-Oxido-11α,15α-dihydroxyprosta-6,(E)-13-dienoic acid methyl ester and 6,9α:6,11α-dioxido-15α-hydroxyprosta-(E)-13-enoic acid methyl ester. Two isomeric forms of prostacyclin (PGI₂), *J. Am. Chem. Soc.,* 100, 2547—2548, 1978.

33. **Nishiyama, H. and Ohno, K.,** Synthesis of 6,7-dehydro-5-oxo-prostaglandin I₁: a stable analog of prostacyclin, *Tetrahedron Lett.,* 3481—3484, 1979.

34. **Sih, J. C. and Graber, D. R.,** Synthesis of (6R)- and (6S)-6(9)-oxy-11,15-dihydroxyprosta-7,13-dienoic acids [(6R)-and (6S)-Δ⁷-PGI₁]: nonidentity with the proposed arachidonic acid metabolite, *J. Org. Chem.,* 43, 3798—3800, 1978.

35. **Ohno, K. and Nishiyama, H.,** 4,5,6,7-Tetradehydro-PGI₁, a stable and potent inhibitor of blood platelet aggregation, *Tetrahedron Lett.,* 3003—3004, 1979.

36. **Gandolfi, C. A. and Gryglewski, R. I.,** 20-Methyl-prostacyclin analogs — two stage screening procedure for biological properties, *Pharmacol. Res. Commun.,* 10, 885—896, 1978.

37. **Nicolaou, K. C., Barnette, W. E., and Magolda, R. L.,** Organoselenium-based synthesis of oxygen-containing prostacyclins, *J. Am. Chem. Soc.,* 103, 3480—3485—1981.

38. **Johnson, R. A. Lincoln, F. H., Nidy, E. G., Schneider, W. P., Thompson, J. L., and Axen, U.,** Synthesis and characterization of prostacyclin, 6-ketoprostaglandin F₁α, prostaglandin I₁, and prostaglandin I₃, *J. Am. Chem. Soc.,* 100, 7690—7705, 1978.

39. **Corey, E. J., Peara, H. L., Szekely, I., and Ishigura, M.,** Configuration of C-6 of 6,9α-oxido-bridged prostaglandins, *Tetrahedron Lett.,* 1023—1026, 1978.

40. **Finch, M. A., Roberts, S. M., and Newton, R. F.,** Synthesis of (±)-6β-prostaglandin I₁ and (±)-6β-decarboxy prostaglandin I₁, *J. Chem. Soc. Perkin Trans. I,* 1312—1316, 1981.

41. **Whittle, B. J. R. and Broughton-Smith, N. K.,** 16-Phenoxy prostacyclin analogs-potent, selective antiulcer compounds, *Prostacyclin,* [Pap. Discuss. Workshop] Vane, J. R. and Bergstrom, S., Eds., Raven Press, New York, 1979, 159—171.

42. **Whittle, B. J. R., Broughton-Smith, N. K., Moncada, S., and Vane, J. R.,** The relative activity of prostacyclin and a stable analogue 6β-PGI₁ on the gastrointestinal and cardiovascular systems, *J. Pharm. Pharmacol.,* 30, 597—599, 1978.

43. **Lin, C. H. and Alexander, D. L.,** Synthesis of 4,4,5,5-tetrahydro and cis-4,5-didehydro prostacyclin analogues, *J. Org. Chem.,* 47, 615—620, 1982.

44. **Sih, J. C., Johnson, R. A., Nidy, E. G., and Graber, D. R.,** Synthesis of the four isomers of 5-hydroxy-PGI₁, *Prostaglandins,* 15, 409—421, 1978.

45. **Bannai, K., Toru, T., Hazato, A., Oba, T., Tanoka, T., Okamura, N., Watanabe, K., and Kurozumi, S.,** Synthesis of new sulfur-containing prostaglandin I₁, *Chem. Pharm. Bull.,* 30, 1102—1105, 1982.

46. **Gandolfi, C. A. and Ceserani, R.,** Prostacyclin analogs: structure-activity relationships, *Prostaglandins Cardiovasc. Dis.,* 183—195, 1981.

47. **Johnson, R. A. and Nidy, E. G.,** Synthesis and stereochemistry of 9-deoxy-5,9-α-epoxyprostaglandins: a series of stable prostacyclin analogues, *J. Org. Chem.,* 45, 3802—3810, 1980.

48. **Johnson, R. A. and Nidy, E. G.,** Synthesis and stereochemistry of stable prostacyclin analogs, in *Chemistry, Biochemistry and Pharmacology of Prostanoids,* Roberts, S. M. and Schienmann, F., Eds., Pergamon Press, Oxford, 1979, 274—285.

49. **Kauffman, G. L., Whittle, B. J. R., Aures, D., and Grossman, M. I.,** Gastric antisecretory and cardiovascular actions of a stable 16-phenoxy prostacyclin analog in the dog, *Adv. Prostaglandin Thormboxane Res.,* 8, 1521—1524, 1980.

50. **Skuballa, W.,** Synthesis of 6,9-homo-prostacyclin a stable prostacyclin analog, *Tetrahedron Lett.,* 21, 3261—3264, 1980.

51. **Nicolaou, K. C., Sipio, W. J., Magolda, R. L., Seitz, S., and Barnette, W. E.,** Total synthesis of carboprostacyclin, a stable and biologically active analogue of prostacyclin (PGI_2), *J. Chem. Soc. Chem. Commun.*, 1067—1068, 1978; **Shibasaki, M., Veda, J., and Ikegami, S.,** New synthetic routes to 9(0)-methanoprostacyclin. A highly stable and biologically potent analog of prostacyclin, *Tetrahedron Lett.*, 433—436, 1979.

52. Carbacyclin, *Drugs of the Future*, 6, 753—760, 1981.

53. **Konishi, Y., Kawamura, M., Iguchi, Y., Arai, Y., and Hayashi, M.,** Facile synthesis of 6a-carba-prostaglandin I_2, *Tetrahedron*, 37, 4391—4399, 1981; **Konishi, Y., Kawamura, M., Arai, Y., and Hayashi, M.,** A synthesis of 9(0)-methanoprostacyclin, *Chem. Lett.*,, 1437—1440, 1979.

54. **Karim, S. M. M., Adaikan, P. G., Lau, L. C., and Tai, M. Y.,** Inhibition of platelet aggregation with intravenous and oral administration of carboprostacyclin in man, *Prostaglandins Med.*, 6, 521—527, 1981.

55. **Morley, J., Page, C. P., Paul, W., Mongelli, N., Cesarini, R., and Gandolfi, C.,** A comparative study of PGI_2 and analogues (FCE 21292 and 21258) in vitro and in vivo, *Prostaglandins, Leukotrienes, Med.*, 11, 391—399, 1983.

56. **Adaikan, P. G., Kottegoda, S. R., Lau, L. C., Tai, M. Y., and Karim, S. M. M.,** Inhibition of platelet aggregation and reversal of vasopressin-induced ECG changes by carboprostacyclin analogue, ONO 41483, in primates, *Prostaglandins, Leukotrienes, Med.*, 9, 307—320, 1982.

57. **Adaikan, P. G., Lau, L. C., Tai, M. Y., and Karim, S. M.,** Inhibition of platelet aggregation with intravenous and oral administration of carboprostacyclin analogue, 15-cyclopentyl-ω-pentanor-5(E)-carba-cyclin *Prostaglandins, Leukotrienes, Med.*, (ONO 41483) in man, 10, 53—64, 1983.

58. **Skuballa, W. and Vorbrüggen, H.,** A new route to 6a-carbacyclins-synthesis of a stable, biologically potent prostacyclin analogue, *Angew. Chem. Int. Ed. Engl.*, 20, 1046—1048, 1981.

59. Ciloprost, *Drugs Future*, 6, 676—677, 1981.

60. **Casals-Stenzel, J., Buse, M., and Losert, W.,** Comparison of the vasopressor action of ZK 36 374, a stable prostacyclin derivative, PGI_2 and PGE_1 with their effect on platelet aggregation and bleeding time in rats, *Prostaglandins, Leukotrienes Med.*, 10, 197—212, 1983.

61. **Vischer, P. and Casals-Stenzel, J.,** Pharmacological properties of ciloprost, a stable prostacyclin analogue, on the gastrointestinal tract, *Prostaglandins, Leukotrienes, Med.*, 9, 517—612, 1982.

62. **Shibasaki, N., Iseki, K., and Ikegami, S.,** Total synthesis of the carbon analog of Δ^6-PGI_1, *Tetrahedron Lett.*, 169—172, 1980.

63. **Sih, J. C.,** Synthesis of 6-membered-ring analogues of 6α-carba-PGI_2, *J. Org. Chem.*, 47, 4311—4315, 1982.

64. **Shimoji, K. and Hayashi, M.,** The synthesis of new aromatic prostacyclin analog, *Tetrhedron Lett.*, 21, 1255—1258, 1980.

65. **Aristoff, P. A. and Harrison, A. W.,** Synthesis of benzindene prostaglandins: a novel potent class of stable prostacyclin analogs, *Tetrahedron Lett.*, 23, 2067—2070, 1982.

66. **Newton, R. F. and Wadsworth, A. H.,** Synthesis of stable prostacyclin analogues from 2,3-disubstituted bicyclo[3.2.0]heptan-6-ones, *J. Chem. Soc. Perkin Trans. I*, 823—830, 1982.

67. **Nakai, H., Arai, Y., Hamanaka, N., and Hayashi, M.,** Synthesis of nitrogen containing prostaglandin I_1 analongs, *Tetrahedron Lett.*, 805—808, 1979.

68. **Bartmann, W., Beck, G., Knolle, J., and Rupp, R.,** Synthesis of stable prostacyclin analogues, *Angew. Chem. Int. Ed. Engl.*, 19, 819, 1980; *Tetrahedron Lett.*, 23, 3647—3650, 1982.

69. **Schölkens, B. A., Gehring, D., Jung, W.,** The orally active thia-imino-prostacyclin HOE 892: antiaggregatory and cardiovascular activities, *Prostaglandins Leukotrienes Med.*, 10, 231—256, 1983.

70. **Nicolaou, K. C., Barnette, W. E., Gasic, G. P., and Magolda, R. L.,** 6,9-Thiaprostacyclin. A stable and biologically potent analogue of prostacyclin (PGI_2), *J. Am. Chem. Soc.*, 99, 7736—7738, 1977.

71. **Shibasaki, M. and Ikegami, S.,** Synthesis of 9(0)-thiaprostacyclin, *Tetrahedron Lett.*, 559—562, 1978.

72. **Nicolaou, K. C., Barnette, W. E., and Magolda, R. L.,** Synthesis of (5Z)- and (5E)-6,9-Thiaprostacyclins, *J. Am. Chem. Soc.*, 103, 3472—3480, 1981.

73. 6,9-Thiaprostacyclin, *Drugs Future*, 4, 220—222, 1979.

74. **Lefer, A. M., Trachte, G. J., Smith, J. B., Barnette, W. E., and Nicolaou, K. C.,** Circulatory and platelet actions of 6,9-thiaprostacyclin (PGI₂-S) in the cat, *Life Sci.*, 25, 259—264, 1964.

75. **Shimoji, K., Arai, Y., and Hayashi, M.,** A new synthesis of 6,9α-thiaprostacyclin, *Chem. Lett.*, 1375—1376, 1978.

76. **Nicolaou, K. C., Barette, W. E., Magolda, R. L.,** Organoselenium-induced ring closures. Sulfur-containing prostacyclins. Stereoselective synthesis of cyclic α,β-unsaturated sulfoxides, sulfones, and sulfides and synthesis of 6,9-sulfoxa-5(E)- and 5(Z)-prostacyclin, 6,9-sulfo-5(E)- and 5(Z)-prostacyclin, 6,9-sulfo-6α- and 6β-4(E)-isoprostacyclin, and 6,9-thiaprostacyclin, *J. Am. Chem. Soc.*, 100, 2567—2570, 1978; Organoselenium-Based Synthesis of Sulfur-Containing Prostacyclins, *J. Am. Chem. Soc.*, 103, 3486—3497, 1981.

77. **Yokomori, H., Torisawa, Y., Shibasaki, M., and Ikegami, S.,** A novel synthesis of the sulfur analog of Δ^6-PGI_1, *Heterocycles*, 18, 251—254, 1981.

78. **Cassidy, F., Moore, R. W., Wotton, G., Baggaley, K. H., Geen, G. R., Jennings, L. J. A., and Tyrrell, A. W. R.**, Synthesis of 12-azaprostacyclin analogues, *Tetrahedron Lett.*, 22, 253—256, 1981.
79. **Bundy, G. L., and Baldwin, J. M.**, The synthesis of nitrogen-containing prostaglandin analogs, *Tetrahedron Lett.*, 1371—1374, 1978.
80. **Smith, H. W., Bach, M. K., Harrison, A. W., Johnson, H. G., Major, N. J., and Wasserman, M. A.**, The synthesis of 6,9-deepoxy-6,9-*N*-phenylimino-$\Delta^{6,8}$-prostagalndin I$_1$, a novel inhibitor of leukotriene C and D synthesis, *Prostaglandins*, 24, 543—546, 1982.
81. **Bach, M. K., Brasher, J. R., Smith, H. W., Fitzpatrick, F. A., Sun, F. F., and McGuire, J. C.**, 6,9-Deepoxy-6,9-(phenylimino)-$\Delta^{6,8}$-prostaglandin I$_1$, (U-60,257), a new inhibitor of leukotriene C and D synthesis: in vitro studies, *Prostaglandins*, 23, 759—771, 1982.
82. **Sun, F. F. and McGuire, J. C.**, Inhibition of human neutrophil arachidonate 5-lipoxygenase by 6,9-deepoxy-6,9-(phenylimino)-$\Delta^{6,8}$-prostaglandin I$_1$ (U-60257), *Prostaglandins*, 26, 211—221, 1983.
83. **Johnson, H. G., McNee, M. L., Bach, M. K., Smith, H. W.**, The activity of a new, novel inhibitor of leukotriene synthesis in Rhesus monkey ascaris reactors, *Int. Arch. Allergy Appl. Immunol.*, 70, 169—173, 1983.
84. **Suzuki, M., Sugiura, S., and Noyori, R.**, Synthesis of a pyrazole prostacyclin, *Tetrahedron Lett.*, 23, 4817—4820, 1982.
85. **Bradbury, R. H. and Walker, K. M.**, Synthesis of analogs of prostacyclin containing a thiazole ring, *Tetrahedron Lett.*, 1335—1338, 1982.
86. **Nicolaou, K. C., Barnette, W. E., and Magolda, R. L.**, 6,9-pyridazaprostacyclin and derivatives: the first "aromatic" prostacyclins, *J. Am. Chem. Soc.*, 101, 766—768, 1979.
87. **Shimoji, K., Arai, Y., Wakatsuka, H., and Hayashi, M.**, Syntheses of new prostacyclin analogs, *Adv. Prostaglandin Thromboxane Res.*, 6, 327—330, 1980.
88. **Lock, J. E., Coceani, F., Hamilton, F., Greenway-Coates, A., and Olley, P. M.**, The pulmonary and vascular effects of three prostaglandin I$_2$ analogs in conscious newborn lambs, *J. Pharmacol. Exp. Ther.*, 215, 156—159, 1980.
89. **See Morton, D. R. Bundy, G. L., and Nishizawa, E. E.**, Five-membered ring-modified prostacyclin analogs, in *Prostacyclin*, Vane, J. R. and Bergstrom, S., Eds., Raven Press, New York, 1978, 43.
90. **Galambos, G., Simonidesz, V., Ivanics, J., Horvath, K., and Kovacs, G.**, Synthesis of stable prostacyclin analogues, *Tetrahedron Lett.*, 24, 1281—1284, 1983.
91. **Radüchel, B.**, An approach to the synthesis of 5-azaprostacyclin, *Tetrahedron Lett.*, 3229—3232, 1983.
92. **Powell, J. R., Kerwin, L. J., McGill, M. H., Haslänger, M., and Fried, J.**, Factors affecting the termination of cardiovascular actions of 10,10-difluoro-13-dehydroprostacyclin, *Prostaglandins*, 25, 457—467, 1983.
93. **Nickolson, R. C. and Vorbrüggen, H.**, Synthesis of stable furan prostacyclin analogs, *Tetrahedron Lett.*, 24, 47—50, 1983.
94. **See Johnson, R. A., Lincoln, F. H., Smith, H. W., Ayer, D. E., Nidy, E. G., Thompson, J. L., Axen, U., Aiken, J. W., Gorman, R. R., Nishizawa, E. E., and Honohan, T.**, Biological Effects of Stable Prostacyclin Analogs, *Prostacyclin*, Vane, J. R. and Bergstrom, S., Eds., Raven Press, New York, 1979, 22.
95. **Baraldi, P. G., Barco, A., Benetti, S., Gandolfi, C. A., Pollini, G. P., and Simoni, D.**, Synthesis of a new sulphur-containing carbaprostacyclin analogue, *Tetrahedron Lett.*, 24, 4871—4874, 1983.
96. **Shibasaki, M., Torisawa, Y., and Ikegami, S.**, Synthesis of 9(0)-methano-$\Delta^{6(9\alpha)}$-PGI$_1$: the highly potent carbon analog of prostacyclin, *Tetrahedron Lett.*, 3493—3496, 1983.
97. **Shibasaki, M., Fukasawa, H., and Ikegami, S.**, Regiocontrolled conversion of α,β-unsaturated ketones to olefins via allylsilanes: synthesis of dl-9(0)-methano-$\Delta^{6(9\alpha)}$-PGI$_1$, *Tetrahedron Lett.*, 3497—3500, 1983.
98. **Collington, E. W., Finch, H., and Wallis, C. J.**, Stereoselective synthesis of a 13-azaprostacyclin methyl ester, *Tetrahedron Lett.*, 3121—3124, 1983.
99. **Wang, C. L. J.**, Azaprostaglandins II. Synthesis of 12-azaprostacyclin analogs, *Tetrahedron Lett.*, 477—480, 1983.
100. **Ivanics, J., Simonidesz, V., Galambos, G., Kormoczy, P., and Kovacs, G.**, Synthesis of prostacyclin analogues via knoevenagel condensation, *Tetrahedron Lett.*, 24, 315—318, 1983.
101. **Aristoff, P. A., Johnson, P. D., and Harrison, A. W.**, Synthesis of 9-substituted carbacyclin analogues, *J. Org. Chem.*, 48, 5341—5348, 1983.
102. **Riefling, B. F. and Radunz, H. E.**, An eight step synthesis of an orally active antihypertensive carbaprostacyclin analogue, *Tetrahedron Lett.*, 24, 5487—5490, 1983.
103. **Collins, J. C., Hess, W. W., and Frank, F. J.**, Dipyridine-chromiun (VI) oxide oxidation of alcohols in dichloromethane, *Tetrahedron Lett.*, 3363—3366, 1968.
104. **Johnson, A. W.**, *Ylid Chemistry*, Academic Press, New York, 1966.
105. **Shibasaki, M., Torisawa, Y., and Ikegami, S.**, A new method for the conversion of aldehydes (RCH$_2$CHO) to acetylenes (RC≡CH) via 1-alkenylstannanes. Application to the synthesis of 9(0)-thia-Δ^6-PGI$_1$, *Tetrahedron Lett.*, 23, 4607—4610, 1982.

106. **Bannai, K., Toru, T., Oba, T., Tanaka, T., Okamura, N., Watanabe, K., Hazato, A., and Kurozumi, S.,** Synthesis of chemically stable prostaglandin analogs, *Tetrahedron,* 39, 3807—3819, 1983.

107. **Barth, H., Lintz, W., Michel, G., Osterloh, G., Seipp, U., and Flohe, L.,** Inhibition of platelet aggregation by intravenous administration of the biochemically stable prostacyclin analogue CG 4203, *Arch. Pharmacol.,* 324(Suppl. 111), R60, 1983.

108. **Sugiura, S., Toru, T., Tanaka, T., Okamura, Hazato, A., Bannai, K., Manabe, K., and Kurozumi, S.,** Total synthesis of stable PGI$_2$ derivatives. Synthesis of 7-hydroxy and 7-fluoro PGI$_2$, *Chem. Pharm. Bull.,* 32, 1248—1251, 1984.

109. **Allan, G., Follenfant, M. J., Lidbury, P., Oliver, P. L., and Whittle, B. J. R.,** Cardiovascular and platelet actions of 9β-methylcarbacyclin (ciprostene), a chemically stable analogue of prostacyclin in the dog and monkey, *Br. J. Pharmac.,* 85, 547—555, 1985.

110. **O'Grady, J., Hedges, A., Whittle, B. J. R., Al-Sinawi, L.A-H., Mekki, Q. A., Burke, C., Moody, S. G., Moti, M. J., and Hassan, S.,** A chemically stable analogue, 9β-methyl carbacyclin, with similar effects to epoprostenol (prostacyclin, PGI$_2$) in man, *Br. J. Clin. Pharmac.,* 18, 921—933, 1984.

111. **Belch, J. J. F., Greer, I., McLaren, M., Sanibadi, A. R., Miller, S., Sturrock, R. D., and Forbes, C. D.,** Effects of intravenous ZK 36,374, a stable prostacyclin analogue, on normal volunteers, *Prostaglandins,* 28, 67—76, 1984.

112. **Yitalo, P., Kaukinen, S., Nurmi, A-K., Seppala, E., Pessi, H., and Vapaatalo, H.,** Effects of a prostacyclin analog iloprost on kidney function, renin-angiotensin and kallikrein-kinin systems, prostanoids and catecholamines in man, *Prostaglandins,* 29, 1063—1071, 1985.

113. **Bugiardini, R., Galvani, M., Ferrini, D., Gridelli, C., Tollemento, D., Mari, L., Puddu, P., and Lenzi, S.,** Myocardial ischemia induced by prostacyclin and iloprost, *Clin. Pharmacol. Ther.,* 38, 101—108, 1985.

114. **Koyama, K. and Kojima, K.,** A Stereoselective total sythesis of a stable prostacyclin analog, *dl*-9(0)-methano-Δ$^{6(9\alpha)}$-prostaglandin I$_1$, *Chem. Pharm. Bull.,* 32, 2866—2869, 1984.

115. **Torisawa, Y., Okabe, H., Shibasaki, M., and Ikegami, S.,** An improved route to (+)-9(0)-methano-Δ$^{6(9\alpha)}$-prostaglandin-I$_1$-(isocarbacyclin), *Chem. Lett.,* 1069—1072, 1984.

116. **Konturek, S. J., Brzozowski, T., Radecki, T., and Pastucki, I.,** Comparison of gastric and intestinal antisecretory and protective effects of prostacyclin and its stable thia-imino-analogue (HOE 892) in conscious rats, *Prostaglandins,* 28, 443—453, 1984.

117. **Ohno, K., Nagase, H., Matsumoto, K., Nishiyama, H., and Nishio, S.,** Stereoselective synthesis of 5,6,7-trinor-4,8-inter-m-phenylene-PGI$_2$ derivatives and their inhibitory activities to human platelet aggregation, *Adv. Prostaglandin, Thromboxane, Leukotriene Res.,* 15, 279—281, 1985.

118. **Sim, A. K., McCraw, A. P., Cleland, M. E., Nishio, S., and Umetsu, T.,** Effect of a stable prostacyclin analogue on platelet function and experimentally-induced thrombosis in microcirculation, *Arzneim. Forsch/Drug Res.,* 35, 1816—1818, 1985.

119. **Mitsunobu, O.,** The use of diethyl azodicarboxylate and triphenylphosphine in synthesis and transformation of natural products, *Synthesis,* 1—28, 1981.

120. **Aristoff, P. A., Harrison, A. W., and Huber, A. M.,** Synthesis of benzopyran prostaglandins, potent stable prostacyclin analogs, via an intramolecular Mitsunobu reaction, *Tetrahedron Lett.,* 25, 3955—3958, 1984.

121. **Amemiya, S., Kojima, K., and Sakai, K.,** Synthesis of (±) 11-oxacarbacyclin, *Chem. Pharm. Bull.,* 32, 805—807, 1984.

122. **Skuballa, A. and Vorbruggen, H.,** Synthesis of ZK-96,480: a chemically and metabolically stable potent prostacyclin analogue, *Adv. Prostaglandin, Thromboxane Leukotriene Res.,* 15, 271—273, 1985.

123. **Sturzebecher, C. S., Habarey, M., Muller, B., Schillinger, E., Schroder, G., Skuballa, W., and Stock, G.,** Prostaglandins and other eicosanoids in the cardiovascular system, *Proc. 2nd. Int. Symp. Nurnberg-Furth,* Karger, Basel, 1985, 485—191.

124. **O'Yang, C., Kertesz, D. J., Kluge, A. F., Kuenzler, P., Li, T.T., Marx, M. M., Bruno, J. J., and Chang, L.,** Synthesis and platelet aggregation inhibition activity of a series of enantiomeric bicyclo [3.2.0] heptane-6-oximinoacetic acids, *Prostaglandins,* 27, 851—863, 1984.

125. **Kluge, A. F., Wu, H. Y., Kertesz, D. J., and O-Yang, C.,** RS-93427, A new potent prostacyclin analogue containing the bicyclo [4.2.0] octane ring system, *6th Int. Conf. on Prostaglandins and Related Compounds,* Florence, Italy, June 1986, Abs. 273B.

126. **Willis, A. L., Smith, D. L., O-Yang, C., Vigo, C., Strosberg, A., Wu, H., Hertesz, D., Kluge, A., and Roszkowski, A. P.,** Pharmacology of RS-93427, an orally active prostacyclin mimetic with potential therapeutic utility in atherosclerotic individuals, *6th Int. Conf. on Prostaglandins and Related Compounds,* Abs. 98B.

127. **Robert, A., Aristoff, P. A., Wendling, M. G., Kimball, F. A., Miller, W. L., and Gorman, R. R.,** Protective and antisecretory properties of a non-diarrheagenic and non-uterotonic prostaglandin analog: U-68,215, *Prostaglandins,* 30, 619—649, 1985.

128. **Aristoff, P. A., Johnson, P. D., and Harrison, A. W.,** Total synthesis of a novel antiulcer agent via a modification of the intramolecular Wadsworth-Emmons-Wittig reaction, *J. Am. Chem. Soc.,* 107, 7967—7974, 1985.
129. **Kecskemeti, V., Stadler, I., and Kormoczy, P.,** Cardiovascular and electrophysiological analysis of a selective vasodilatory prostacyclin analogue, 5-nitro PGI₁, *Biomed. Biochem. Acta,* 43, 5287—5290, 1984.
130. **Baraldi, P. G., Barco, A., Benetti, S., Gandolfi, C. A., Pollini, G. P., Pollo, E., and Simoni, D.,** Synthesis of sulphur containing carbaprostacyclin analogues, *Gazz. Chim. Ital.,* 114, 177—183, 1984.

Table 4
PGF ANALOGS

No.	Systematic name	Structure	Mol. formula (mol. wt.)	Synthesis	Biological actions
1	$9\alpha,11\alpha,15\alpha$-Trihydroxy-15$\beta$-methyl-prosta-4-*cis*-13-*trans*-dienoic acid		$C_{21}H_{36}O_5$ (368)	Corey type[23]	Hamster antifertility (s.c.); $4 \times$ $PGF_{2\alpha}$,[68] resistant to metabolic degradation of α-chain by β-oxidation[4]
2	Ethyl $9\alpha,11\alpha,15\alpha$-trihydroxyprosta-5-*cis*-13-*trans*-dienoate		$C_{22}H_{38}O_5$ (382)	Reaction of $PGF_{2\alpha}$ with ethyl iodide in the presence of diisopropylethylamine (Hunig's base)[1]	Hamster antifertility (s.c.); $2 \times PGF_{2\alpha}$[1]
3	(+)1-Dimethylamino-$9\alpha,11\alpha,15\alpha$-trihydroxyprosta-5-*cis*-13-*trans*-prostadiene		$C_{22}H_{41}NO_3$ (367)	Reaction of $PGF_{2\alpha}$ with dimethylamine followed by lithium aluminum hydride reduction[2]	Specific antagonist of $PGF_{2\alpha}$-mediated increases in mean pulmonary arterial pressure and mean systemic arterial pressure (rats, i.v.); compound alone has no effect on these parameters;[3] competitively antagonized $PGF_{2\alpha}$-mediated contractile response on isolated gerbil colon; compound alone has no effect on the muscle[2]
4	Methyl $9\alpha,11\alpha,15\alpha$-trihydroxyprosta-13-*trans*-en-5-ynoate		$C_{21}H_{34}O_5$ (366)	Prepared by bromination, dehydrobromination of $PGF_{2\alpha}$ tribenzoate methyl ester[5]	Methyl ester less active than natural compound on gerbil colon and in hamster antifertility assays[5]
5	Methyl 5-oxa-$9\alpha,11\alpha,15\alpha$-trihydroxy-17-phenyl-ω-trinorprosta-13-*trans*-enoate		$C_{23}H_{34}O_6$ (406)		Fertility control: injection of the compound (7.5 mg, i.m.) plus Meteneprost (Cpd. 26, Table 5, 0.5 mg) in rhesus monkeys on day 28 of fertile cycle terminated pregnancy in all cases (10/10); mild anorexia observed in one monkey; the two agents together were more effective than each singly[64]

#	Compound	Structure	Formula (MW)	Preparation	Activity
6	Methyl 5-fluoro-9α,11α,15α-trihy-droxyprosta-5-cis-13-trans-dienoate		$C_{21}H_{35}FO_5$ (386)	From the addition product of the lithium enolate of methyl 2-fluoro-6-hydroxyhexanoate (protected as 2-methoxy-2-methylethyl ether) and a Corey type aldehyde with THP protecting groups at C-9,11,15[6]	
7	5α- or 5β,9α,11α,15α-Tetrahydroxyprosta-6-trans-13-trans-dienoic acid		$C_{20}H_{34}O_6$ (370)	Photooxygenation of PGF$_{2α}$ 11,15-diacetate methyl ester, chromatographic separation and hydrolysis[7]	5β-Isomer was 0.002 × PGF$_{2α}$ for contraction of isolated hamster fundus[7]
8	6α- or 6β,9α,11α,15α-Tetrahydroxyprosta-4-trans-13-trans-dienoic acid		$C_{20}H_{34}O_6$ (370)	Part of mixture obtained with compound 7[7]	
9	5,6-Dinor-4,7-inter-o-phenylene-9α,11α,15α-trihydroxyprosta-13-trans-enoic acid		$C_{24}H_{36}O_5$ (404)	Conjugate addition (cuprate) of protected ω-chain to fully elaborated 11-hydroxylated cyclopentenone[8]	Inactive in vitro on standard smooth muscle assays[8]
10	Methyl 9α,11α-15ξ-trihydroxy-15ξ-meth-ylprosta-4,5,13-trans-trienoate (prostalene)		$C_{22}H_{36}O_5$ (380)	Allene unit introduced by addition of dilithio-1-hydroxy-1-hydroxy-4-pentyne to Corey type lactol, acetylation, and reaction with lithium dimethyl copper[9]	Synchronizes estrus in mares at dose of 1.5 mg (s.c.); no side effects such as sweating or diarrhea and no affect on heart or respiration rates[10]

Table 4 (continued)
PGF ANALOGS

No.	Systematic name	Structure	Mol. formula (mol. wt.)	Synthesis	Biological actions
11	9β-Hydroxy-ω-trinor-prostanoic acid (Rosaprostol)		$C_{18}H_{34}O_3$ (298)		Inhibited gastric acid secretion in pylorus ligated rats with ED_{50} = 130 mg/kg (i.p.); inhibited pentagastrin induced gastric acid secretion in Heidenhain pouch dogs at 100 mg/kg (p.o.); protected against acute gastric lesions induced in rats by aspirin, indomethacins 0.6 N HCl, and ethanol, with ED_{50} values of 135, 132, 73, and 132 mg/kg, respectively No cardiovascular activity and no effect on uterine motility and GI transit[11] In duodenal ulcer patients, 500 mg (q.i.d.) orally for 6 weeks was nearly equivalent to cimetidine in healing (78.9 vs. 90%); see Reference 116, p. 99
12	9α,11α-Dihydroxy-prosta-5,6,13-*trans*-trienoic acid		$C_{20}H_{32}O_4$ (336)	From product of reaction of acetylenic acetate (protected primary alcohol at C-1) with lithium 1-pentynylbutyl copper[16]	
13	(+) 9α-Fluoro-11α,15α-dihydroxyprosta-5-*cis*-13-*trans*-dienoic acid		$C_{20}H_{33}FO_4$ (356)	Reaction of methyl 9β-hydroxy-11α,15α-*bis*-tetra-hydropyranyloxy-prosta-5-*cis*-13-*trans*-dienoate with diethyl-(2-chloro-1,1,2-tri-fluoroethyl)amine (fluoramine reagent) and subsequent deprotection[26]	Hamster antifertility (s.c.): 0.02 × $PGF_{2\alpha}$; gerbil colon; 0.5 × PGE_2 Hamster uterus; 1 × $PGF_{2\alpha}$[26]

No.	Compound	Structure	Formula (MW)	Synthesis	Biological activity
14	(+) 9β-Fluoro-11α,15α-dihydroxy-prosta-5-cis-13-trans-dienoic acid		$C_{20}H_{33}FO_4$ (356)	Reaction of $PGF_{2\alpha}$-11,15-*bis*-tetrahydropyranyl ether methyl ester and fluoramine reagent and subsequent deprotection[26]	Bronchial dilator in guinea pig; 4 × PGE_2 (i.v.); 2.5 × PGE_2 (aerosol)[26] Upper airway irritation in man (aerosol);[27] gerbil colon; 8 × PGE_2; hamster uterus; 0.3 × $PGF_{2\alpha}$[26]
15	(−) 9β-Chloro-11α,15α-dihydroxy-prosta-5-cis-13-trans-dienoic acid		$C_{20}H_{33}ClO_4$ (373)	Reaction of $PGF_{2\alpha}$-11-tetrahydropyranyl ether 15-acetate methyl ester with carbon tetrachloride/triphenylphosphine and subsequent deprotection[26]	Bronchial dilator in guinea pig; 0.5 × PGE_2 (i.v.); gerbil colon; 1 × PGE_2; Hamster uterus; 0.4 × $PGF_{2\alpha}$[26]
16	9β,11α,15α-Trihydroxy-16β-methyl-prosta-5-cis-en-13-ynoic acid		$C_{21}H_{34}O_5$ (366)	Inversion of 9α-hydroxy compound via a Mitsunobu[66] reaction[12]	Inhibited histamine-induced bronchial spasms in guinea pigs (i.p.); 5—6 × PGE_2 Inhibited gastric acid secretion in pylorus ligated rats; 0.6 × PGE_2 Inhibited stress ulcers in rats; 0.7 × PGE_2[12]
17	Methyl 9α,11α,15α-trihydroxy-10α-fluoroprosta-5-cis-13-trans-dienoate		$C_{21}H_{35}FO_5$ (386)	Corey-like process with introduction of the 10α-fluoro group at the bicyclic lactone stage[24]	
18	Methyl 9α,11α,15α-trihydroxy-10β-fluoroprosta-5-cis-13-trans-dienoate		$C_{21}H_{35}FO_5$ (386)	Reaction of 10α,11α-epoxy-$PGF_{2\alpha}$ with potassium fluoride in hot ethylene glycol[25]	
19	(+) 9α,15α-Dihydroxy-11β-fluoro-prosta-5-cis-13-trans-dienoic acid		$C_{20}H_{33}FO_4$ (356)	Reaction of $PGF_{2\alpha}$-9-tetrahydropyranyl ether 15-acetate methyl ester with fluoramine reagent and subsequent deprotection[26]	Hamster antifertility (s.c.); 0.4 × $PGF_{2\alpha}$; hamster uterus; 0.3 × $PGF_{2\alpha}$[26]

Table 4 (continued)
PGF ANALOGS

No.	Systematic name	Structure	Mol. formula (mol. wt.)	Synthesis	Biological actions
20	Methyl 9α,11α,12α,15α-tetrahydroxyprosta-5-*cis*-13-*trans*-dienoate		$C_{21}H_{36}O_6$ (384)	Via a "Corey-like" bicyclic lactone with 12α-hydroxy group in place[28]	Hamster antifertility (s.c.); 1 × PGF$_{2\alpha}$; Essentially inactive on gerbil colon and hamster uterus assays[28]
21	(+) 9α,11α,15α-Tri-hydroxy-12α-fluoro-prosta-5-*cis*-13-*trans*-dienoic acid		$C_{20}H_{33}FO_5$ (372)	From (−) 7-fluoro-spiro[bicyclo[2.2.1]hept-5-ene-2,2′-[1,3]dioxolane]-7-methanol[13]	Hamster antifertility assay; 25 × PGF$_{2\alpha}$(s.c.); hamster uterine concon-traction; 0.65 × PGF$_{2\alpha}$; gerbil colon; 0.28 × PGF$_{2\alpha}$[13]
22	(+) Methyl 9α,11α,15α-trihydroxy-12α-fluoroprosta-5-*cis*-enoate		$C_{21}H_{37}FO_5$ (388)	From (−) 7-fluorospiro[bicyclo[2.2.1]hept-5-ene-2,2′-[1,3]di-oxolane]-7-methanol[14]	2 × PGF$_{2\alpha}$ as an antifertility agent (s.c.) in hamsters; 0.014 and 0.004 × PGE$_{2\alpha}$ in hamster uterine contraction and gerbil colon assays, respectively[14]
23	Methyl 9α,11α,15α-trihydroxy-12α-hy-droxymethylprosta-5-*cis*-13-*trans*-dienoate		$C_{22}H_{38}O_6$ (398)	From 2-*exo*-bromo-7-*syn*-methoxycarbonylbicyclo [2.2.1]heptan-5-one ethylene ketal[15]	Hamster antifertility (s.c.): inactive at 125 μg/hamster[15]
24	(+) Methyl 9α,11α,15α-trihydroxy-12α-fluoromethylprosta-5-*cis*-13-*trans*-dienoate		$C_{22}H_{32}FO_5$ (400)	From (1α,4α,5α,7R)-5-bromo-7-(fluoro-methyl)-7-formyl-spiro[bicyclo[2.2.1] heptane-2,2′-[1,3] dioxolane][67]	Inactive in hamster antifertility assay at 25 μg/hamster (s.c.)[67]
25	(+) 9α,11α,15β-Tri-hydroxy-11-aza-12-keto-16-(4-fluoro-phenoxy)-ω-tetranor-prosta-5-*cis*-enoic acid		$C_{21}H_{28}FNO_6$ (409)	Oxidation of Corey lac-tone aldehyde 11-ace-tate to carboxylic acid followed by a Curtius reaction α-Chain then attached in normal way followed by introduction of ω-chain by acylation[65]	Hamster antifertility (s.c.); 1 × PGF$_{2\alpha}$[65]

No.	Name	Structure	Formula	Synthesis	Activity
26	(+) Methyl 9α,11α, 15β-trihydroxy-14-fluoroprosta-5-*cis*-13-*trans*-dienoate		$C_{21}H_{13}FO_5$ (384)	From (E)-enone derived from (−)-spiro[bicyclo[2.2.1]-hept-5-ene-2,2′-[1,3]dioxolane] 7-methanol and dimethyl 1-fluoro-2-oxoheptylphosphonate[17,18]	Hamster antifertility assay; 0.5 × $PGF_{2\alpha}$ (s.c.); Hamster uterine contraction ; 0.025 × $PGF_{2\alpha}$; gerbil colon; 0.003—0.004 × $PGF_{2\alpha}$[17,18]
27	(+) Methyl 9α,11α, 15α-trihydroxy-14-fluoroprosta-5-*cis*-13-*cis*-dienoate		$C_{21}H_{13}FO_5$ (384)	From (E)-enone derived from (−)-spiro[bicyclo[2.2.1] hept-5-ene-2,2′-[1,3]dioxolane]-7-methanol and dimethyl 1-fluoro-2-oxoheptylphosphonate[18]	Hamster antifertility assay 3—4 × $PGF_{2\alpha}$ (s.c.); hamster uterine contraction ; 0.36—1.28 × $PGF_{2\alpha}$; gerbil colon; 0.15—0.26 × $PGF_{2\alpha}$[18]
28	9α,11α,15α-Trihydroxy-14-phenyl-prosta-5-*cis*-13-*trans*-dienoic acid		$C_{26}H_{18}O_5$ (430)	Via cuprate addition of protected ω-chain to fully elaborated cyclopentenone[29]	0.001 × $PGF_{2\alpha}$ in various typical pharmacological assays[29]
29	9α,11α,14ξ-Trihydroxyprosta-5-*cis*-enoic acid		$C_{20}H_{36}O_5$ (356)	Modified Corey type synthesis[30]	Hamster uterus; <0.02 × PGE_2[30]
30	9α,11α,16ξ-Trihydroxy-20-methyl-prosta-5-*cis*-13,14-trienoic acid		$C_{21}H_{34}O_5$ (366)	One of four diasterioisomeric racemates derived from reaction of lithium dimethylcopper with acetate obtained from addition product of lithium salt of oct-1-yn-3-ol tetrahydroxypyranyl ether and 11-protected bicyclic Corey aldehyde[31]	Hamster antifertility (s.c.): 0.2 × $PGF_{2\alpha}$[31]
31	9α,11α,15α-Trihydroxy-15β-methyl-prosta-5-*cis*-13-*trans*-dienoic acid (Carbaprost)		$C_{21}H_{34}O_5$ (366)	Oxidation of $PGF_{2\alpha}$ with DDQ, followed by silylation, addition of methylmagnesium bromide, and deprotection[61]	Induction of 1st and 2nd trimester abortion; incidence of GI side effects high (about 50%), but the severity thereof can be reduced with anti-emetic and antidiarrheal agents; induction of cervical dilation (3rd and 4th trimester)[19]

Table 4 (continued)
PGF ANALOGS

No.	Systematic name	Structure	Mol. formula (mol. wt.)	Synthesis	Biological actions
32	9α,11α,15α-Trihydroxyprost-5-*cis*-13-*trans*-trienoic acid-1,15-olide		$C_{20}H_{32}O_5$ (352)	Lactonization of PGF$_{2\alpha}$ 9,11-*bis*-THP ether using dipyridyldisulfide-triphenylphosphine procedure[22]	Hamster antifertility (s.c.); 2 × PGF$_{2\alpha}$; gerbil colon; 0.001 × PGF$_{2\alpha}$; hamster uterus; 0.007 × PGF$_{2\alpha}$[21]
33	9α,11α,15α-Trihydroxy-15-(4-trifluoromethylphenyl)-ω-pentanor-5-*cis*-13-*trans*-dienoic acid		$C_{23}H_{27}F_3O_5$ (440)	Corey type[32]	Hamster antifertility (s.c.); 50 × PGF$_{2\alpha}$[32]
34	Methyl 9α,11α,15α-trihydroxy-15-benzo[b]-furan-2-yl-ω-pentanorprosta-5-*cis*-13-*trans*-dienoate		$C_{24}H_{30}O_6$ (414)	Addition of 2-lithio benzo[b]furan to ω-hexanor-15-aldehydo PGF$_{2\alpha}$ derivative[52]	Rat antifertility (s.c.); 50—100 × PGF$_{2\alpha}$[52]
35	Methyl 9α,11α,15ξ-trihydroxy-15,17-methano-17-ξ-phenoxy-ω-trinorprosta-5-*cis*-13-*trans*-dienoate		$C_{25}H_{34}O_6$ (430)	Modified Corey synthesis[33]	Rat antifertility; 5 × PGF$_{2\alpha}$ methyl ester[33]
36	9α,11α,15α-Trihydroxy-16,16-dimethyl-prosta-5-*cis*-13-*trans*-dienoic acid		$C_{22}H_{36}O_5$ (380)	Corey synthesis; introduction of α-chain first[20]	Hamster antifertility (s.c.); 1.2 × PGF$_{2\alpha}$; isolated gerbil colon; 0.6 × PGF$_{2\alpha}$; hamster uterus; 1.8 × PGF$_{2\alpha}$[21]
37	Methyl 9α,11α,15α-trihydroxy-(E)-16-fluoromethylene-prosta-5-*cis*-13-*trans*-dienoate		$C_{22}H_{35}FO_5$ (398)	Corey synthesis[37]	Rat uterine contractility (i.v.); 4 × PGF$_{2\alpha}$; mouse diarrhea (p.o.); 1 × PGF$_{2\alpha}$[37]

No.	Structure	Name	Formula	Synthesis	Activity
38		Methyl 9α,11α,15α-trihydroxy-16-cyclohexyl-ω-tetranorprosta-5-cis-13-ynoic acid (alfaprostol)	$C_{22}H_{34}O_5$ (378)		In Hanoverian mares, 2—4 mg (i.v.) induced heat and ovulation; used to synchronize estrus but not ovulation;[34] in sows, 2—3 mg (i.m.) induced parturition; reduced incidence of stillbirths and piglet losses during first 10 days of life[35]
39		9α,11α,15α-Trihydroxy-16-(3-chlorophenoxy)-ω-tetranorprosta-5-cis-13-trans-dienoic acid (estrumate)	$C_{22}H_{29}ClO_6$ (425)	Corey synthesis[36]	Synchronization of estrus in cattle and swine; induction of parturition in sows[36]
40		9α,11α,15α-Trihydroxy-16-(3-trifluoromethyl-phenoxy)-ω-tetranorprosta-5-cis-13-trans-dienoic acid (equimate)	$C_{23}H_{29}F_3O_6$ (458)	Corey synthesis[36]	Induction of estrus in mares[36]
41		9α,11α,15α-Trihydroxy-16-(3-thienyloxy)-ω-tetranorprosta-5-cis-13-trans-dienoic acid (tiaprost)	$C_{20}H_{28}O_6S$ (396)	Corey synthesis[43]	Induction of parturition, synchronization of estrus in cattle and horses; induction of parturition in pigs with concomitant decrease in mastitis, metritis, agalactia (MMA) syndrome; treatment of pyometra in cattle[44]
42		Methyl 9α,11α,15α-trihydroxy-16-(3-chlorophenoxy)-ω-tetranorprosts-2-trans-5-cis-13-trans-trienoate (delprostenate)	$C_{23}H_{29}ClO_6$ (437)	Corey type with introduction of α-chain first Double bond at C-2 introduced by phenylselenation and then selenoxide elimination[38]	Rat antifertility (s.c.); 200 × $PGF_{2\alpha}$;[38] in cows, 800 µg in 2 doses (i.m.), 10 days apart, caused estrus in 100% of animals and could be used to synchronize estrus; in cows 50 µg (i.m.), on day 112—113 of gestation, caused parturition in 80% of animals within 36 hr; in mares, 250 µg induced estrus and fertilization thereafter was effective in 63% of animals[39]

Table 4 (continued)
PGF ANALOGS

No.	Systematic name	Structure	Mol. formula (mol. wt.)	Synthesis	Biological actions
43	Methyl 9α,11α,15α-trihydroxy-16-(3-trifluoromethyl)phenoxy)-ω-tetranorprosta-2-*trans*-5-*cis*-13-*trans*-dienoate (ONO-995)		$C_{24}H_{29}F_3O_6$ (470)	Synthesis analogous to that used for *m*-chloro compound (no. 42)[38]	Rat antifertility (s.c.); 100 × PGF$_{2\alpha}$,[38] in rabbits, i.v. administration increased rhythmical movement of the trigonal muscle and resistance of bladder neck; may be useful for treatment of neurogenic bladder[40,41]
44	9α,11α,15α-Trihydroxy-16-(3-chlorophenoxy)-ω-tetranor-4-*cis*-13-*trans*-dienoic acid		$C_{22}H_{27}ClO_6$ (423)	Modified Corey synthesis with attachment of α-chain first[42]	Hamster antifertility (i.v.); 0.05 μg (minimum effective dose); this compound is equiactive to estrumate (no. 39)[42]
45	Methyl 9α,11α,15α-trihydroxy-16-phenoxy-ω-tetranorprosta-4,5,-13-*trans*-trienoate (fenprostalene)		$C_{23}H_{30}O_6$ (402)	Allene unit introduced by addition of dilithio-4-pentynoic acid to Corey lactol containing protected ω-chain; the product was esterified, acetylated, and then reacted with lithium dimethyl copper; deprotection gave fenprostalene[45]	Induction of abortion in feedlot cattle; synchronization of estrus in cattle[46,62,63]
46	Methyl 9α,11α-*bis*-acetylamino-11α-hydroxy-16-(3-chlorophenoxy)-ω-tetranorprosta-5-*cis*-13-*trans*-dienoate		$C_{27}H_{37}ClN_2O_6$ (521)	From a PGF intermediate with intact upper side chain and unprotected α-hydroxyl groups at C-9,11; reaction sequence involved inversion of hydroxyl groups via a Mitsunobu reaction followed by reaction of the dimesylate with azide ion[47]	Inactive as luteolytic agent[47]

No.	Name	Molecular formula (MW)	Synthesis	Biological activity
47	11α,15α-Dihydroxy-16-(3-chlorophenoxy)-ω-tetranor-prosta-5-cis-9,13-trans-trienoic acid	$C_{22}H_{27}ClO_5$ (407)		Isolated rat uterus; 2.2 × $PGF_{2\alpha}$. Rat antifertility (i.v.); ≥ 40 × $PGF_{2\alpha}$
48	9α,15α-Dihydroxy-16-(3-thienyloxy)-ω-tetranor-prosta-5-cis-13-trans-dienoic acid	$C_{20}H_{28}O_5S$ (380)	From *trans*-2-methoxy-carbonyl-3-tetrahydro-pyranyl-oxymethylcyclopenta-none[49]	Hamster antifertility (s.c.); ~100 × $PGF_{2\alpha}$. Luteolysis in heifers (i.m.); 0.5 mg (threshold dose)[48]. Hamster antifertility; 25 × $PGF_{2\alpha}$[49]
49	9α,11α,15α-Trihydroxy-13-thia-16-(3-chlorophenoxy)-ω-tetranorprosta-5-cis-enoic acid	$C_{21}H_{29}ClO_6S$ (445)		Hamster antifertility (s.c.); 50 × $PGF_{2\alpha}$. Threshold dose for luteolysis in heifers (i.m.); ≥ 2.5 mg; single 15 mg dose (i.m.) causes luteolysis and synchronization of estrus in lactating cows[48]
50	9α,11α-Dihydroxy-15-keto-16-phenoxy-ω-tetranor-prosta-5-cis-13-trans-dienoic acid 15-ethylene ketal	$C_{24}H_{32}O_7$ (432)	Modified Corey synthesis[50]	Rat abortifacient (s.c.); 10-30 × $PGF_{2\alpha}$ contraction of isolated rat uterus; 10 × $PGF_{2\alpha}$; guinea pig ilium; 0.1 × $PGF_{2\alpha}$; used to synchronize estrus in heifers and cows[50]
51	Methyl 9α,11α,15α-trihydroxy-17-(2-chlorophenyl)-ω-trinorprosta-5-cis-13-trans-16-ynoate	$C_{24}H_{29}ClO_5$ (433)	Addition of the lithium acetylide to the protected ω-hexanor-15-aldehyde[52]	Rat antifertility (s.c.); 40—80 × $PGF_{2\alpha}$[52]
52	Methyl 9α,11α,15α-trihydroxy-17-(2-furyl)-ω-trinor-prosta-5-cis-13-trans-16-ynoate	$C_{22}H_{28}O_6$ (388)	Corey type[51]	Hamster antifertility (s.c.); 50 × (±) $PGF_{2\alpha}$; isolated guinea pig uterus; 0.05 × (±) $PGF_{2\alpha}$[51]

Table 4 (continued)
PGF ANALOGS

No.	Systematic name	Structure	Mol. formula (mol. wt.)	Synthesis	Biological actions
53	Methyl 9α,11α,15α-trihydroxy-16ξ,17ξ-methano-17-phenyl-ω-trinorprosta-5-cis-13-trans-dienoate		$C_{25}H_{34}O_5$ (414)	Corey type[51]	Hamster antifertility (s.c.); 50 × (±) PGF$_{2α}$; isolated guinea pig uterus; 0.1 × (±) PGF$_{2α}$[51]
54	Methyl 9α,11α,15α-trihydroxy-17-keto-17-phenyl-ω-trinorprosta-5-cis-13-trans-dienoate		$C_{24}H_{32}O_6$ (416)	Addition of lithium enolate of acetophenone to protected hexanor-15-aldehyde[52]	Rat antifertility (s.c.); 40 × PGF$_{2α}$[52]
55	9α,11α,15α-Trihydroxy-17,17-dimethylprosta-5-cis-13-trans-dienoic acid		$C_{22}H_{38}O_5$ (382)	Corey synthesis; introduction of α-chain first[20]	Hamster infertility (s.c.); 1.9 × PGF$_{2α}$; isolated gerbil colon; 0.14 × PGF$_{2α}$; isolated hamster uterus; 0.11 × PGF$_{2α}$[21]
56	9α,11α,15α-Trihydroxy-18,18,20-trimethylprosta-5-cis-13-trans-dienoic acid		$C_{23}H_{40}O_5$ (396)	Corey synthesis; introduction of α-chain first[20]	Hamster antifertility (s.c.); 4.6 × PGF$_{2α}$; gerbil colon; 0.9 × PGF$_{2α}$; hamster uterus; 0.8 × PGF$_{2α}$[21]
57	9α,11α,15α-Trihydroxy-18-cyano-ω-bisnorprosta-5-cis-13-trans-dienoic acid		$C_{19}H_{29}NO_5$ (351)	Corey type[53]	
58	9α,11α,15α-Trihydroxy-20-N,N-pentamethylene-carboxamidomethyl)-prosta-5-cis-13-trans-dienoic acid		$C_{27}H_{45}NO_6$ (479)	Corey type[53]	
59	Methyl 9α,16α,-dihydroxy-12-iso-11α,13α-methanoepoxy-13-thia-20-methyl-prosta-5-cis-enoate		$C_{22}H_{38}O_5S$ (414)	From aucubin[54]	Hamster antifertility; 0.1 × PGF$_{2α}$; isolated hamster fundus; 0.002 × PGF$_{2α}$[54]

No.	Name	Structure	Formula (MW)	Synthesis	Biological activity
60	Methyl 9α,11,15-trihydroxy-12,15-dimethyleneprosta-5-*cis*-dienoate		$C_{23}H_{40}O_5$ (396)	From cyclopentane-1,3-dione[55]	No affect on i.v. adminstration to pregnant rats at 200 γ/kg; no affect on blood pressure in rabbits or dogs at 50 γ/kg (i.v.)[55]
61	Ethyl 9α,11α,15α-trihydroxy-9a-*homo*-prosta-5-*cis*-13-*trans*-dienoate		$C_{23}H_{40}O_5$ (396)	From Diels-Alder adduct of 1,4-diacetoxy-butadiene and dimethylfumarate[56]	
62	Methyl 9α,11α,15α-trihydroxy-11a-*homo*-11a-fluoroprosta-5-*cis*-11a, 13-*trans*-trienoate		$C_{22}H_{35}FO_5$ (398)	Addition of chlorofluorocarbene to Corey dienone followed by ring expansion and subsequent elaboration of the bicyclic product[57]	Much less active (<0.01) than $PGF_{2\alpha}$ in several standard assays[57]
63	9α,15α-Dihydroxy-11-nor-prosta-5-*cis*-13-*trans*-dienoic acid		$C_{19}H_{30}O_4$ (322)	From the adduct of cyclopentadiene and dichloroketene[58,59]	
64	(+) Methyl 9α-hydroxy-methyl-11-nor-15α-hydroxyprosta-5-*cis*-13-*trans*-dienoate		$C_{21}H_{36}O_4$ (352)	From 11,12-epoxy derivative of PGF_2 methyl ester[60]	Hamster antifertility (s.c.); 0.1 × $PGF_{2\alpha}$[60]
65	Methyl 9β-chloro-11α,15α-dihydroxy-16,16-dimethyl-prosta-5-*cis*-13-*trans*-dienoate (Nocloprost)		$C_{23}H_{39}ClO_4$ (414)		Antisecretory and cytoprotective activity in rats comparable to that of 16,16-dimethyl PGE_2 but with a much lower diarrhea induction potential; see Reference 69 pp. 99—100

Table 4 (continued)
PGF ANALOGS

No.	Systematic name	Structure	Mol. formula (mol. wt.)	Synthesis	Biological actions
69	(+)Methyl 9a,11a,15a-trihydroxy-16ß-aminoprosta-5-cis-13-/ra«.s-dienoate		C21H35NO5 (385)	Modified Corey synthesis[70]	Acetic acid salt not deactivated by 15-hydroxyprostaglandin dehydrogenase; abortifacient in mice (week three, ED50 = 0.07 (µg/kg, s.c; 250 x PGF2a), hamsters (week one, ED50 = 0.2 (xg/day/animal, s.c; 90 x PGF2a) and rats (ED50 = 165 fxg/day/animal, s.c; 7 x PGF2a) with no enteropooling (rats, 20 mg/kg, s.c.) or diarrhea (mice, 5 mg/kg, s.c.) effects[70]

REFERENCES

1. **Morozowich, W., Osterling, T. O., Miller, W. L., and Douglas, S. C.,** Prostaglandin prodrugs. II. New method for synthesising prostaglandin C_1-aliphatic esters, *J. Pharm. Sci.,* 68, 836—838, 1979.

2. **Maddox, Y. T., Ramwell, P. W., Shriner, C. S., and Corey, E. J.,** Amide and 1-amino derivatives of F prostaglandins as prostaglandin antagonists, *Nature (London),* 273, 549—552, 1978.

3. **Stinger, R. B., Fitzpatrick, T. M., Corey, E. J., Ramwell, P. W., Rose, J. C., and Kot, P. A.,** Selective antagonism of prostaglandin $F_{2\alpha}$-mediated vascular responses by *N*-dimethylamino substitution of prostaglandin $F_{2\alpha}$, *J. Pharm. Exp. Ther.,* 200, 521—525, 1982.

4. **Hansson, G.,** Metabolism of two $PGF_{2\alpha}$ analogues in primates; 15(S)-15-methyl-Δ^4-*cis*-$PGF_{1\alpha}$ and 16,16-dimethyl-Δ^4 *cis*-$PGF_{1\alpha}$- *Prostaglandins,* 18, 745—771, 1979.

5. **Lin, C. H., Stein, S. J., and Pike, J. E.,** The synthesis of 5,6-acetylenic prostaglandins, *Prostaglandins,* 11, 377—380, 1976.

6. **Nakai, H., Hamanka, N., Miyake, H., and Hayashi, M.,** Synthesis of 5-fluoroprostaglandins, *Chem. Lett.,* 1499—1502, 1979.

7. **Nishiyama, H. and Ohno, K.,** The synthesis of 6,7-didehydro-5-hydroxy and 4,5-didehydro-6-hydroxy-prostaglandin $F_{1\alpha}$; photosensitized oxygenation of prostaglandin $F_{2\alpha}$;, *Chem, Lett.,* 661—664, 1979.

8. **Buckler, R. T., Ward, F. E., Hartzler, H. E., and Kurchacova, E.,** The synthesis and biological activity of benzo[5,6]prostaglandins A_2, E_2 and $F_{2\alpha}$, *Eur. J. Med. Chem.,* 12, 463—465, 1977.

9. **Crabbé, P. and Carpio, H.,** Synthesis of allenyl prostaglandins, *J. Chem. Soc. Chem. Commun.,* 904—905, 1972.

10. Prostalene, *Drugs of the Future,* 2, 755—760, 1977.

11. **Valcavi, U., Caponi, R., Brambilla, A., Palmira, M., Monoja, F., Bernini, R., Musanti, R., and Fumagalli, R.,** Gastric, antisecretory, antiulcer and cytoprotective properties of 9-hydroxy-19,20-bis-nor-prostanoic acid in experimental animals, *Arzneim. Forsch. Drug. Res.,* 32, 657—663, 1982.

12. **Gandolfi, C., Fumagalli, A., Pellegata, R., Doria, G., Ceserani, R., and Franceschini, J.,** Prostaglandins. X. An improved route for synthesis of $PGF_{2\beta}$ derivatives, *Il Farmaco Ed. Sci.,* 31, 649—654, 1976.

13. **Grieco, P. A., Owens, W., Wang, G.-L., Williams, E., Schillinger, W. J., Hirotsu, K., and Clardy, J.,** Fluoroprostaglandins; synthesis and biological evaluation of the methyl esters of (+)-12-fluoro-, (−)-*ent*-12-fluoro-(+)-15-*epi*-fluoro- and (−)-*ent*-15-epi-12-fluoroprostaglandin $F_{2\alpha}$, *J. Med. Chem.,* 23, 1072—1077, 1980.

14. **Grieco, P. A. and Takigawa, T.,** Synthesis and biological evaluation of the methyl esters of (+)-12-fluoro-13,14-dihydroprostaglandin $F_{2\alpha}$ and (+)-15-*epi*-12-Fluoro-13,14-dihydroprostaglandin $F_{2\alpha}$, *J. Med. Chem.,* 24, 839—843, 1981.

15. **Grieco, P. A., Wang, C.-L., and Okuniewicz, F. J.,** Total synthesis of 12-hydroxymethylprostaglandin $F_{2\alpha}$ methyl ester, *J. Chem. Soc. Chem. Commun.,* 939—940, 1976.

16. **Baret, P., Barreiro, E., Greene, A. E., Luche J.-L., Teixeira, M.-A., and Crabbé, P.,** Synthesis of new allenic prostaglandins, *Tetrahedron,* 35, 2931—2938, 1979.

17. **Grieco, P. A., Yokoyama, Y., Nicolaou, K. C., Barnette, W. E., Smith, J. B., Ogletree, M., and Lefer, A. M.,** Total synthesis of 14-fluoroprostaglandin $F_{2\alpha}$ and 14-fluoroprostacyclin, *Chem. Lett.,* 1001—1004, 1978.

18. **Grieco, P. A., Schillinger, W. J., and Yokoyama, Y.,** C(14)-fluorinated prostaglandins: synthesis and biological evaluation of the methyl esters of (+)-14-fluoro-, (+)-15-*epi*-14-fluoro-, (+)-13-(E)-14-fluoro-, and (+)-13-(E)-15-*epi*-fluoroprostaglandins $F_{2\alpha}$, *J. Med. Chem.,* 23, 1077—1083, 1980.

19. 15-Methyl $PGF_{2\alpha}$, *Drugs of the Future,* 6, 652—653, 1981.

20. **Anderson, N. H. Subramanian, N., Inamoto, S., Picker, D. H., Ladner, D. W., McCrae, D. A., Lin, B., and De, B.,** Omega chain methylated analogs of $PGF_{2\alpha}$ and PGE_2, *Prostaglandins,* 22, 809—830, 1981.

21. **Anderson, N. H., Inamoto, S., Subramanian, N., Picker, D. H., Ladner, D. W., De, B., Tynan, S. S., Eggerman, T. L., Harker, L. A., Robertson, R. P., Oien, H. G., and Rao, C. V.,** Molecular basics for prostaglandin potency III. Tests of the significance of the "hairpin conformation" in biorecognition phenomena, *Prostaglandins,* 841—856, 1981.

22. **Anderson, N. H., Inamoto, S., and Subramanian, N.,** Synthesis of macrolide prostaglandin analogs, *Prostaglandins,* 831—840, 1981.

23. **Nidy, E. G. and Johnson, R. A.,** Synthesis of 15-methyl-cis-Δ^4-prostaglandins, *J. Org. Chem.,* 45, 1121—1125, 1980.

24. **Grieco, P. A., Williams, E., and Sugahara, T.,** Ring-fluorinated prostaglandins; total synthesis of (+)-10α-fluoroprostaglandin $F_{2\alpha}$ methyl ester, *J. Org. Chem.,* 44, 2194—2199, 1979.

25. **Grieco, P. A., Sugahara, T., Yokoyama, Y., and Williams, E.,** Fluoroprostaglandins: synthesis of (±)-10β-fluoroprostangladin $F_{2\alpha}$ methyl ester, *J. Org. Chem.,* 2189—2193, 1979.

26. **Arróniz, C. E., Gallina, J., Martinez, E., Muchowski, J. M., and Velarde, E.,** Synthesis of ring halogenated prostaglandins, *Prostaglandins,* 16, 47—65, 1978.

27. **Muchowski, J. M.,** Synthesis and bronchial dilator activity of some novel prostaglndin analogues, in *Chemistry, Biochemistry and Pharmacological Activity of Prostaglandins,* Roberts, S. M. and Scheinmann, F., Eds., Pergamon Press, Oxford, 1979, 39—60.

28. **Grieco, P. A., Yokoyama, Y., Withers, G. P,, Okuniewicz, F. J., and Wang, C. L. J,** C-12-substituted prostaglandins: synthesis and biological evaluation of ($\pm$)-12-hydroxyprostaglandin $F_{2\alpha}$ methyl ester, *J. Org. Chem.,* 43, 4178—4182, 1979.

29. **Buckler, R. T. and Garling, D. L.,** Synthesis of 14-phenylprostaglandins E_1, A_1 and $F_{2\alpha}$, *Tetrahedron Lett.,* 21, 2257—2260, 1978.

30. **Plattner, J. J. and Gager, A. H.,** Synthesis of optically active 14-hydroxyprostaglandins, *Tetrahedron Lett.,* 2479—2482, 1977.

31. **Fried, J. H., Muchowski, J. M., and Carpio, H.,** Synthesis of lower chain allenic prostaglandins, *Prostaglandins,* 14, 807—811, 1977.

32. **Binder, D., Bowler, J., Crossley, N. S., Hutton, J., King, R. R., Lilly, T. J., and Senior, M. W.,** Synthesis and biological activity of 15-arylprostaglandins, *Prostaglandins,* 15, 773—778, 1978.

33. **Niwa, H. and Kurono, M.,** Synthesis of 15,17-methylene-prostaglandins, *Chem. Lett.,* 23—26, 1979.

34. **Martin, J. C., Klug, E., Merkt, H., Himmler, V., and Jöchle, W.,** Luteolysis and cycle synchronization with a new prostaglandin analog for artificial insemination in the mare, *Theriogenology,* 16, 433—446, 1981.

35. **Cerne, F. and Jöchle, W.,** Clinical evaluation of new prostaglandin analog in pigs. I. Control of parturition and of the MMA-syndrome, *Theriogenology,* 459—467, 1981.

36. **Binder, D., Bowler, J., Brown, E. D., Crossley, N. S., Hutton, J., Senior, M., Slater, L., Wilkinson, P., and Wright, N. C. A.,** 16-Aryloxprostaglandins: a new class of potent luteolytic agent, *Prostaglandins,* 6, 87—90, 1974.

37. **Kosuge, S., Nakai, N., and Kurono, M.,** Synthesis of 16-fluoromethylene-prostaglandin $F_{1\alpha}$, *Prostaglandins,* 18, 737-743, 1979.

38. **Hayashi, M., Arai, Y., Wakatsuka, H., Kawamura, M., Konishi, Y., Tsuda, T., and Matsumoto, K.,** Prostaglandin analogues possessing antinidatory effects, II. Modification of the α-chain, *J. Med. Chem.,* 23, 525—535, 1980.

39. Delprostenate, *Drugs Future,* 6, 17—19, 1981.

40. ONO-995, *Drugs Future,* 5, 499—501, 1980.

41. ONO-995, *Drugs Future,* 6, 654, 1981.

42. **Bowler, J., Brown, E. D., Crossley, N. S., Heaton, D. W., Lilley, T. J., and Rose, N.,** Double bond isomers of cloprostenol, *Prostaglandins,* 17, 789—800, 1979.

43. **Bartmann, W., Beck, G., Lerch, U., Teufel, H., and Schölkens, B.,** Luteolytic prostaglandins synthesis and biological activity, *Prostaglandins,* 301—311, 1979.

44. Tiaprost, *Drugs Future,* 8, 141—144, 1983.

45. *Drugs Future,* 7, 812, 1982.

46. **Herschler, R. C.,** Estrus synchronization and conception rates in beef heifers using fenprostalene in both single- and double-injection programs, *Agri-practice,* 4, 28—31, 1983.

47. **Swain, M. L. and Turner, R. W.,** Synthesis of 9α,11α-diacylamino-9,11-dideoxyprostaglandin $F_{2\alpha}$ analogues, *J. Chem. Soc. Chem. Commun.,* 840—841. 1981.

48. **DeVries, H. and Feenstra, H.,** The selection of prostaglandin $F_{2\alpha}$ analogues for luteolysis in cattle, *J. Vet. Pharmacol. Ther.,* 2, 223—229, 1979.

49. **Buendia, J. and Schalbar, J.,** Synthese de nouvelles desoxy-11 prostaglandines $F_{2\alpha}$, *Tetrahedron Lett.,* 4499—4502, 1977.

50. **Skuballa, W., Raduchel, B., Loge, O., Elger, N., and Vorbrüggen, H.,** 15,15-Ketals of natural prostaglandins and prostaglandin analogues. Synthesis and biological activities, *J. Med. Chem.,* 21, 443—447, 1978.

51. **Fletcher, D. G., Gibson, K. H., Moss, M. R., Sheldon, D. R., and Walker, E. R. H.,** Synthesis and biological activity of 16,17-configurationally-rigid-17-aryl-18,19,20-prostaglandins, *Prostaglandins,* 12, 493—500, 1976.

52. **Hayashi, M., Miyake, H., Kori, S., Tanouchi, T., Wakatsuka, H., Arai, Y., Yamamoto, T., Kajiwara, I., Konishi, Y., Tsuda, T., and Matsumoto, K.,** Prostaglandin analogues possessing antinidatory effects. I. Modification of the α-chain, *J. Med. Chem.,* 23, 519—524, 1980.

53. **Disselnkötter, H., Lieb, F., Oediger, H., and Wendisch, D.,** Synthese von prostaglandin-analoga, *Liebigs Ann. Chem.,* 150—166, 1982.

54. **Ohno, K. and Naruto,** Synthesis of novel prostanoids having a cyclopenta[c]furan structure with a hemithioacetal side chain from aucubin, *Chem. Lett.,* 175—176, 1980.

55. **Hamanaka, N., Nakai, H., and Kurono, M.,** Synthesis of dℓ-12,15-Ethylene-13,14-dihydro-prostaglandin-$F_{2\alpha}$ methyl ester, *Bull. Chem. Soc. Jpn.,* 53, 2327—2329, 1980.

56. **Eggelte, T. A., de Koning, H., and Huisman, H. O.**, Synthesis of six-membered ring analogues of prostaglandin F$_{2\alpha}$, *Chem. Lett.*, 433—436. 1977.

57. **Muchowski, J. M., and Verlarde, E.**, Synthesis of six-membered prostaglandin analogues, *Prostaglandins*, 10, 297—302, 1975.

58. **Greene, A. E., Deprés, J. P., Meana, M. C., and Crabbé, P.**, Total synthesis of 11-nor prostaglandins, *Tetrahedron Lett.*, 3755—3758. 1976.

59. **Reuschling, D., Kuhlein, K., and Linkies, A.**, Synthese von cyclobutanprostaglandin, *Tetrahedron Lett.*, 17—18, 1977.

60. **Guzmán, A., Muchowski, J. M., and Vera, M.**, Synthesis of cyclobutanoprostaglandins, *Chem. Ind. (London)*, 884—885, 1975.

61. **Bundy, G. L., Lincoln, F. H., Nelson, N. A., Pike, J. A., and Schneider, W. P.**, Novel prostaglandin syntheses, *Ann. N.Y. Acad. Sci.*, 180, 76—90, 1971.

62. **Herschler, R. C., Kent, J. S., and Tomlinson, R. V.**, New prostaglandins: present study and future, in *Veterinary Pharmacology and Toxicology*, MTP Press, Lancaster, England, 1983, 213—220.

63. **Tomlinson, R. V., Spires, H. R., and Kent, J.S.**, Absorption, excretion and tissue residue in feedlot heifers injected with the synthetic prostaglandin, fenprostalene, *J. Anim. Sci.*, 57(Suppl. 1), 377, 1983.

64. **Wilks, J. W.**, Pregnancy interception with a combination of prostaglandins: studies in monkeys, *Science*, 221, 1407—1409, 1983.

65. **Favara, D., Guzzi, U., Ciabatti, R., Battaglia, F., Depaoli, A., Gallico, L., and Galliani, G.**, Synthesis and antifertility activity of 13-aza 14-oxo-prostaglandins, *Prostaglandins*, 25, 311—320, 1983.

66. **Mitsunobu, O.**, The use of diethyl azodicarboxylate and triphenylphosphine in synthesis and transformation of natural products, *Synthesis*, 1—28, 1981.

67. **Grieco, P. A. and Vedana, T. R.**, C(12)-substituted prostaglandins. An enantiospecific total synthesis of (+)-12-(fluoromethyl)prostaglandin F$_{2\alpha}$ methyl ester, *J. Org. Chem.*, 48, 3497—3502, 1983.

68. **Tarpley, W. G. and Sun, F. F.**, Metabolism of *cis*-Δ^4-15(S)-15-methylprostaglandin F$_{1\alpha}$ methyl ester in the rat, *J. Med. Chem.*, 21, 288—291, 1978.

69. **Garay, G. L. and Muchowski, J. M.**, Agents for the treatment of peptic ulcer disease, *Ann. Rep. Med. Chem.*, 20, 93—105, 1985.

70. **Ambrus, G., Cseh, G., and Toth-Sarudy, E.**, Synthesis and biological properties of 16(S)-amino-PGF$_{2\alpha}$ methyl ester, *Prostaglandins*, 29, 303—312, 1985.

Table 5
PGE ANALOGS

No.	Systematic name	Structure	Mol. formula (mol. wt.)	Synthesis	Biological Actions
1	*i*-Propyl 9-keto-11α,15α-dihydroxy-prosta-5-*cis*-13-*trans*-dienoate		$C_{23}H_{40}O_5$ (396)	Reaction of PGE_2 with 2-iodopropane in the presence of Hunig's base[1]	Hamster antifertility (s.c.); $1 \times PGE_2$[1]
2	N-Acetyl-9-oxo-11α,15α-prosta-5-*cis*-13-*trans*-dienoamide (CP-27,987)		$C_{22}H_{35}NO_5$ (393)	Corey type[2]	Bronchial dilator in asthmatics (aerosol) with long duration of activity and potency ~10 × isoproterenol; side effects included increased heart rate, increase in systolic blood pressure at 35 μg but decreased at 105 μg, and increase in diastolic blood pressure at 6 μg Bronchial dilation was about the same for all doses (6—140 μg) after 1.5 hr[3]
3	9-Keto-11α-15α-dihydroxyprosta-13-*trans*-13-en-1-ol (TR 4161)		$C_{20}H_{36}O_4$ (340)	PGE_1 converted into oxime, reduced with LAH, and the de-protected with sodium nitrite in acetic acid;[4] also by cuprate addition of protected ω-chain to fully elaborated, protected cyclopentanone[7]	Bronchial selective bronchial dilator; in guinea pigs, equipotent to PGE_1 in anesthetized, histamine challenged animals (aerosol); equipotent to PGE_1 in reducing basal resistance (i.v.); reduced airway resistance (2.9 × PGE_1) and increased dynamic compliance (0.4 × PGE_1) in guinea pigs (aerosol); inhibited histamine-induced bronchial constriction in guinea pigs by oral route (2.7, i.v. PGE_2); caused 100—1000 times less tracheal-bronchial irritation than PGE_2 in restrained conscious cats (aerosol)[5]

4	9-Keto-15α- and 15β-hydroxy-16,16-dimethylprosta-13-*trans*-en-1-ol		$C_{22}H_{40}O_4$ (368)	From *trans*-2-(1-hydroxyheptyl)-3-formylcyclopentanone ethylene ketal[8]	Relaxed human bronchial muscle at 0.001—100 µg/mℓ, whereas PGE_2 was mainly a contractant in this dose range[6]
5	*N*-Acetyl-9-keto-11α,15α-dihydroxy-16,16-dimethylprosta-5-*cis*-13-*trans*-dienoamide		$C_{24}H_{39}NO_5$ (421)	Corey type[2]	Uterine stimulant (i.v.) in rats; 3.8 × PGE_1; much less effective, 0.1 × PGE_2, in reversing morphine-induced diarrhea in mice (p.o.)[8] Inhibition of pentagastrin-induced gastric acid secretion in anesthetized rats (i.v.): 3.3 × PGE_2; isolated guinea pig uterus; 1 × PGE_2; inhibition of histamine-induced bronchial constriction in guinea pigs aerosol; 0.24 × PGE_2 Hypotensive in anesthetized dogs (i.v.): 0.025 × PGE_2; diarrhea induction in mice (i.v.); 0.0008 × PGE_2[2]
6	9-Keto-15α-hydroxy-16,16-diemthyl-2-*trans*-4-*trans*-13-*trans*-prostatreinoic acid		$C_{22}H_{34}O_4$ (362)		Protected female rats against cysteamine induced duodenal ulcers (p.o., ED_{50} = 1 mg/kg); 5 × PGE_2; reduced basal gastric acid secretion in pylorus ligated rats (p.o.) by 71% at 100 µg/kg[14]
7	9-Keto-16α- and 16β-hydroxy-20-methyl-prosta-2-*trans*-14-*trans*-dienoic acid		$C_{21}H_{34}O_4$ (350)	From ethyl cyclopentanone-2-carboxylate[15]	Isolated rat stomach muscle contraction; 1 × PGA_2[14]

Table 5 (continued)
PGE ANALOGS

No.	Systematic name	Structure	Mol. formula (mol. wt.)	Synthesis	Biological Actions
8	Methyl 9-keto-15α-acetoxy-16,16-dimethyl-18-oxaprosta-2-*trans*-13-*trans*-dienoate (HR-601)		$C_{24}H_{38}O_6$ (422)	From 2-ethoxycarbonyl-cyclopentanone[123]	Decreased mean arterial blood pressure by >20% in anesthetized dogs (i.v.) at 10 μg/kg; decreased systemic blood pressure in conscious renal hypertensive dogs (p.o.) by 15—25% at 0.1 mg/kg with a duration of >5 hr[124]
9	Methyl 9-keto-11α,15α-dihydroxy-16,16-dimethyl-prosta-2-*trans*-13-*trans*-dienoate (Gemeprost)		$C_{23}H_{38}O_5$ (394)	Corey type[9]	Highly effective in termination of 1st, 2nd and 3rd trimester pregnancies (vaginal pessaries) with low incidence of side effects which included fever, diarrhea, mild nausea, and vomiting[10-13]
10	9-Keto-11α,15α-dihydroxy-17α,20-dimethylprosta-2-*trans*-13-*trans*-dienoic acid (ONO-1206)		$C_{22}H_{36}O_5$ (380)	From *cis-anti-trans*-1-acetoxy-2-(6-methoxycarbonylhexyl)-3-formyl-4-tetrahydropyranyloxycyclopentane Unsaturation at C-2 introduced by phenylselenation and oxidative elimination[16]	May be useful in treatment of ischemic heart disease and peripheral circulating disorders Oral administration (20 and 30 μg t.i.d.) for 2 weeks to patients with thromboembolic disorders produced marked inhibition of platelet aggregation, increased platelet cAMP levels, and decreased (slight) platelet adhesiveness; BP was lowered, but HR was unaffected; patients with deep vein thrombosis and angina pectoris showed improvement in symptoms[16]

No.	Compound	Structure	Formula (MW)	Synthesis	Biological activity
11	9-Keto-11α,15α-dihydroxy-17ξ-ethylprosta-2-*trans*-13-*trans*-dien-oic acid (ONO 747)		$C_{22}H_{36}O_5$ (380)	Modified Corey synthesis[19]	In rabbits; 100—1000 µg/kg (p.o.) suppressed vasopressin-induced ST depression on ECG;[17] also observed in baboons and monkeys (bolus i.v. doses of 5—15 µg/kg)[18] Hypotensive in SH and normotensive rats (100—300 µg/kg, p.o.) with long duration of activity (>5 hr, < 24 hr);[17] in baboons, 14 × PGE_2 in lowering arterial BP (i.v.)[18] Intracoronary injection (dogs) caused pronounced increase in coronary blood flow with no change in BP, HR, myocardial oxygen consumption, or redox potential; decreased resistance in coronary vessels in dogs (i.v.), like nitroglycerin, but of longer duration (about 20 min)[17] Decreased platelet adhesiveness in guinea pigs (p.o.; 1000 × PGE_2) and inhibited platelet aggregation induced by collagen or ADP in guinea pigs (100—300 µg/kg, p.o.) for ≥ 4 hr; prolonged bleeding time in guinea pigs (p.o.); inhibited thrombocytopenia induced by infusion of collagen or ADP in guinea pigs (p.o.)[17] Inhibited ADP-induced platelet aggregation in PRP of human, rat, rabbit, and dog (16, 12.2, and 2 × PGE_1, respectively); oral administration to rat caused potent inhibition of aggregation 0.5 hr after dosing[19]

Table 5 (continued)
PGE ANALOGS

No.	Systematic name	Structure	Mol. formula (mol. wt.)	Synthesis	Biological Actions
12	(−)3-Oxa-4,5,6-trinor-3,7-inter-*m*-phenylene-9-keto-11α,15α-dihydroxyprosta-13-*trans*-enoic acid		$C_{22}H_{30}O_6$ (390)	Upjohn-type synthesis from 6-*endo*-hept-*cis*-1-enylbicyclo[3.1.0]hexan-3-one,[20] or from photoadduct of benzaldehyde and bicyclo[3.1.0]hex-2-en-6-*endo* carboxaldehyde neopentylglycol acetal;[21] also a combination Corey-Stork synthesis[22]	Bronchial selective bronchial dilator Inhibited ADP-induced aggregation of HPRP (30 × PGE₁);[21] methyl ester inhibited ADP-induced platelet aggregation ex vivo for up to 45 min when given orally to rats (20 mg/kg);[20] methyl ester resistant to 15-hydroxyprostaglandin dehydrogenase (0.1 × PGE₂)[20]
13	3-Thia-9-keto-15α-hydroxyprosta-13-*trans*-enoic acid		$C_{19}H_{32}O_4$ (356)	Cuprate addition of ω-chain to fully elaborated cyclopentanone[23]	Inhibited bronchial constriction in guinea pigs induced by serotonin, histamine, and acetylcholine with potency 0.8, 0.14, and 0.65 × PGE₁, respectively Inhibited bronchial constriction induced by pilocarpine in dogs (aerosol) by 50% at 1.6—3.2 μg/kg with a duration of action much greater than 1 hr[23]
14	Methyl 9-keto-11α,15α-dihydroxy-16-phenoxy-ω-tetra-norprosta-4,5,13-*trans*-trienoate (Enprostil)		$C_{23}H_{28}O_6$ (400)	Allene moiety introduced by addition of dilithio-4-pentynoic acid to Corey lactol containing protected ω-chain; the product was then esterified, acetylated, and reacted with lithium dimethyl copper	Long lasting antisecreotry and anti-ulcer agent[24] Profound inhibitory activity on stimulated acid secretion in rat, cat, dog, and rhesus monkey (e.g. in *dimaprit* stimulated gastric fistula cats, ID₅₀ = 2.5 μg/kg) and persistent (up to 14 hr in the rat, 200—400 ×

Removal of protecting groups at C-9,11,15, selective silylation, oxidation, and desilylation gave the desired product[24]

PGE$_2$ in blocking experimentally induced gastric and duodenal ulcers and much more potent than natural PG's in stimulating mucus secretion; rate of onset and degree of acid inhibition in the rat is greater p.o. than i.d.; in Heidenhain pouch dog, food stimulated acid secretion inhibited at lower dose when agent is deposited in the pouch (ID$_{50}$ = 0.9 μg/kg) than the stomach (ID$_{50}$ = 7 μg/kg) In healthy humans, single oral dose (35 or 70 μg) showed similar efficacy to cimetidine (600 mg) for suppression of food stimulated gastric acid secretion, but the duration was much longer (83—94% for 8 hr); gastrin levels were depressed (30—40%) in these individuals whereas cimetidine augmented gastrin release; in healthy volunteers, 3.5 μg (b.i.d.) conferred significant protection against aspirin induced (650 mg, q.i.d.) gastric mucosal damage; in clinical trials, 35 or 70 μg (b.i.d.) for 4 weeks was superior to placebo in healing duodenal ulcers and 35 μg (b.i.d.) for 6 weeks was equivalent to cimetidine (400 mg, b.i.d.); digestive system complaints were more frequent for the prostaglandin; see Reference 133, p. 99 and Reference 134.

Table 5 (continued)
PGE ANALOGS

No.	Systematic name	Structure	Mol. formula (mol. wt.)	Synthesis	Biological Actions
15	(−)9-Keto-11α,15α-dihydroxyprosta-13-*trans*-en-4-ynoic acid		$C_{20}H_{30}O_5$ (350)	Jones[130] oxidation of 1,9-diol-11-15,*bis*-THP derivative[25]	Hamster antifertility; $<PGE_2$; hypotensive in rat; $1 \times PGE_1$; isolated gerbil colon; $<PGE_{26}$
16	(−)Methyl 9-keto-11α,15α-dihydroxy-prosta-13-*trans*-en-5-ynoate		$C_{21}H_{32}O_5$ (364)	From corresponding F derivative[26]	Inactive as hypotensive in SH rat at doses (i.p) up to 10 mg/kg Inactive on smooth muscle assays[27]
17	5,6-Bisnor-4,7-inter-*o*-phenylene-9-keto-11α,15α-dihydroxy-prosta-13-*trans*-enoic acid		$C_{24}H_{34}O_5$ (400)	Cuprate addition of protected ω-chain to fully elaborated cyclopentenone[27]	Inhibited histamine-induced bronchial constriction in anesthetized (i.v., $1.7 \times PGE_2$) and conscious (aerosol, $0.07 \times PGE_2$) guinea pigs[28]
18	5ξ,6ξ,10α,11α-*Bis*-methano-9-keto-15α-hydroxyprosta-13-*trans*-enoic acid (R-45,661)		$C_{22}H_{34}O_4$ (362)		
19	6,9-Diketo-11α,15α-dihydroxyprosta-13-*trans*-enoic acid		$C_{20}H_{32}O_6$ (368)	Metabolite of 6-keto-PGF$_{1α}$. Chromic acid oxidation of 6-keto-PGF$_{1α}$-11, 15-*bis*-THP and subsequent deprotection[29]	Potent stimulation of renin release ($5 \times PGI_2$)[30] and reduction of renal vascular resistance ($3 \times PGE_2$) in dogs (i.v. infusion, nonfiltering β receptor blocked kidneys)[30-32] Caused strong dilation of pulmonary vascular bed, decreased systemic arterial pressure, increased cardiac output in anesthetized cats (i.v.)[33]

No.	Structure	Formula (MW)	Method of synthesis	Pharmacology
20	Methyl 5ξ-fluoro-6,9-diketo-11α,15α-dihydroxyprosta-13-*trans*-enoate	$C_{21}H_{33}FO_6$ (400)	From the addition product of the lithium enolate of methyl 2-fluoro-6-hydroxyhexanoate (protected as 2-methoxy-2-methylethyl ether) and a Corey type aldehyde with THP protecting groups at C-9,11,15.[37]	Slightly less potent as vasodepressor in anesthetized dogs (i.v.) than PGI_2. Inhibited ADP-induced platelet aggregation (HPRP); $\leq 0.05 \times PGI_2$. Stimulated platelet cAMP; $\sim 0.03 \times PGI_2$. Desaggregatory agent (ADP induced); $\sim 0.01 \times PGI_2$. Inhibited thrombus formation in vivo in dog (anesthetized, i.v.) coronary arteries; $0.1 \times PGI_2$.[34] Increased intraocular pressure on topical or intravitreal administration with a potency greater tha PGI_2 or PGE_2.[35,36] For a recent review on 6-keto-PGE_1, see Ref. 129. Inhibition of stress ulcer in rat; $10 \times PGE_1$; isolated rat uterus; $10 \times PGE_1$[37]
21	Methyl 7,9-diketo-11α,15ξ-dihydroxy-15ξ-methylprosta-13-*trans*-enoate (TE-1226)	$C_{22}H_{36}O_6$ (396)		Inhibited indomethacin-induced ulcers in rats (p.o.); $2 \times PGE_2$. Diarrhea induction; $\sim PGE_2$; in vitro smooth muscle stimulation i.e., rabbit aorta, guinea pig trachia, rat stomach, rat colon; $\sim 0.01 \times PGE_2$.[38]

Table 5 (continued)
PGE ANALOGS

No.	Systematic name	Structure	Mol. formula (mol. wt.)	Synthesis	Biological Actions
22	(+)Methyl 7-thia-9-keto-15α-hydroxy-prosta-13-*trans*-enoate		$C_{20}H_{34}SO_4$ (370)	Reaction of 2,3-epoxycyclopentanone with methyl 6-mercaptohexanoate and then cuprate addition of protected optically active ω-chain to cyclopentanone thus formed[39]	Protected against formation of indomethacin induced ulcers; contraction of isolated guinea pig aortic strip; $2 \times PGF_{2\alpha}$[39]
23	Methyl 7,13,-*bis*-thia-9-keto-15ξ-hydroxy-15ξ-methylprostanoate		$C_{20}H_{36}O_5S_2$ (420)	Reaction of methyl 6-mercaptohexanoate with suitably protected hydroxylated-2,3-epoxy-cyclopentanone followed by conjugate addition of lower side chain to the resulting enone[40]	Inhibited gastric acid secretion with minimal side effects[40]
24	Methyl 8β-phenylthio-9-keto-15α-hydroxy-prosta-5-*cis*-13-*trans*-dienoate		$C_{27}H_{38}O_4$ (458)	Cuprate addition of protected ω-chain to cyclopentenone followed by trapping with diphenyl disulfide and subsequent alkylation with methyl 7-bromohept-5-(*trans*)-enoate[41]	
25	(−)9-Keto-11α,15α-dihydroxyprosta-5-*cis*-13-*trans*-dienoic acidethylene ketal		$C_{22}H_{36}O_6$ (396)	Direct ketalization of PGE_2[42]	Bioequivalent to PGE_2 in rhesus monkey (p.o.); cleaved in stomach to PGE_2 without significant formation of PGA_2[42]

No.	Name	Structure	Formula	Synthesis	Activity
26	9-Methylene-11α,15α-dihydroxy-16,16-di-methylprosta-5-cis-13-trans-dienoic acid (meteneprost)		$C_{23}H_{38}O_4$ (378)	Reaction of *N,S*-dimethyl-(S)-phenylsulfoximine, bromomagnesium salt with 16,16-dimethyl PGE$_2$ methyl ester *bis*-trimethylsilyl ether and subsequent treatment with aluminum amalgam and acetic acid[43]	Induction of 1st and 2nd trimester abortions; frequency of side effects for early first trimester menses induction very low[43,44] Administration of meteneprost (0.5 mg) in combination with methyl 5-oxa-9α,11α-dihydroxy-17-phenyl-ω-trinorprosta-13-*trans*-enoate (cpd. 5, Table 4; 7 mg i.m.) to rhesus monkeys on day 28 of fertile cycle terminated pregnancies in all cases (10/10) with milk anorexia being the only side effect (1/10) observed;[45] the 2 agents together were more effective than each singly
27	9-Keto-10,10-dimethyl-11α,15α-dihydroxy-prosta-13-*trans*-enoic acid		$C_{22}H_{38}O_5$ (382)	From 2,2-dimethylcyclo-pentane-1,3-dione[46,47]	Contraction of isolated rat uterus; $0.001 \times PGF_{2\alpha}$[46]
28	9-Keto-10ξ-methyl-15α-hydroxyprosta-13-*trans*-enoic acid		$C_{21}H_{36}O_4$ (352)	From 2-(7-hydroxyheptyl)-3-cyanocyclopentanone[48]	
29	9-Keto-10,10-dimethyl-15α-hydroxyprosta-5-cis-13-*trans*-dienoic acid		$C_{22}H_{36}O_4$ (364)	Cuprate addition of protected ω-chain to 5,5-dimethylcyclopent-2-en-1-one and subsequent alkylation with the iodide methyl ester corresponding to α-chain[49]	Highly potent inhibitor of gastric acid secretion in the dog and powerful hypotensive agent in the rat[49]
30	Methyl-9-keto-10α,15α-dihydroxyprosta-13-*trans*-enoate		$C_{21}H_{36}O_5$ (368)	Cuprate addition of protected ω-chain to fully elaborated cyclopentanone[50]	Stimulation of pregnant rat (anesthetized) uterus (i.v.); $0.1 \times PGE_1$ Caused short-lived vasodepression in anesthetized cat (i.v.); $0.25 \times PGE_1$

Table 5 (continued)
PGE ANALOGS

No.	Systematic name	Structure	Mol. formula (mol. wt.)	Synthesis	Biological Actions
31	Methyl 9,10-diketo-15α-hydroxyprosta-13-*trans*-enoate		$C_{21}H_{34}O_5$ (366)	Oxygenation of 10α-hydroxy compound under alkaline conditions; exists as 8,9-enol form[51]	Caused bronchial constriction in anesthetized cats (i.v.); 1 × PGE_1[50]
32	(+)Methyl 9-keto-10α,11α-methano-15α-hydroxyprosta-5-*cis*-13-*trans*-dienoate		$C_{22}H_{34}O_4$ (362)	Cyclization of mesylate of 11α-hydroxymethyl derivative[52]	
33	9-Keto-10,11-benzo-15α-hydroxyprosta-5-*cis*-13-*trans*-dienoic acid		$C_{24}H_{32}O_4$ (384)	Modified Corey type synthesis[53]	Inhibited gastric acid secretion; hypotensive activity[53]
34	Methyl 9-keto-11α,15α-dimethoxy-prosta-5-*cis*-13-*trans*-dienoate		$C_{23}H_{38}O_5$ (394)	Reaction of PGE_2 methyl ester with a large excess of ethereal diazomethane in the presence of silica gel[54]	
35	9-Keto-11α-mercapto-15α-hydroxyprosta-5-*cis*-13-*trans*-dienoic acid		$C_{20}H_{32}O_4S$ (368)	Reaction of PGA_2 with sodium sulfide in aqueous THF[55]	Neuroleptic activity in mice and rats qualitatively similar to clozapine[55]

| 36 | (−)-9-Keto-11α-(2-hy-droxyethylthio)-15α-hydroxyprosta-5-cis-13-trans-dienoic acid | | $C_{22}H_{36}O_5S$ (412) | Conjugate addition of mercaptoethanol to (−) PGA_2.[56] | Potent bronchial dilator in animal models, but mainly causes bronchial constriction in human asthmatics; thus, in guinea pigs (i.v.) it is 0.03—0.14 × isoproterenol in protection against bronchospasms induced by 5-hydroxytryptamine, histamine, or acetyl choline;[56,57] in dogs (i.v.) bronchoconstricted with pilocarpine it is ≈ isoproterenol, but with much greater duration of activity and shows no cardiovascular effects; in chronic bronchospasms induced by sulfur dioxide in dogs (aerosol); ~0.5 × PGE_1; aerosol administration to human asthmatics elicited variable responses but bronchoconstriction was most frequent; isolated carbachol contracted human bronchus; constriction; isolated carbachol constricted dog bronchus; relaxation.[57] |
| 37 | Methyl 9-keto-11α,15α-dihydroxy-11β-methyl]prosta-5-cis-13-trans-dienoate | | $C_{22}H_{36}O_5$ (380) | Addition of trimethylaluminum to Corey lactone containing 11-ketone-15-propionate and subsequent elaboration to E derivative via the corresponding $F_{2\alpha}$ compound[58] | |

Table 5 (continued)
PGE ANALOGS

No.	Systematic name	Structure	Mol. formula (mol. wt.)	Synthesis	Biological Actions
38	9-Keto-11α,16,16-tri-methyl-15α-hydroxy-prosta-5-cis-13-trans-dienoic acid (RO-21-6937; Trimeprostil)		$C_{23}H_{38}O_4$ (378)	Corey type from bicyclic lactone with 11-methyl group in place[59]	Antisecretory agent; inhibited gastric acid production (50—65%) in conscious, gastric fistula rats either i.v. (20—30 μg/kg) or orally (10 μg/kg) with significant effects lasting up to 1.3 hr[59] Prevented formation of duodenal and gastric ulcers in rats and guinea pigs on oral administration; does not appear to be cytoprotective[60] In clinical trials, 750 μg (q.i.d.) given orally, was less effective (62%) than cimetidine (200 mg, t.i.d. and 400 mg at night, 90%) in healing duodenal ulcers after 4 weeks and use thereof was associated with more side effects (nausea, vomiting) and a greater relapse rate (1 year post-treatment); Reference 133, pp. 98—99
39	9-Keto-11,11-dimethyl-15α-hydroxyprosta-5-cis-13-trans-dienoic acid		$C_{22}H_{36}O_4$ (364)	Corey type from bicyclic lactone with both 11-methyl groups in place[61]	Antiulcer activity nearly equal to PGE_1[61]

No.	Structure	Name	Formula (MW)	Method	Activity
40		9-Keto-11α-phenyl-15α-hydroxyprosta-13-*trans*-enoic acid	$C_{26}H_{38}O_4$ (414)	Addition of lithium di-phenylcopper to 7-meth-oxybicyclo[3.2.0]-hepta-3,6-dien-2-one and subsequent elaboration[62]	Inhibited histamine-induced bronchial constriction in guinea pigs by i.v. (5 × PGE$_2$) and aerosol (0.3 × PGE$_2$) routes; essentially inactive in mildly asthmatic patients (aerosol)[72]
41		9-Keto-11α,12α-difluoromethano-15α-hydroxyprosta-5-*cis*-13-*trans*-dienoic acid	$C_{21}H_{30}F_2O_4$ (384)	Addition of difluorocarbene to Corey dienone and subsequent elaboration[71]	Inhibited histamine-induced bronchial constriction in guinea pigs by i.v. (80 × PGE$_2$) and aerosol (10 × PGE$_2$) routes; inhibited methocholine-induced increase in airway resistance in anesthetized rhesus monkeys by aerosol route (1.0—0.5 × PGE$_1$) with a duration of action similar to PGE$_1$ and precipitated milk upper airway irritation and tachycardia[132]
41a		9-Keto-11α,12α-methano-15α-hydroxy-prosta-5-*cis*-13-*trans*-dienoic acid	$C_{21}H_{32}O_4$ (348)	Reaction of Corey dienone with dimethylsulfoxonium methylide and subsequent elaboration[132]	
42		9-Keto-12α-methyl-15α-hydroxyprosta-5-*cis*-13-*trans*-dienoic acid	$C_{21}H_{34}O_4$ (350)	From 2-ethoxycarbonyl cyclopentanone;[81] also, from 2-methoxycarbonyl-methyl-3-methyl-3-benzyloxymethylcyclopentanone[82]	Methyl ester shows no activity on isolated gerbil colon or hamster uterus; also inactive in hamster antifertility assay at 200 µg (s.c.)[83]
43		9-Keto-15α-hydroxy-15β-ethynylprosta-5-*cis*-13-*trans*-dienoic acid	$C_{22}H_{32}O_4$ (362)	Cuprate addition of ω-chain to fully elaborated cyclopentenone,[63] or by addition of acetylenic Grignard reagent to 9,15-diketo compound in which 9-ketone is protected as methyl enol ether[64]	Equipotent to PGE$_2$ as a bronchial dilator in the guinea pig[63]

Table 5 (continued)
PGE ANALOGS

No.	Systematic name	Structure	Mol. formula (mol. wt.)	Synthesis	Biological Actions
44	9-Keto-15α-hydroxy-prosta-13-*trans*-enoic acid (AY 23,578)		$C_{20}H_{34}O_4$ (338)	From 2-methoxycarbonyl-cyclopetanone;[65] for other syntheses see Ref. 66	Bronchial dilator in animal models with potency 0.1—1 × PGE_2,[66] and depending on animal model, ≤potency of doxaprost (Cpd. 45) but with shorter duration of action[68]
45	9-Keto-15α-hydroxy-15α-methylprosta-13-*trans*-enoic acid (doxaprost)		$C_{21}H_{36}O_4$ (352)	From addition product of methylmagnesium chloride to 9-hydroxy-15-oxo derivative[67]	Bronchial dilator in animal models, with potency ≥ that of 11-deoxy PGE_1 (see above), but with a longer duration of action[68]
46	9-Keto-15ξ-hydroxy-15ξ-methylprostanoic acid (deprostil)		$C_{21}H_{38}O_4$ (354)	Reaction of 9,15-diketo-methyl ester with methyl-magnesium iodide, followed by saponification;[69] for other syntheses see Ref. 70	Orally active antiulcer-antisecretory agent; in pylorus ligated rats, ED_{50} for gastric acid secretion was 3 mg/kg. (p.o.) In rats, ED_{50} for inhibition of indomethacin induced ulcers was 0.11 mg/kg (p.o.); in starved rats, ED_{50} = 140 mg/kg for diarrhea induction[70]
47	Methyl 9-keto-13ξ-hydroxyprosta-14-*trans*-enoate		$C_{21}H_{36}O_4$ (352)	From allylic rearrangement product of 8-ethoxycarbonyl-11-deoxy-PGE_1 ethyl ester[73]	Appreciable spasmolytic and bronchial spasmolytic activity[73]
48	9-Keto-15α-hydroxy-16ξ-methylprosta-5-*cis*-13-*trans*-dienoic acid		$C_{21}H_{34}O_4$ (350)	Cuprate addition of protected ω-chain to fully elaborated cyclopentanone[74]	Inhibited bronchial constriction in guinea pigs (i.v.) induced by serotonin, histamine, and acetylcholine with potency of 0.8, 6.5, and 2 × PGE; caused initial brief spasmogenic events in guinea pigs and in pilocarpine bronchoconstricted dogs (aerosol)[74]

49	9-Keto-15α-hydroxy-16β-methylprosta-5-cis-en-13-ynoic acid (FCE 20700)	$C_{21}H_{32}O_4$ (348)	Corey type[75]	Inhibited gastric acid secretion in pylorus ligated rats (ED_{50} = 65 µg/kg s.c. and 3607 µg/kg p.o.). Protected against stress-induced lesions of gastric mucosa in rats (ED_{50}; 20.7 µg/kg s.c. and 182 µg/kg p.o.); less potent in this assay than 16,16-diemthyl PGE_2 (ED_{50} = 14.2 µg/kg s.c. and 18.9 µg/kg p.o.), but also less effective (0.02 × 16,16-dimethyl PGE_2) in enteropooling;[75] see also Reference 133, p. 99
50	9-Keto-15α-hydroxy-16ξ-methylprosta-13-trans-enoic acid	$C_{21}H_{36}O_4$ (352)	From *trans*-2-(7-hydroxyheptyl)-3-formylcyclopentanone ethylene ketal[76]	Inhibited pentagastrin-stimulated gastric acid secretion in the rat (p.o.); 0.8 × PGE_2; inhibited indomethacin-induced ulcers in rats (p.o.); >0.2 × PGE_2[76]
51	9-Keto-15α-hydroxy-16-phenoxy-ω-tetranorprosta-13-trans-enoic acid (M + B 28, 767)	$C_{22}H_{30}O_5$ (374)	From *trans*-2-(1-hydroxyheptyl)-3-formylcyclopentanone ethylene ketal[77]	Inhibited indomethacin-induced ulcers in rats (p.o.); ED_{50} = 3 µg/kg; reversed morphine-induced constipation in mice at 1000—1400 µg/kg (p.o.) indicative of low diarrhea induction tendency[77] In the rat, it is 60 × more potent (p.o.) than 16,16-dimethyl PGE_2 for inhibition of pentagastrin stimulated acid secretion but less potent (0.3) in prevention of indomethacin induced ulcers; much less uterotonic (rat, s.c.) and diarrheagenic (mouse, p.o.) than 16,16-dimethyl PGE_2; see Reference 133, p. 99

Table 5 (continued)
PGE ANALOGS

No.	Systematic name	Structure	Mol. formula (mol. wt.)	Synthesis	Biological Actions
					Powerful constrictor of isolated rabbit aorta (EC_{50} = 2 μM) and mesenteric artery (EC_{50} = 0.2 μM), but less active than U-46619; induced irreversible aggregation of HPRP (0.2 × R-46619) and rat PRP (0.9 × U-46619) in vitro; i.v. bolus administration of low doses (1—20 μg/kg) to anesthetized rats caused immediate thrombocytopenia (1 × U-46619)[135]
52	2-Decarboxy-2-(tetrazol-5-yl)-9-keto-11-deoxy-15α-hydroxy-16-phenyl-ω-tetranorprostane	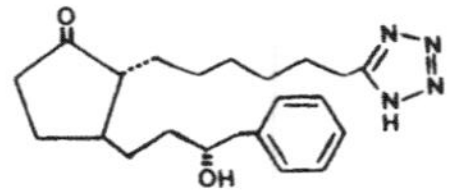	$C_{22}H_{32}N_4O_2$ (384)	Modified Corey synthesis[93]	Potent long-lasting tissue selective antihypertenisve agent; oral administration to conscious renal hypertensive dogs caused a profound reduction in BP (100 × PGE_2) with a duration of 3 hr; inhibited ADP-induced aggregation of HPRP (30 × PGE_2)[93]
53	9-Keto-15α-hydroxy-16,16-trimethylene-prosta-13-*trans*-enoic acid	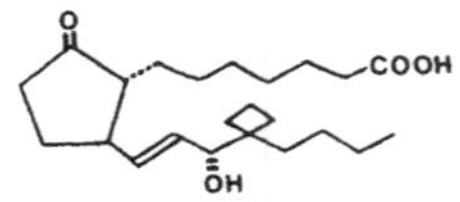	$C_{23}H_{38}O_4$ (378)	Cuprate addition of protected ω-chain to fully elaborated cyclopentanone[78]	Inhibited bronchial constriction in guinea pigs induced by histamine (0.8 × PGE_2) and serotonin (1.6 × PGE_2) Potent and long-lived inhibitor of pilocarpine-induced bronchial constriction in dogs (aerosol), but produced short-lived pulmonary hypertension at effective doses)[78]

No.	Name	Structure	Formula (MW)	Preparation	Pharmacology
54	Methyl 9-keto-16ξ-hydroxyprosta-5-*cis*-14-*trans*-dienoate		$C_{21}H_{34}O_4$ (350)	From 2-allylcyclopent-2-enone[79]	
55	9-Keto-15α-hydroxy-20-*N,N*-dimethylaminoprosta-13-*trans*-enoic acid		$C_{22}H_{39}NO_4$ (381)	From *trans*-2-(6-methoxycarbonylhexyl)-3-formylcyclopentanone ethylene ketal[80]	No inhibition of gastric acid secretion in rats (i.p.) in doses up to 5 mg/kg. No protection against ulcers produced by pylorus ligature in rat. Transient increase in arterial blood pressure at doses of 1—2 mg/kg (i.v., rat?). No effect on isolated guinea pig intestine[80]
56	9-Keto-11α,15α-dihydroxy-13-thia-15β-methylprostanoic acid 4-(benzoylamino)phenyl ester (EM 33,290)		$C_{33}H_{45}NO_6S$ (583)	Conjugate addition of ω-chain to fully elaborated cyclopentanone[84]	Antisecretory and cytoprotective; basal acid output in healthy humans decreased by 75 and 65% after single doses (intragastric) of 3 and 2 mg; no antisecretory effect at 1 mg; single 50 or 250 μg doses protected human gastric mucosa against injurious effects of aspirin or sodium taurocholate; the antisecretory effect was of short duration[84,85]
57	9-Keto-1ξ-hydroxy-20-methylprosta-13,14-dienoic acid		$C_{21}H_{34}O_4$ (350)	Reaction of corresponding acetylenic acetate with lithium dimethylcopper[86]	
58	9-Keto-11α,15β-dihydroxy-15α-methylprosta-5-*cis*-13-*trans*-dienoic acid (arbaprostil)		$C_{21}H_{34}O_5$ (366)	Oxidation of PGE$_{2α}$ with DDQ to the 15-ketone, followed by *tris* silylation and reaction with methylmagnesium bromide. Desilylation, reaction with diazomethane, selective silylation of 11-OH, chromium trioxide oxidation, and de-protection gave arbaprostil[87]	Antisecretory, antiulcer agent;[88] the inactive 15β (15R) serves as a prodrug for the active 15α-hydroxy (15S) isomer when given orally because of the acid-promoted isomerization in the gut[69]

Table 5 (continued)
PGE ANALOGS

No.	Systematic name	Structure	Mol. formula (mol. wt.)	Synthesis	Biological Actions
59	Methyl 9-keto-11α-hydroxy-15-thiaprosta-5-cis-enoate 15α-and 15β-sulfoxides		$C_{20}H_{34}O_5S$ (386)	Modified Corey synthesis[90]	
60	9-Keto-11α,15α-dihydroxy-15-(dihydrobenzpyran-2-yl)-ω-pentanorprosta-5-cis-13-trans-dienoic acid		$C_{24}H_{30}O_6$ (414)	Corey synthesis[91]	Contraction of isolated guinea pig uterus; 1.3 × PGE_2; diarrhea induction in mice (i.v.); 0.001 × PGE_2[91]
61	9-Keto-11α,15α-dihydroxy-16-phenyl-ω-tetranorprosta-5-cis-enoic acid		$C_{22}H_{30}O_5$ (374)	Modified Corey synthesis[92]	Potent hypotensive (20 × PGE_2; i.v., anesthetized dogs) Weak bronchial dilator (0.1 × PGE_2; conscious guinea pig, aerosol, histamine-induced), and uterine stimulant (0.01 × PGE_2; isolated guinea pig uterus) Not a substrate for 15-hydroxy PG dehydrogenase[92]
62	N-Methanesulfonyl-9-keto-11α,15α-dihydroxy-16-phenoxy-ω-tetranorprosta-5-cis-13-transdienoamide (sulprostone)		$C_{23}H_{31}NO_7S$ (465)	Modified Corey synthesis[94]	Termination of 1st[96] 2nd[95] and 3rd trimester[97] pregnancy; very low incidence of side effects
63	1-Hydroxymethyl-1,9-diketo-11α,15α-dihydroxy-16-phenoxy-ω-tetranorprosta-5-cis-13-trans-diene (CL-116,069)		$C_{23}H_{30}O_6$ (402)	Conjugate addition of ω-chain to fully elaborated cyclopentanone[98]	Long acting (>1 hr) nasal decongestant in dogs (i.v. or topical); blood pressure not affected and tolerance did not develop on long-term administration (30 days). Threshold doses; i.v., 0.1 μg/kg; topical, 0.08—4 μg[99,100].

64	(−) 9-Keto-11α,15α-dihydroxy-16β-methyl-20-methoxy-prosta-5-*cis*-13-*trans*-dienoic acid (YPG-209)		$C_{22}H_{36}O_6$ (396)	Corey synthesis[101]	Bronchial dilator; inhibited histamine-induced bronchial constriction in guinea pigs by i.v. ($230 \times$ PGE$_2$) and oral (ED$_{50}$ = 7 μg/kg) routes. Diarrhea induction in guinea pigs (p.o.) very low (ED$_{20\%}$ = 140 μg/kg)[102]
65	9-Keto-11α,15α-dihydroxy-16,16-dimethyl-prosta-5-*cis*-13-*trans*-dienoic acid		$C_{22}H_{36}O_5$ (380)	Corey synthesis[103]	Antisecretory, antiulcer agent with potent cytoprotective activity in animals and in man;[104,105] see also Reference 136, pp. 93—94
66	Methyl 9-keto-11α,15α-dihydroxy-16α-methoxy-16β-methylprosta-13-*trans*-enoate (MDL 646)		$C_{23}H_{40}O_6$ (412)	Cuprate addition of protected ω-chain to fully elaborated cyclopentanone[106]	Antisecretory agent; inhibited histamine and bethanechol-induced gastric acid secretion in gastric fistula Heidenhain pouch dogs by 60 and 75% at 50 μg/kg (intragastric); no inhibition of pentagastrin-induced secretion; similar but less dramatic results were observed in cats; in anesthetized rats histamine and pentagastrin-induced acid secretion were inhibited with ID$_{50}$ = 4.5 and 18.3 μg/kg, respectively (i.v. infusion)[107] In rats, more effective p.o. than i.v. as a cytoprotective agent and less potent than natural PGE's as a luteolytic agent in hamsters (p.o.); in healthy humans, a single oral dose of 1000 μg reduced gastric acid output by 36% 2.5 hr post-dose; see Reference 133, p. 99

Table 5 (continued)
PGE ANALOGS

No.	Systematic name	Structure	Mol. formula (mol. wt.)	Synthesis	Biological Actions
67	Methyl 9-keto-11α,15α-dihydroxy-16,15-dimethylene-prosta-5-cis-13-trans-dienoate		$C_{23}H_{36}O_5$ (392)	Corey synthesis[108]	Inhibited pentagastrin-induced gastric acid secretion in rats; 80 × PGE_2. Diarrhea induction in mice (p.o.); 0.1 × 16-(R)-methyl PGE_2[108]
68	9-Keto-11α,15α-dihydroxy-16,16-trimethyl-leneprosta-5-cis-13-trans-dienoic acid		$C_{23}H_{38}O_5$ (394)	Cuprate addition of protected lower side chain to fully elaborated cyclopentanone derivative[78]	Inhibited bronchial constriction in anesthetized guinea pigs (i.v.) induced by serotonin, histamine, and acetylcholine by 0.85, 1.3 and 2.3 × PGE_1, respectively. Long-lasting protection to pilocarpine bronchoconstricted dogs (aerosol) but causes short-lived pulmonary hypertension (spasms) at effective doses[78]
69	Methyl 9-keto-11α,15α-dihydroxy-16,16-(2ξ-fluorotrimethylene)-prosta-13-trans-enoate		$C_{24}H_{39}FO_5$ (426)	Selective catalytic reduction of α-chain of $F_{2α}$ derivative which was in turn obtained by a Corey synthesis; the $F_{1α}$ derivative was converted into the PGE_1 derivative by standard techniques[109]	Contraction of pregnant rat uterus; 5 × PGE_1. Diarrhea induction in mice (p.o.), 0.5 × PGE_1[109]

No.	Name	Structure	Formula (MW)	Synthesis	Activity
70	Methyl 9-keto-11α,15α-dihydroxy-16-methyleneprosta-5-*cis*-13-*trans*-dienoate		$C_{22}H_{34}O_5$ (378)	Corey synthesis[108,110]	Antisecretory, antiulcer; inhibited pentagastrin-induced acid secretion (20 × PGE_2, i.v. infusion) and indomethacin-induced or cold water immersion stress ulcers (50% reduction at 1 µg/kg p.o.) iun rats; in man, it was <50% as emetic as 16,16-dimethyl PGE_2[111]
71	(−) Methyl 9-keto-11α,16α-dihydroxy-16β-methylprosta-13-*trans*-enoate (SC 30, 249)		$C_{22}H_{38}O_5$ (382)	Cuprate addition of protected optically active ω-chain to fully elaborated optically active cyclopentanone[112]	Antisecretory agent; inhibited histamine induced gastric acid secretion in Heidenhain pouch and gastric fistula dogs by i.v. (14 × PGE_1) and intragastric (48 × PGE_1) routes, respectively Completely elminated meal-stimulated gastric secretion at 1 µg/kg i.v. Although histamine-induced gastric acid secretion is inhibited in Heidenhain pouch dogs, gastric mucosal blood flow is not reduced[113]
72	9-Keto-11α,16ξ-dihydroxy-16ξ-trifluoromethylprosta-13-*trans*-enoic acid		$C_{21}H_{31}F_3O_5$ (420)	Cuprate addition of protected ω-chain to fully elaborated cyclopentanone[114]	Inhibited bronchial dilation induced by histamine, serotonin, or acetylcholone in anesthetized guinea pigs (i.v.) with a potency ~4 × PGE_1[114]
73	Methyl 9-keto-11α,16ξ-dihydroxy-16ξ-vinylprosta-5-*cis*-13-*trans*-dienoate (CL 115,347; Viprostadil)		$C_{22}H_{34}O_5$ (378)	Cuprate addition of ω-chain to fully elaborated protected cyclopentanone[131]	Topical application of acid or methyl ester to SH rats caused a rapid (3—5 min), profound (25%), and long-lasting (≥24 hr) hypotensive effect; similar effects but of shorter duration were observed on oral or i.v. administration[115,131]

Table 5 (continued)
PGE ANALOGS

No.	Systematic name	Structure	Mol. formula (mol. wt.)	Synthesis	Biological Actions
					Also active in conscious, normotensive rats, deoxycorticosterone-salt induced hypertensive rats, and aorta-coarcted hypertensive rats, at 1 mg/kg orally; lowered mean arterial BP in 2-kidney 1-clip Goldblatt renal hypertenisve dogs orally and s.c. and was accompanied by tachycardia; lowered systolic BP of conscious rhesus monkeys, on oral administration maximally at 0.1 and 0.2 mg/kg with only slight tachycardia.[115] In a double blind trial in Reynaud's phenomenon vs. placebo, patients were given 1 mg of the PG transdermally for 6 weeks; the treated group had fewer and shorter spasm attacks and better healing of ulcers; blood supply was improved as were hand temperatures; platelet count increased but initial decreases in platelet aggregation were not maintained to the end of the study[137]

No.	Name	Structure	Formula (MW)	Synthesis	Activity
74	1-Hydroxymethyl-1,9-diketo-11α,16ξ-dihydroxy-16ξ-methyl-prosta-13-*trans*-ene (CL 115,574)		$C_{22}H_{38}O_5$ (382)	Cuprate addition of ω-chain to protected fully elaborated cyclopentanone[116]	Potent inhibitor of histamine-induced gastric acid secretion in dogs; complete inhibition observed at 20 µg/kg (p.o.) which is 1/50 dose where overt side effects are observed; the 20 µg dose causes 100% inhibition for 1 hr with 30% inhibition of secretion still present at 4 hr[117] inhibited pentagastrin-stimulated gastric acid secretion in fasted males on i.v. infusion; peak inhibition was at 30 min after infusion and was 39, 50, and 51% for doses of 500, 750, and 1000 µg, respectively. Increases in mucus secretion of 115, 162, and 208% were observed at these doses[118] In healthy individuals, maximally effective dose for inhibition of pentagastrin stimulated gastric acid secretion was 750 µg; this dose elicited substantial and subtained stimulation of mucus secretion into the gastric juice[138]
75	(−) Methyl 9-keto-11α,15α-dihydroxy-16,18-methanoprosta-13-*trans*-dienoate		$C_{22}H_{36}O_5$ (380)	Cuprate addition of protected ω-chain to fully elaborated cyclopentenone[132]	Relaxed isolated guinea pig and human tracheal muscles; 10 × PGE_1 Contracted isolated rat uterus; ~1 × PGE_1 Hypotensive in anesthetized cat (i.v.); 1 × PGE_1; inhibited gastric acid secretion in pylorus-ligated rats (s.c.); 1 × PGE_1; inhibited collagen-induced platelet (HPRP) aggregation; 1 × PGE_1[120]

Table 5 (continued)
PGE ANALOGS

No.	Systematic name	Structure	Mol. formula (mol. wt.)	Synthesis	Biological Actions
76	9-keto-11α,15α-dihydroxy-17ξ-methyl-prosta-13-*trans*-enoic acid (ONO-358)		$C_{21}H_{36}O_5$ (368)	Modified Corey synthesis[125]	Coronary vasodilator; increased coronary blood flow in the dog with potency much greater than PGE_1 but with shorter duration of action[125]
77	9-Keto-11α,16β-dihydroxy-17,17-dimethyl-prosta-13-*trans*-en-1-ol (TR-4752)		$C_{22}H_{40}O_4$ (368)	Cuprate addition of protected ω-chain to fully elaborated cyclopentenone[118,119]	Relaxed isolated guinea pig trachea but less potent than PGE_2; also relaxed isolated human tracheal muscle; aerosol administration to conscious cats caused induction of severe cough response[6]
78	9-Keto-11α,15α,19β-trihydroxyprosta-13-*trans*-trienoic acid		$C_{20}H_{34}O_6$ (370)	Cuprate addition of fully elaborated ω-chain as *cis*-isomer to fully elaborated cyclopentanone; the product, which had 13-*cis*-14β-hydroxy structure, was subjected to inversion of both centers to the natural stereochemistry via the phenylsulfenate ester[121]	Bronchial dilator inhibited histamine-induced bronchial constriction in conscious rhesus monkeys by i.v. or aerosol administration with a potency ~isoproterenol, but with a somewhat longer duration of action; there was little evidence of cardiovascular effects or upper airway irritation[122]
79	9-Keto-11α,15α-dihydroxy-20-methyl-prosta-13-*trans*-enoic acid		$C_{21}H_{36}O_5$ (368)	Corey type with attachment of α-chain first[126]	Inhibited histamine-induced gastric acid secretion in Heidenhain pouch dogs (i.v.): $1.1 \times PGE_1$ Hypotensive in anesthetized cats (i.v.): $0.68 \times PGE_1$ Contracted isolated gerbil colon; $0.54 \times PGE_1$[126]

No.	Name	Structure	Synthesis	Pharmacology
80	9-Keto-11α,15α-dihydroxy-20-isopropylidenprosta-5-*cis*-13-*trans*-dienoic acid (CS 412)		Corey synthesis[127]	Bronchial dilator; inhalation of 50 μg caused a significant increase in specific airway conductance in bronchial asthmatics for at least 0.5 hr; the specific airway conductance increased with increasing doses from 50 — 200 μg, but larger doses caused no further increase; neither tracheal irritation nor hypotensive effects were observed[128] Inhibited ADP induced aggregation of rabbit PRP in vitro (IC_{50} = 1 ng/mℓ; 3.5 × PGE_1)[139]
		$C_{23}H_{36}O_5$ (392)		
81	Methyl 7-thia-9-keto-11α,15α-dihydroxy-17α,20-dimethyl-prosta-13-*trans*-enoate		Cuprate addition of optically active protected ω-chain to fully elaborated optically active enone[139]	Antisecretory and cytoprotective agent; inhibited histamine stimulated gastric acid secretion in Heidenhain pouch dogs (ED_{50} = 40 μg/kg, p.o.) and in gastric fistula dogs (MED = 100 μg/kg, i.g.); inhibited ethanol induced gastric lesions in the rat (ED_{50} = 2 μg/kg, p.o.) at dose ≤100 times MED for gastric acid secretion[140]
		$C_{22}H_{38}O_5S$ (414)		
82	9-Keto-11α,16ξ-dihydroxy-16ξ-methyl-prosta-13-*trans*-en-1-ol (Rioprostol)			At 300 and 600 μg/kg oral doses, reduced basal acid (54 and 88%) and pepsin (86 and 68%) secretion and pentagastrin stimulated acid (44 and 59%) and pepsin (40 and 67%) secretion in normal subjects;[141] in normal subjects, 300 μg (q.i.d.) completely protected against the endoscopically observed injurious effects of aspirin (975 mg, q.i.d.)[142]
		$C_{22}H_{38}O_4$ (366)		

Table 5 (continued)
PGE ANALOGS

No.	Systematic name	Structure	Mol. formula (mol. wt.)	Synthesis	Biological Actions
83	Methyl 9-keto-11α, 16ξ-dihydroxy-16ξ-methylprosta-4-*cis*-13-*trans*-dienoate (Enisoprost)		$C_{22}H_{36}O_5$ (380)	Cuprate addition of protected ω-chain to fully elaborated cyclopentenone[143]	Antisecretory agent; inhibited acid secretion in the Heidenhain pouch dog by i.v. (ID_{50} = 0.5 μg/kg) and oral (ED_{50} = 0.3—0.6 μg/kg) route with a longer duration of action (>5 hr, p.o.) than Misoprostol (Compound 84); less diarrheagenic than Misoprostol[144]
84	Methyl 9-keto-11α, 16ξ-dihydroxy-16ξ-methylprosta-13-*trans*-enoate (Misoprostol)		$C_{22}H_{38}O_5$ (382)	Cuprate addition of protected ω-chain to fully elaborated cyclopentenone[145]	Antisecretory agent; histamine induced gastric acid secretion in Heidenhain pouch dog (ID_{50} = 0.16 μg/kg, i.v.; 48 × PGE_1) and gastric fistula dog (ED_{50} = 5.5 μg/kg, i.g.); greatly reduced tendency to induce diarrhea in fasted rats (ED_{50} = 366 μg/kg, p.o.)[146] In clinical trials 200 μg (q.i.d.) given orally was similar in efficacy to 300 mg cimetidine q.i.d. and was considerably more effective than placebo in healing duodenal ulcers after 4 weeks treatment; diarrhea was observed in 10—13% of the patients usually in the initial stages of the trials; see Reference 133, p. 98

85 9-Keto-11α,15α,
dihydroxy-16,16-
dimethyl-1-oxaprosta-
5-*trans* dienoic acid
(HOE 260)

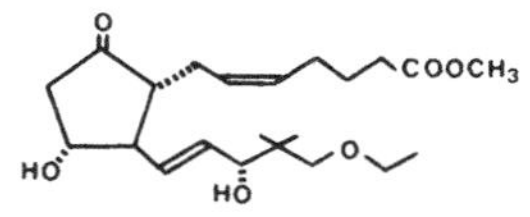

$C_{21}H_{34}O_6$ (382)

Potent, long lasting antisecretory and cytoprotective agent; in Heidenhain pouch dog inhibited histamine ($ED_{50} = 1.75$ µg/kg, i.v.) and heptagastrin ($ED_{50} = 3.69$ µg/kg, i.v.) stimulated gastric acid secretion; in the rat, prevented gastric lesions produced by 80% ethanol ($ID_{50} = 17$ µg/kg, p.o.) but caused enteropooling only at much higher doses ($ED_{50} \simeq 1$ mg/kg, p.o.)[147]

REFERENCES

1. **Morozowich, W., Oesterling, T. O., Miller, W. L., and Douglas, S. C.,** Prostaglandin prodrugs II: new method of synthesizing prostaglandin C_1-aliphatic esters, *J. Pharm. Sci.,* 68, 836—838, 1979.
2. **Schaaf, T. K., and Hess, H. J.,** Synthesis and biological activity of carboxyl-terminus modified prostaglandin analogues, *J. Med. Chem.,* 22, 1340—1346, 1979.
3. **Spector, S. L. and Ball, R. C.,** The effect of varying doses of prostaglandin E_1 analog bronchodilator in asthmatic patients, *Ann. Allergy,* 38, 302, 1977.
4. **Pike, J. E., Lincoln, F. H., and Schneider, W. P.,** Prostanoic acid chemistry, *J. Org. Chem.,* 34, 3552—3557, 1969.
5. **Gardiner, P. J., Copas, J. L., Schneider, C., and Collier, H. O. J.,** 2-Decarboxy-2-hydroxymethyl prostaglandin E_1 (TR4161), a new prostaglandin bronchodilator of low tracheobronchial irritancy, *Prostaglandins,* 19, 349—370, 1980.
6. **Gardiner, P. J. and Collier, H. O. J.,** Specific receptors for prostaglandins in airways, *Prostaglandins,* 819—841, 1980.
7. **Kluender, H. C. and Peruzzotti, G. P.,** Synthesis and biological activity of some PGE_1 carbinols, *Tetrahedron Lett.,* 2063—2066, 1977.
8. **Broughton, B. J., Caton, M. P. L., Christmas, A. J., Coffee, E. C. J., and Hambling, D. J.,** Uterine stimulant action of some ω-chain modified (±)-11-deoxyprostaglandins, *Prostaglandins,* 22, 53—64, 1981.
9. **Suga, H., Konishi, Y., Wakatsuka, H., Miyake, H., Kori, S., and Hayashi, M.,** Synthesis of 16,16-dimethyl-*trans*-Δ^2-PGE_1 methyl ester (ONO-802), *Prostaglandins,* 15, 907—912, 1978.
10. **Karim, S. M. M., Ragnam, S. S., and Ilancheran, A.,** Menstrual induction with vaginal administration of 16,16-dimethyl *trans*-Δ^2-PGE_1 methyl ester (ONO 802), *Prostaglandins,* 14, 615—1977.
11. Gemeprost, *Drugs Future,* 6, 54, 1981.
12. **Takagi, S., Yoshida, T., Ohya, A., Tsubata, K., Sakata, H., Fujii, K. T., Iizuka, S., Tochigi, B., Tochigi, M., and Mochigi, A.,** The abortifacient effect of 16,16-dimethyl-*trans*-Δ^2-PGE_1 methyl ester, a new prostaglandin analogue, on mid-trimester pregnancies and long-term follow-up observations, *Prostaglandins,* 23, 591—601, 1982.
13. **Sakamoto, S., Satoh, K., Nishiya, I., Kunimoto, K., Chimura, T., Oda, T., Takeuchi, S., Satoh, R., Iizuka, T., Kobayashi, S., Takagi, S., Yoshida, T., Tomoda, Y., Ninagawa, T., Kurachi, K., Tanizawa, O., Tojo, S., Mochizuki, M., Maeda, K., Tominaga, Y., Torigoe, T., Koresawa, M., Taki, I., and Hamada, T.,** Abortifacient effect and uterine cervix-dilating action of 16,16-dimethyl *trans* Δ^2 PGE_1 methyl ester (ONO 802) in the form of a vaginal suppository (a randomized, double-blind, controlled study in the second trimester of pregnancy), *Prostaglandins, Leukotrienes, Med.,* 9, 349—361, 1982.
14. **Bickel, M.,** Prevention of duodenal ulcers in the rat by a synthetic prostaglandin analog and natural prostaglandin E_2, *IRCS Med. Sci. Libr. Compend.,* 7, 409, 1979; *Chem. Abstr.,* 91, 187075a, 1979.
15. **Bartmann, W., Beck, G., Kunstmann, R., Lerch, V., and Teufel, H.,** Darstellung von 15-Hydroxy-9-oxo-prosta-2-(E)-13-(E)-dien-säure und von 16-Hydroxy-9-oxo-20-homo-prosta-2-(E)-14-(E)-dien-säure, *Tetrahedron Lett.,* 2563—2566, 1977.
16. ONO-1206, *Drugs of the Future,* 7, 116—118, 1982.
17. **Tsuboi, T., Hatano, N., Nakatsuji, K., Fujitami, B., Yoshida, K., Schimizu, M., Kawasaki, A., Sakata, M., and Tsuboshima, M.,** Pharmacological evaluation of OP 1206, a prostaglandin E_1 derivative, as an antianginal agent, *Arch. Int. Pharmacodyn.,* 247, 89—102, 1982.
18. **Kottegoda, S. R., Adaikan, P. G., and Karim, S. M. M.,** Reversal of vasopressin-induced coronary vasoconstriction by a PGE_1 analogue (ONO-1206) in primates, *Prostaglandins, Leukotrienes, Med.,* 8, 343—348, 1982.
19. ONO-747, *Drugs of the Future,* 2, 320—322, 1977.
20. **Nelson, N. A., Jackson, R. W., Au, A. T., Wynalda, D. J., and Nishizawa, E. E.,** Synthesis of dℓ-4,5,6-trinor-3,7-inter-*m*-phenylene-3-oxoprostaglandins including one which inhibits platelet aggregation, *Prostaglandins,* 10, 795—806, 1975.
21. **Morton, D. R. and Morge, R. A.,** Total synthesis of 3-oxa-4,5,6-trinor-3,7-*inter*-*m*-phenylene prostaglandins. I. Photochemical approach, *J. Org. Chem.,* 43, 2093—2101, 1978.
22. **Morton, D. R. and Thompson, J. L.,** Total synthesis of 3-oxa-4,5,6-trinor-3,7-*inter*-*m*-phenylene prostaglandins. II. Conjugate addition approach, *J. Org. Chem.,* 43, 2102—2106, 1978.
23. **Skotnicki, J. S., Schaub, R. E., Weiss, M. H., and Dessy, F.,** Prostaglandins and congeners. XVI. Synthesis and bronchodilator activity of dℓ-11-deoxy-3-thiaprostaglandins, *J. Med. Chem.,* 20, 1662—1665, 1977.
24. RS-84,135, *Drugs of the Future,* 7, 812—814, 1982.
25. **Lin, C. H. and Stein, S. J.,** The synthesis of 4,5-acetylenic prostaglandins, *Synth. Comm.,* 6, 503—508, 1976.
26. **Lin, C. H., Stein, S. J., and Pike, J. E.,** The synthesis of 5,6-acetylenic prostaglandins, *Prostaglandins,* 11, 377—380, 1976.

27. **Buckler, R. T., Ward, F. E., Hartzler, H. E., and Kurchacova, E.**, The synthesis and biological activity of benzo[5,6]prostaglandins A_2, E_2 and $F_{2\alpha}$, *Eur. J. Med. Chem.*, 12, 463—465, 1977.

28. **Yamaguchi, T., Sakai, K., Yusa, T., and Yamazaki, M.**, The bronchodilator activity of 20-isopropylidene prostaglandin E_2 (CS-412), *Prostaglandins*, 20, 521—532, 1980.

29. 6KPGE$_2$, *Drugs Future*, 5, 348, 1980.

30. **Jackson, E. K., Herzer, W. A., Zimmerman, J. B., Branch, R. A., Oates, J. A., and Gerkens, J. F.**, 6-Keto-prostaglandin E_1 is more potent than prostaglandin I_2 as a renal vasodilator and renin secretagogue, *J. Pharmacol. Exp. Ther.*, 216, 24—27, 1981.

31. **Lee, W. H., McGiff, J. C., Householder, R. W., Sun, F. F., and Wong, P. Y. K.**, Inhibition of platelet aggregation by 6-keto-prostaglandin $F_{1\alpha}$ (6-KPGF$_{1\alpha}$), *Fed. Proc.*, 38, 419, 1979.

32. **Wong, P. Y. K., McGiff, J. C., Sun, F. F., and Lee, W. H.**, 6-Keto-prostaglandin F_1 inhibits the aggregation of human platelets, *Eur. J. Pharmacol.*, 60, 245—252, 1979.

33. **Hyman, A. L. and Kadowitz, P. J.**, Vasodilator actions of prostaglandin 6-keto-E$_1$ in the pulmonary vascular bed, *J. Pharmacol. Exp. Ther.*, 213, 468—472, 1950.

34. **Miller, O. V., Aiken, J. W., Shebuski, R. J., and Gorman, R. R.**, 6-Keto-prostaglandin E$_1$ is not equipotent to prostacyclin (PGI$_2$) as an antiaggregatory agent, *Prostaglandins*, 20, 391—400, 1980.

35. **Kulkarni, P. S. and Srinivasan, B. D.**, The effect of intravitreal and topical prostaglandins on intraocular inflammation, *Invest. Opthamol. Vis. Sci.*, 23, 383—392, 1982.

36. 6-KPGE$_1$, *Drugs Future*, 8, 651—652, 1983.

37. **Nakai, H., Hamanaka, N., Miyake, H., and Hayashi, M.**, Synthesis of 5-fluoroprostaglandins, *Chem. Lett.*, 1499—1502, 1979.

38. **Hoshina, K., Kawashima, H., Kurozumi, S., and Hashimoto, Y.**, Inhibition of indomethacin-induced gastric ulceration in the rat by 7-oxo-prostaglandin E$_1$ analogs, *Yakugaku Zasshi*, 101, 247—253, 1981; *Chem. Abstr.*, 94, 202990z, 1981.

39. **Kurozumi, S., Toru, T., Kobayashi, M., and Hashimoto, Y.**, Synthesis of E-type 7-thiaprostaglandins, *Chem. Lett.*, 331—334, 1977.

40. **Kurozumi, S., Toru, T., Kobayashi, M., and Hashimoto, Y.**, Synthesis of new 7,13-dithiaprostaglandins, *Synth. Comm.*, 7, 169—177, 1977.

41. **Kurozumi, S., Toru, T., Tanaka, T., Kibayashi, M., Miura, S., and Ishimoto, S.**, Prostaglandin chemistry. VII. A synthesis of new 8-phenylthio-11-deoxyprostaglandins, *Chem. Lett.*, 4091—4094, 1976.

42. **Cho, M. J., Bundy, G. L., and Biermacher, J. J.**, Prostaglandin prodrugs. V. Prostaglandin E$_2$ ethylene ketal, *J. Med. Chem.*, 20, 1525—1527, 1977.

43. Meteneprost, *Drugs Future*, 6, 686—687, 1981.

44. **Green, K., Vesterqvist, O., Bydeman, M., Christenssen, M., and Bergström, S.**, Plasma levels of 9-deoxo-16,16-dimethyl-9-methylene-PGE$_2$ in connection with its development as an abortifacient, *Prostaglandins*, 24, 451—466, 1982.

45. **Wilks, J. W.**, Pregnancy interception with a combination of prostaglandins: studies in monkeys, *Science*, 221, 1407—1409, 1983.

46. **Hamon, A., Lacoume, B., Pasquet, G., and Pilgrim, W. R.**, Synthesis of (±)- and 15-epi-(±)-10,10-dimethylprostaglandin E$_1$, *Tetrahedron Lett.*, 211—214, 1976.

47. **Plantema, O. G., de Koning, H., and Huisman, H. O.**, Synthesis of (±)-10,10-dimethylprostaglandin E$_1$ methyl ester and its 15-epimer, *J. Chem. Soc. Perkin Trans. I*, 304—308, 1978.

48. **Burton, T. S., Caton, M. P. L., Coffee, E. C. J., Parker, T., Stuttle, K. A. J., and Watkins, G. L.**, Prostaglandins. IV. Synthesis of (±)-11-deoxyprostaglandins from 2-(ω-hydroxyheptyl)cyclopent-2-enones, *J. Chem. Soc. Perkin Trans. I*, 2550—2556, 1976.

49. **Peret, A. G., Nakamoto, H., Ishizuka, N., Aburatani, M., Nakahashi, K., Sakamoto, K., and Takeuchi, T.**, Prostaglandin analogs modified at the 10 and 11 positions, *Tetrahedron Lett.*, 3933—3936, 1979.

50. **Caton. M. P. L., Darnbrough, G., Hambling, D. J., Jordan, R., Parker, T.**, Prostaglandins. V. Synthesis and Pharmacology of (±)-11-deoxy-10-hydroxyprostaglandin E$_1$ methyl ester, *Prostaglandins*, 18, 569—576, 1979.

51. **Caton, M. P. L., Darnbrough, G., and Parker, T.**, Prostaglandins, VI. Base-catalysed acetoxidation of α-hydroxycyclopentanones and the synthesis of 9,10-diketoprostanoic acids, *Tetrahedron Lett.*, 21, 1685—1686, 1981.

52. **Guzman, A. and Muchowski, J. M.**, Synthesis of 10,11-methylene prostaglandins of unambiguous stereochemistry, *Chem. Ind. (London)*, 790, 1975.

53. **Sugie, A., Shimomura, H., Katsube, J., and Yamamoto, H.**, Synthesis of benzo-type prostaglandin analogs, *Tetrahedron Lett.*, 2759—2762, 1977.

54. **Ohno, K., Nishiyama, H., and Nagase, H.**, A mild methylation of alcohols with diazomethane catalyzed by silica gel, *Tetrahedron Lett.*, 4405—4406, 1979.

55. SHPGE$_2$, *Drugs Future*, 4, 739—740, 1979.

56. **Floyd, M. B., Schaub, R. E., Suita, G. J., Skotnicki, J. S., Grudzinskas, C. V., and Weiss, M. J.,** Prostaglandins and congeners. XXII. Synthesis of 11-substituted derivatives of 11-deoxyprostaglandins E$_1$ and E$_2$. Potent bronchodilators, *J. Med. Chem.*, 23, 903—913, 1980.

57. **Birnbaum, J. E., Birkhead, N. C., Oronsky, A. L., Dessy, F., Rihoux, J. R., and Van Humbeeck, L.,** Bronchodilator activity of a PGE$_2$ analog in animals and in man, *Prostaglandins*, 21, 457—469, 1981.

58. **Lin, C. H.,** Synthesis of 11-methyl prostaglandins, *Chem. Ind. (London)*, 994—995, 1976.

59. RO-21-6937/000, *Drugs Future*, 6, 42—43, 1981.

60. RO-21-6937/000, *Drugs Future*, 8, 72, 1983.

61. **Inoue, K., Ide, J., and Sakai, K.,** Synthesis of 11-methyl- and 11,11-dimethylprostaglandins via the Wittig reaction of α-dienol, *Bull. Chem. Soc. Jpn.*, 51, 2361—2365, 1978.

62. **Greene, A. E., Teixeira, M. A., Barreiro, E., Cruz, A., and Crabbé, P.,** The total synthesis of prostaglandins by the tropolone route, *J. Org. Chem.*, 47, 2553—2564, 1982.

63. **Kao, W. and Strike, D. P.,** Synthesis of (±)-11-deoxy-15-ethynyl prostaglandins, *Prostaglandins*, 16, 467—471, 1978.

64. **Buendia, J., Nierat, J., and Vivat, M.,** Synthèse de prostaglandines alkylées en 15, *Bull. Soc. Chim. Fr.*, 2, 614—622, 1979.

65. **Bagli, J. and Bogri, T.,** Prostaglandins. V. Utility of the Nef reaction in the synthesis of prostanoic acids. A total synthesis of (±)-11-deoxy-PGE$_1$, -PGE$_2$, and their C-15 epimers, *Tetrahedron Lett.*, 3815—3817, 1972.

66. AU 23,578, *Drugs Future*, 1, 201—209, 1976.

67. **Bagli, J. F., Greenberg, R., Abraham, N. H., and Pelz, K.,** Prostaglandins. IX. Synthesis of (±)-15-methyl-11-deoxy PGE$_1$ (doxaprost), a potent bronchodilator — and its C-15-epimer, *Prostaglandins*, 11, 981—986, 1976.

68. **Greenberg, R., Smorong, K., and Bagli, J. F.,** A comparison of the bronchodilator activity of (±) 11-deoxy prostaglandin E$_1$ with its 15-methyl analogue (doxaprost), *Prostaglandins*, 961—980, 1976.

69. **Bagli, J. F., Bogri, T., and Sehgal, S. N.,** Prostaglandins. VII. 15-Methyl-15-hydroxy-9-oxoprostanoic acid — a potent orally active antisecretory agent, *Tetrahedron Lett.*, 3329—3332, 1973.

70. Deprostil, *Drugs Future*, 1, 282—285, 1976.

71. **Crabbé, P. and Cervantes, A.,** Synthesis of difluoromethylene-prostaglandins, *Tetrahedron Lett.*, 1319—1321, 1973.

72. **Muchowski, J. M.,** Synthesis and bronchial dilator activity of some novel prostaglandin analogues, in *Chemistry, Biochemistry and Pharmacology of Prostanoids*, Roberts, R. M. and Scheinmann, F., Eds., Pergamon Press, Oxford, 1979, 45, 57, 58.

73. **Bartmann, W., Beck, G., Kunstmann, R., Lerch, V., and Teufel, H.,** Synthese von 13-Hydroxy-9-oxo-14-(E)-prostensäure Durch Allylumlagerung aus 8-Äthoxycarbonyl-11-desoxy-PGE$_1$, *Tetrahedron Lett.*, 3879—3882, 1976.

74. **Skotnicki, J. S., Schaub, R. E., Bernady, K. F., Siuta, G. J., Poletto, J. F., and Dessy, F.,** Prostaglandins and congeners. IV. Synthesis and bronchodilator activity of dℓ-11-Deoxy-15- or 16-alkylprostaglandins, *J. Med. Chem.*, 20, 1551—1557, 1977.

75. K-11,550 *Drugs Future*, 6, 780—782, 1981.

76. **Banerjee, A. K., Broughton, B. J., Burton, T. S., Caton. M. P. L., Christmas, A. J., Coffee, E. C. J., Crowshaw, D., Heazell, M. A., Stuffle, K. A. F., and Watkins, G. L.,** Synthesis and gastrointestinal pharmacology of some 15- and 16-modified (±)-11-deoxyprostaglandins, *Prostaglandins*, 16, 541—554, 1978.

77. **Banerjee, A. K., Broughton, . J., Burton, T. S., Caton, M. P. L., Christmas, A. J., Coffee, E. C. J., Crowshaw, K., Hardy, C. J., Heazell, M. A., Paltreyman, M. N., Parker, T., Saunders, L. C., and Stuttle, K. A. J.,** Synthesis and antiulcer activity of 16-phenoxy analogues of (±)-11-deoxyprostaglandin E$_1$, *Prostaglandins*, 22, 167—182, 1981.

78. **Skotnicki, J. J., Schaub, R. E., Weiss, M. J., and Dessy, F.,** Prostaglandins and congeners. XIV. Synthesis and bronchodilator activity of dℓ-16,16-trimethyleneprostaglandins, *J. Med. Chem.*, 20, 1042—1050, 1977.

79. **Pirillo, D. and Traverso, G.,** Synthesis of (±)-8,12-*trans*-Δ^5-en-*cis*-9-keto-14,15-en-*trans*-16-R,S-hydroxyprostanoic acid methyl ester, *Il Farmaco Ed. Sci.*, 31, 468—470, 1976.

80. **Troconoi, G., d'Atri, G., and Scolastico, C.,** Synthesis and biological activity of modified amino prostaglandins, *Prostaglandins*, 13, 1067—1071, 1977.

81. **Nakamura, N. and Sakai, K.,** A synthesis of 12-substituted prostaglandins, *Tetrahedron Lett.*, 2049—2052, 1976.

82. **Godoy, L., Guzman, A., and Muchowski, J. M.,** Synthesis of 11-deosoxy-12-methylprostaglandins, *Chem. Lett.*, 327—330, 1975.

83. **Rooks, W. H. and Tomolonis, A. J.,** Unpublished data.

84. EMD-33,290, *Drugs Future*, 8, 585—586, 1983.

85. **Dammann, H. G., Müller, P., Heun, J., and Simon, B.,** Säurehemmung und Vorhinderung des Transmukösen Potentialdifferenz-Abfalls durch das Thiaprostaglandin EMD-33,290 Beim Henschem, *Schweiz, Med. Wochenschr.,* 112, 829—831, 1982.

86. **Baret, P., Barreiro, E., Greene, A. E., Luche, J.-L., Teixeira, M. A., and Crabbé, P.,** Synthesis of new allenic prostanoids, *Tetrahedron,* 35, 2931—2938, 1979.

87. **Yankee, E. W. and Bundy, G. L.,** (15S)-15-Methylprostaglandins, *J. Am. Chem. Soc.,* 94, 3651—3652, 1972.

88. Arbaprostil, *Drugs Future,* 8, 80S, 1983.

89. **Robert, A. and Yankee, E. W.,** Gastric antisecretory effect of 15(R)-15-methyl PGE₂, methyl ester and of 15(S)-15-methyl PGE₂, methyl ester (38707), *Proc. Soc. Exp. Biol. Med.,* 148, 1155—1158, 1975.

90. **Plattner, J. J. and Gager, A. H.,** Synthesis of optically active 15-thiaprostaglandins, *Tetrahedron Lett.,* 1629—1632, 1977.

91. **Schaaf, T. K., Johnson, M. R., Constantine, J. W., Bindra, J. S., Hess, H. J., and Elger, W.,** Structure-activity studies of configurationally rigid arylprostaglandins, *J. Med. Chem.,* 26, 328—334, 1983.

92. **Johnson, M. R., Schaaf, T. K., Constantine, J. W., and Hess, H. J.,** Structure activity studies leading to a tissue selective hypotensive prostaglandin analog, 13,14-dihydro-16-phenyl-ω-tetranor PGE₂, *Prostaglandins,* 20, 515—520, 1980.

93. **Schaaf, T. K., Johnson, M. R., Eggler, J. F., Bindra, J. S., Constantine, J. W., and Hess, H. J.,** Hypotensive prostalgandin structure activity relationships: 11-deoxy-16-aryl-ω-tetranor prostaglandins, in *Advances in Prostaglandin, Thromboxane and Leukotriene Research,* Vol. 2, Samuelsson, B. Paoletti, R., and Ramwell, P., Eds., Raven Press, New York, 1983, 313—318.

94. Sulprostone, *Drugs Future,* 3, 59—61, 1978.

95. **Karim, S. M. M., Choo, H. T., Lim, A. L., Yeo, K. C., and Ratnam, S. S.,** Termination of second trimester pregnancy with intramuscular administration of 26-phenoxy-ω-17,18,19,20-tetranor PGE₂ methylsulfonylamide, *Prostaglandins,* 15, 1063—1068, 1978.

96. Sulprostone Sh-B-286, *Drugs Future,* 5, 53—54, 1980.

97. Sulprostone nalador, *Drugs Future,* 7, 64, 1982.

98. CL-116-069, *Drugs Future,* 796—798, 1982.

99. **Jackson, R. T. and Birnbaum, J. E.,** Nasal vasoconstrictor activity of a novel PGE₂ analog, *Prostaglandins,* 21, 1015—1024, 1981.

100. **Jackson, R. T. and Birnbaum, J. E.,** A comparison of a synthetic prostaglandin and xylometazoline hydrochloride as nasal decongestants, *Otolaryngol. Head Neck Surg.,* 90, 595—597, 1982.

101. **Iwamoto, H., Inuka, N., Yanagisawa, I., Ishii, Y., Tamura, T., Shiozaki, T., Takagi, T., Tomioka, K., and Murakami, M.,** Studies on prostaglandins, VI. Synthesis of 16(S)-methyl-20-methoxy-PGE₂ (YPG-209) having oral bronchodilator activity and its analog, *Chem. Pharm. Bull.,* 28, 1422—1431, 1980.

102. **Tomioka, K., Terai, M., and Haeno, H.,** Oral bronchodilator effect of 16(S)-methyl-20-methoxy-PGE₂ (YPG-209) in guinea pigs, *Arch. Int. Pharmacodyn.,* 226, 224—234, 1977.

103. **Magerlein, D. W., Ducharme, W. E., Magee, W. E., Miller, W. L., Robert, A., Weeks, J. R.,** Synthesis and biological properties of 16,16-alkylprostaglandins, *Prostaglandins,* 4, 143—145, 1973.

104. 16,16-Dimethyl PGE₂, *Drugs Future,* 5, 317—318, 1981.

105. 16,16-Diemthyl PGE₂, *Drugs Future,* 8, 450—452, 1983.

106. **Hallett, W. A., Wissner, A., Grudzinskas, C. V., Partridge, R., Birnbaum, J. E., and Weiss, M. J.,** Prostaglandins and congeners. XIII. The synthesis of dℓ-erythro-16-methoxyprostaglandins, *Prostaglandins,* 13, 409—415, 1977.

107. **Scarpignato, C., Spina, G., Signorini, G. C., and Bertaccini, G.,** Action of MDL-646, a new synthetic prostaglandin, on gastric acid secretion of some experimental animals, *Res. Commun. Chem. Pathol. Pharmacol.,* 39, 211—228, 1983.

108. **Miyake, K., Iguchi, S., Kori, S., and Hayashi, M.,** Synthesis of 16-methylene-prostaglandins, and 16,16-ethano-prostaglandins, *Chem. Lett.,* 211—214, 1976.

109. **Nakai, H., Hamanaka, N., and Kurono, M.,** Synthesis of 16,16-(2-fluorotrimethylene)-prostaglandins and 16,16-(2,2-difluorotrimethylene)-prostaglandins, *Chem. Lett.,* 63—66, 1979.

110. **Tamura, T., Inukai, N., Iwamoto, H., Yanagisawa, I., Ishii, Y., Takagi, T., Tomioka, K., and Murakami, M.,** Studies on prostaglandins. VI. Synthesis of prostaglandins derivatives possessed ylidene group on C-16, *Rep. Yamanouchi Res. Lab.,* 4, 16—29, 1980.

111. 16-Methylene-PGE₂ methyl ester, *Drugs Fugure,* 2, 247—249, 1977.

112. SC 30,249, *Drugs Future,* 6, 106—107, 1981.

113. **Dajani, E. Z., Driskill, D. R., Bianchi, R. G., Phillips, E. L., Woods, E. M., Colton, D. G., Collins, P. W., and Pappo, R.,** Effect of a prostaglandin analogue. SC-30249, on canine gastric secretion, *Drug Dev. Res.,* 3, 339—347, 1983.

114. **Chan, S. M. L. and Grudzinskas, C. V.,** Prostaglandins and congeners. XXVII. Synthesis of biologically active 16-halomethyl derivatives of 15-deoxy-16-hydroxyprostaglandin E₂, *J. Med. Chem.,* 45, 2278—2282, 1980.

115. **Chan, P. S., Cervoni, P., Ronsberg, M. A., Accomando, R. C., Quirk, G. J., Scully, P. A., and Lipchuck, L. M.**, Antihypertensive activity of dℓ-15-deoxy-16-hydroxy-16(α/β)-vinyl prostaglandin E$_2$ methyl ester (CL 115,347), a new orally and transdermally long-acting antihypertensive agent, *J. Pharmacol. Exp. Ther.*, 226, 726—732, 1983.

116. **Wissner, A., Birnbaum, J. E., and Wilson, D. E.**, Prostaglandins and congeners. XXV. Inhibition of gastric acid secretion. Replacement of the carboxylate moiety of a prostaglandin with a hydroxymethylketo functional group, *J. Med. Chem.*, 23, 715—717, 1980.

117. **Wilson, D. E., Scruggs, W., and Birnbaum, J. E.**, A new prostaglandin E$_1$ analogue (CL-115,574), II. Effects of gastric acid secretion in the dog, *Prostaglandins*, 22, 971—978, 1981.

118. CL-115,574, *Drugs Future*, 7, 793—795, 1982.

119. TR-4752, *Drugs Future*, 6, 303—305, 1981.

120. **Arndt, H. C., Gardiner, P. J., Hong, E., Kleuader, H. C., Meyers, C., and Woessner, W. D.**, The synthesis and biological activity of ω-pentanor-15-alkylcyclobutyl-PGE$_1$ analogs, *Prostaglandins*, 16, 67—77, 1978.

121. **Lüthy, C., Konstantin, P., and Untch, K.**, Total synthesis of dℓ-19-hydroxyprostaglandin E$_1$ and dℓ-13-*cis*-15-*epi*-19-hydroxyprostaglandin E$_1$, *J. Am. Chem. Soc.*, 100, 6211—6217, 1978.

122. **Garay, G. L. and Weissberg, R. W.**, Unpublished data.

123. **Bartmann, W., Beck, G., Lerch, U., and Teufel, H.**, Synthese des 15α-acetoxy-16,16-dimethyl-9-oxo-18-oxa-(2E,13E)-2,13-prostadiensäure-methylesters, *Liebigs Ann. Chem.*, 1739—1748, 1978.

124. **Beck, G., Bartmann, W., Lerch, U., Teufel, H., and Schölkens, B.**, Antihypertensive activity of 16,16-dimethyl-oxa-alkyl-prostaglandins of the PGA$_2$, PGE$_2$ and trans-Δ^2-11-deoxy-PGE$_1$ series: structure-activity relationships, *Prostaglandins*, 20, 153—169, 1980.

125. ONO-358, *Drugs Future*, 3, 536—538, 1978.

126. **Dajani, E. Z., Rozek, L. F., Sanner, J. H. and Miyano, M.**, Synthesis and biological evaluation of ω-homologues of prostaglandin E$_1$, *J. Med. Chem.*, 19, 1007—1010, 1976.

127. CS 142, *Drugs Future*, 7, 152—153, 1982.

128. **Murao, M., Uchiyama, K., Shida, A., Sakai, K., Yusa, T., and Yamaguch, T.**, Studies on 20-isopropylidene PGE$_2$ as a new bronchodilator, *Adv. Prostaglandin Thromboxane Res.*, 7, 985—988, 1980.

129. **Moore, P. K. and Griffiths, R. J.**, 6-Keto-prostaglandin-E$_1$, *Prostaglandins*, 26, 509—517, 1983.

130. **Bowden, K., Hielbron, I. M., Jones, E. R. K., and Weedon, B. C. L.**, Researches on acetylenic compounds. I. The preparation of acetylenic ketones by oxidation of acetylenic carbinols and glycols, *J. Chem. Soc.*, 39—45, 1946.

131. **Birnbaum, J. E., Cervoni, P., Chan, P. S., Chan, S. M. L., Floyd, M. B., Grudzinskas, C. V., and Weiss, M. J.**, Prostaglandins and congeners. XXIX. (16RS)-($\pm$)-16-Hydroxy-16-vinyl prostaglandin E$_2$, an orally and transdermally active hypotensive agent of prolonged duration, *J. Med. Chem.*, 25, 492—494, 1982.

132. **Guzman, A., Velarde, E., Davis, R., Garay, G. L., Muchowski, J. M., Rooks, W. H., and Tomolonis, A.J.**, 11-Deoxy-11α,12α-methanoprostaglandin E$_2$, a potent, short acting bronchial dilator, *J. Pharm. Sci.*, 75, 307—312, 1986.

133. **Garay, G. L. and Muchowski, J. M.**, Agents for the treatment of peptic ulcer disease, *Ann. Rep. Med. Chem.*, 20, 93—105, 1985.

134. Protective and therapeutic effects of gastrointestinal prostaglandins. Enprostil: A new modality. Proceedings of a symposium, *Am. J. Med.*, 81(2A), 1—88, 1986.

135. **Bannerjee, A. K., Tuffin, D. P., and Walker, J. L.**, Pharmacological effects of ($\pm$)-11-deoxy-16-phenoxyprostaglandin E$_1$ derivatives in the cardiovascular system, *Br. J. Pharmac.*, 84, 71—80, 1985.

136. **Bays, D. E. and Stables, R.**, Agents for the treatment of peptic ulcer disease, *Ann. Rep. Med. Chem.*, 18, 89—98, 1983.

137. **Belch, J. J. F., Shaw, B., Sturrock, R. D., Madhok, R., Lieberman, P., and Forbes, C. D.**, Double blind trial of CL115,347, a transdermally absorbed prostaglandin E$_2$ analogue, in treatment of Reynaud's phenomenon, *The Lancet*, 1180—1183, 1985.

138. **Wilson, D. E., Hulya, M. D., and Levendoglu, M. D.**, A new PGE$_1$ analogue (CL115,574) III. Effects on gastric acid and mucus secretion in man, *Prostaglandins*, 28, 5—11, 1984.

139. **Tanaka, T., Okamura, N., Bannai, K., Hazato, A., Sugiura, S., Manabe, K., Kamimoto, F., and Kurozumi, S.**, Synthesis of 7-thiaprostaglandin E$_1$ congeners: potent inhibitors of platelet aggregation, *Chem. Pharm. Bull.*, 33, 2359—2385, 1985.

140. **Shriver, D. A., Rosenthale, M. E., Kleunder, H. C., Schut, R. N., McGuire, J. L., and Hong, E.**, Pharmacology of rioprostol, a new gastric cytoprotective/antisecretory agent, *Arzneim. Forsch/Drug Res.*, 35, 839—845, 1985.

141. **Demol, P., Wingender, W., Weihrauch, T. R., and Graefe, K. H.**, Inhibition of gastric secretion in man by rioprostol, a new synthetic methyl prostaglandin E$_1$, *Arzneim. Forsch/Drug Res.*, 861—863, 1985.

142. **Detweiler, M. D., Harrison, C. A., Rollins, D. E., Tolman, K. G., McCormack, G. H., and Simon, D. M.,** Effect of rioprostol on aspirin induced gastrointestinal mucosal changes in normal volunteers, *Gastroenterol.,* 86, 1062, 1984.

143. **Collins, P. W., Gasiecki, A. F., Decktor, D. L., Dajani, E. Z., Woods, E. M., and Bianchi, R. C.,** Alpha chain derivatives of 16-hydroxy prostaglandins, *19th Medicinal Chemical Symposium of the American Chemical Society,* Tucson, Arizona, June 1984, pp. 102—112.

144. **Collins, P. W., Dajani, E. Z., Pappo, R., Gasiecki, A. F., Bianchi, R. G., and Woods, E. M.,** Synthesis and gastric antisecretory properties of 4,5-unsaturated derivatives of 15-deoxy-16-hydroxy-16-methylprostaglandin E_1, *J. Med. Chem.,* 26, 786—790, 1983.

145. **Collins, P. W., Dajani, E. Z., Driskill, D. R., Bruhn, M. S., Jung, C. J., and Pappo, R.,** Synthesis and gastric antisecretory properties of 15-deoxy-16-hydroxyprostaglandin E analogues, *J. Med. Chem.,* 20, 1152—1159, 1977.

146. **Collins, P. W., Pappo, R., and Dajani, E. Z.,** Chemistry and synthetic development of misoprostol, *Dig. Dis. Sci.,* 30 (Nov. Suppl.), 114S-117S, 1985.

147. **Bickel, M., Bartmann, W., Beck, G., Lerch, U., Schleyerbach, R., and Scholkens, B. A.,** Gastrointestinal and cardiovascular effects of the prostaglandin analogue HOE 260, *Biomed. Biochim. Acta,* 43 5235—5238, 1984.

Table 6
HETEROCYCLIC PG ANALOGS

No.	Systematic name	Structure	Mol. formula (mol. wt.)	Synthesis	Biological actions
1	8-Aza-9-keto-15α-hydroxyprosta-13-*trans*-enoic acid		$C_{19}H_{33}NO_4$ (339)	From pyroglutamic acid[1-5,52]	Prolonged inhibition of histamine induced bronchial constriction in guinea pig by i.v. (5 × PGE$_2$) and aerosol routes (1 × PGE$_2$), the most active enantiomer having the 12R,15S configuration[5] Inhibited gastric acid secretion[3,4] and stress ulcers (rat);[52] substrate for 15-hydroxy PG dehydrogenase[2]
2	8-Aza-9-keto-15α-hydroxyprosta-5-*cis*-13-*trans*-dienoic acid		$C_{19}H_{31}NO_4$ (337)	From pyroglutamic acid[2,7,8]	Inhibited histamine-induced bronchial constriction in guinea pig by i.v. route (0.007 × PGE$_2$);[6] substrate for 15-hydroxy PG dehydrogenase[2]
3	8-Aza-9-keto-15α- and 15β-hydroxyprosta-5-ynoic acid		$C_{19}H_{31}NO_4$ (337)	From pyroglutamic acid[9]	Contracted isolated human bronchial muscle[9]
4	8-Aza-9-keto-15α-hydroxyprostanoic acid		$C_{19}H_{35}NO_4$ (341)	Catalytic reduction of compound *1*,[10] or by total synthesis not involving pyroglutamic acid[11,12]	Prolonged inhibition of histamine-induced bronchial constriction in guinea pig by i.v. (2.5 × PGE$_2$) and aerosol routes (0.3 × PGE$_2$); stimulated formation of cAMP (mouse ovary) 8-fold at 25 μg/mℓ (cf. PGE$_1$; 60-fold at 25 μg/mℓ)[12]
5	9-Keto-15α- and 15β-hydroxyprost-13-*trans*-ene		$C_{19}H_{35}NO_2$ (309)	From pyroglutamic acid[13]	Mild inhibitor of platelet aggregation[13]
5a	8-Aza-9-keto-15α- and 15β-hydroxy-16-(3-trifluoromethyl-phenoxy)-ω-tetra-norprosta-13-*trans*-enoic acid		$C_{22}H_{28}F_3NO_5$ (413)	From pyroglutamic acid[54]	Inactive as antifertility agent in hamsters[54]

6	8-Aza-9-keto-16ξ-hydroxy-16ξ-methyl-prosta-13-*trans*-enoic acid		$C_{20}H_{35}NO_4$ (353)	From methyl pyroglutamate[14]	Cytoprotective; 30% at 2 mg/kg (p.o.). Inhibited histamine-induced bronchoconstriction; 15% at 50 μg/kg (i.p.)[14]
7	7-Keto-8-aza-15α- and 15β-hydroxy-prosta-13-*trans*-enoic acid		$C_{19}H_{33}NO_4$ (339)	From L-2-hydroxy-methylpyrrolidine[15]	Mild inhibitor of platelet aggregation[15]
8	8-Aza-15-hydroxy-prosta-9,11,13-*trans*-trienoic acid		$C_{19}H_{31}NO_3$ (321)	From pyrrole-2-carboxaldehyde[16]	
9	2-(3-Hydroxy-*trans*-oct-1-enyl)-3-(6-carboxyhexyl)-indole		$C_{23}H_{33}NO_3$ (371)	From methyl 3-(6-methoxcarbonylhexyl)-indole-2-carboxylate[17]	Contracted isolated guinea pig ileum; 0.002 × PGE₁, Contracted isolated rat fundus; 0.0007 × PGE₂. Stimulated cAMP formation in rat liver homogenates; 0.002 × PGE₁, but had only partial agonist activity since partially inhibited PGE₁ stimulated cAMP generation[17]
10	9-Aza-9-ethoxycarbonyl-11α, 15α-dihydroxyprosta-13-*trans*-enoic acid		$C_{22}H_{39}NO_6$ (413)	From N-ethoxycarbonyl-glycinate and diethyl 2-decendioate[18]	
11	9,11-Diketo-10-aza-15α-hydroxyprosta-5-*cis*-enoic acid		$C_{19}H_{31}NO_5$ (353)	From product of alkylation of diethyl allylmalonate with ethyl 2-bromodecanoate[19]	Inhibited aggregation of HPRP induced by ADP; 0.1 × PGE₁[19]
12	9-Keto-10-aza-10-methyl-15α-hydroxy-prosta-13-*trans*-enoic acid		$C_{20}H_{35}NO_4$ (353)	From methyl 1-methyl-2-pyrrolidone-4-carboxylate[20,21]	Strong bronchial dilator and spasmogenic activity[20]
13	Methyl 9-keto-10-aza-10-*i*-propyl-11α- and 11β-methoxy-15α-hydroxyprosta-13-*trans*-enoate		$C_{24}H_{43}NO_5$ (425)	From 1-*i*-propyl-3-(6-methoxycarbonyl-hexyl)-*trans*-4-hydroxy-methyl-2-pyrrolidone[22]	Prostaglandin-like activity[22]

Table 6 (continued)
HETEROCYCLIC PG ANALOGS

No.	Systematic name	Structure	Mol. formula (mol. wt.)	Synthesis	Biological actions
14	10-Keto-11-aza-11H-15α-hydroxyprosta-13-*trans*-enoic acid		$C_{20}H_{35}NO_4$ (353)	From *trans*-4-(6-ethoxy-carbonylhexyl)-5-ethoxy-carbonyl-2-pyrrolidone[23]	Hypotensive in anesthetized rats (i.v.); 0.13 × PGE$_2$.[23]
15	9-Keto-10-aza-10-H-11-nor-15α-hydroxy-prosta-13-*trans*-enoic acid		$C_{18}H_{31}NO_4$ (325)	From 3-vinylazetidin-2-one[28]	
15a	10-aza-11α-homo-15-hydroxyprosta-5-*cis*-Δ^{8,12}, Δ^{11,11a}, 13-*trans*-pentaenoic acid		$C_{20}H_{29}NO_3$ (331)	From 3-bromo-4-formylpyridine[55]	Thromboxane synthesis inhibitor (IC$_{50}$ = 3 µM; washed human platelets)[55]
16	Methyl 9-keto-12-aza-15α- and 15β-hydroxyprostanoate		$C_{22}H_{41}NO_4$ (383)	From dimethyl 2-aminononanedioate[24]	Contracted isolated rat fundus strip at 50 ng/mℓ; contracted rat uterus muscle[24]
17	9,11-Diketo-12-aza-15α- and 15β-hydroxyprostanoic acid		$C_{19}H_{33}NO_5$ (355)	From dimethyl[25] or diethyl[26] 2-aminononanedioate	Same profile in isolated smooth muscle assays as PGEs; hypotensive in rate (i. a.), 0.001 × PGE$_2$, with duration of response being somewhat longer than PGE$_2$; Smooth muscle activity not lost after passage through superfused guinea pig lungs[26]
18	Methyl 12-aza-15α- and 15β-hydroxy-prosta-9-enoate		$C_{20}H_{37}NO_3$ (339)	From Δ^3-purroline[27]	
19	8,10-Diaza-9-keto-10H-15α-hydroxy-prosta-13-*trans*-enoic acid		$C_{18}H_{32}N_2O_4$ (340)	From methyl 3-benzyloxy-ycarbonylimidazolidin-2-one-4-carboxylate[29]	Mild bronchial dilator activity[29]

No.	Name	Structure	Formula (MW)	Synthesis	Activity
20	8,10-Diaza-9,11-diketo-10H-15α-hydroxyprostanoic acid		$C_{18}H_{32}N_2O_5$ (356)	From ethyl 7-aminoheptanoate[30]	Inhibited aggregation of HPRP induced by ADP; 0.025 × PGE$_1$[30]
21	8,11-Diaza-15-hydroxyprosta-9,11,13-*trans*-enoic acid		$C_{18}H_{28}N_2O_5$ (320)	From 1-(6-methoxycarbonyl)hexyl)hexyl-2-hydroxymethyl-imidazole[31]	No useful activity in vitro in or in vivo[31]
22	8,12-Diaza-9,11-diketo-15-hydroxy-15-cyclohexyl-ω-pentanorprostanoic acid		$C_{19}H_{32}N_2O_5$ (368)	From product of reaction of ethyl 7-bromoheptanoate with ethyl carbazate[32]	Inhibited aggregation of HPRP induced by ADP, 0.01 × PGE$_1$[32]
23	9,12-Diaza-15-hydroxyprosta-8,10,13-*trans*-trienoic acid		$C_{18}H_{28}N_2O_3$ (320)	Self catalysed addition of 2-(6-methoxycarbony-lhexyl)-imidazole to 1-octyn-3-one followed by borohydride reduction and saponification[33]	Hypotensive in animals with efficacy ~ PGE$_2$[33]
24	(+) 9,11-Diketo-10,12-diaza-10H-15α-hydroxy-15-cyclohexyl-ω-pentanorprostanoic acid (BW 245C)		$C_{19}H_{32}N_2O_5$ (368)	From (S)-2-aminononanedoic acid 9-ethyl ester[30,34] The 8S, 15R configuration was supported by an X-ray analysis[34]	Inhibitor of ADP-induced platelet aggregation in vitro and in vivo the potency of which is species dependent In vitro for PRP of human, horse, sheep, rabbit, and rat, it has 0.2, 2.6, 0.03, and 0.003 times the potency of PGI$_2$[35] The inhibition is probably mediated by PGD$_2$ receptors since it parallels that observed for PGD$_2$ in the above species[35] and because the compound has a high affinity for human platelet PGD$_2$ receptors, but not for other PG receptors;[36] in vivo, in the anesthetized rabbit, dog, or monkey (i.v. infusion), ex vivo ADP-induced platelet aggregation was inhibited with potencies 0.08, 0.04, and 0.05 times PGI$_2$

Table 6 (continued)
HETEROCYCLIC PG ANALOGS

No.	Systematic name	Structure	Mol. formula (mol. wt.)	Synthesis	Biological actions
					The activity was of short duration (<30 min) by i.v. infusion (rabbit), but prolonged (>5 hr) by the oral route; there was a substantial drop in the BP and increase in HR in the dog and the monkey but not in the rabbit (i.v. infusion);[35] as a vasodepressor in the anesthetized dog, monkey, and rabbit (i.v. infusion), its potency was 0.5, 0.1, and 0.02 × PGI_2.[35] In conscious SH rats (i.v. bolus) potency was ~0.1 × PGE_2, but duration of hypotensive response was much greater (>50 min); HR was increased at hypotensive doses (max. 31% at 250 μg/kg)[36]
					Compound not inactivated by passage through the lungs (rat);[35] i.v. infusion in normal humans caused inhibition (ex vivo) of ADP-induced platelet aggregation, vasodilation, and HR increase comparable to PGI_2, but with a longer duration of action; except for nasal congestion, side effects were like those observed with PGI_2 (e.g., nausea, restlessness, headache)[37]
					In double blind study vs placebo, repeated oral dosing (150 μg q.i.d. for 6 days) in normal volunteers produced significantly higher incidence of headaches, facial flushing, nasal stuffiness and abdominal discomfort than placebo, but platelet aggregation was not inhibited[57]

No.	Name	Structure	Formula (MW)	Source	Activity
25	9,11-Diketo-10,12-diaza-10H-homo-15α-hydroxy-15-cyclohexyl-ω-pentanorprostanoic acid		$C_{20}H_{34}N_2O_5$ (382)	From diethyl-2-aminononanedioate[32]	Inhibited aggregation of HPRP induced by ADP; 0.1 × PGE_1.[32]
26	8,10,12-Triaza-9,11-diketo-10H-15-hydroxy-15-cyclohexyl-ω-pentanorprostanoic acid		$C_{18}H_{31}N_3O_5$ (369)	From product of reaction of ethyl 7-bromoheptanoate with ethyl carbazate[32]	Inhibited aggregation of HPRP induced by ADP; 0.13 × PGE_2.[32]
26a	8,10,12-Triaza-9,11-diketo-5-thia-10,15,16-trimethylprostanoic acid		$C_{19}H_{35}N_3O_5$ (417)	From 4-methylurazole[56]	Inhibited serotonin-induced bronchial constriction in the anesthetized guinea pig by i.v. (ED_{50} = 0.55 µg/kg; 0.8 × PGE_2); prolonged (>4 hr) inhibition of histamine-induced asphyxic collapse in conscious guinea pigs p.o. at 1.0 mg/kg[56]
27	9-Keto-10-oxa-15α-hydroxyprosta-13-*trans*-enoic acid		$C_{19}H_{32}O_5$ (340)	From 1,9-nonanedioic acid;[37] also from 3-(6-carboxyhexyl)-4-hydroxymethylbutyrolactone[39]	Contracted isolated diestrous rat uterus at 0.1 µg/mℓ[38]
28	9-Keto-10-oxa-15α-hydroxyprosta-5-*cis*-13-*trans*-dienoic acid		$C_{19}H_{30}O_5$ (338)	From 3-(6-methoxycarbonyl-hex-*cis*-2-enyl)-4-hydroxy-methylbutyrolactone[40]	
29	10-Oxa-15-hydroxyprosta-8,11,13-*trans*-trienoic acid		$C_{19}H_{30}O_4$ (322)	From Diels-Alder addition product of 3-phenyloxazole with methyl 9-formylnon-8-yne[41]	
30	Methyl 9-keto-10-thia-11-nor-15α-hydroxyprosta-5-*cis*-13-*trans*-dienoate		$C_{19}H_{30}O_4S$ (354)	From methyl 4,4-dimethoxy-3-oxobutyrate[42]	Transient hypotensive activity in anesthetized rabbits (i.v.)[42]
31	9β-Methyl-10-thia-11-nor-15α-hydroxyprosta-13-*trans*-enoic acid-10,10-dioxide		$C_{19}H_{34}O_3S$ (342)	From *trans*-2-hydroxymethyl-3-(6,methoxycarbonylhexyl)-4-methylthietane sulfoxides[43]	Strong constriction of smooth muscles (like TBX_2), but no effect on platelet aggregation[43]

Table 6 (continued)
HETEROCYCLIC PG ANALOGS

No.	Systematic name	Structure	Mol. formula (mol. wt.)	Synthesis	Biological actions
32	8-Aza-9-keto-10-oxa-15α-hydroxyprosta-13-*trans*-enoic acid		$C_{18}H_{31}NO_5$ (341)	From (±) serine[44]	Mild hypotensive and bronchial dilator activity in dogs at 50 mg/kg. Accelerated platelet aggregation induced by ADP or collagen at 10^{-6} M[44]
33	9-Aza-11-oxa-15-hydroxyprosta-$\Delta^{8,12\omega}$,13-*trans*-trienoic acid		$C_{18}H_{29}NO_4$ (323)	From methyl 4-(6-methoxycarbonylhexyl)oxazole-5-carboxylate[45]	
34	9-Oxa-11-aza-15-hydroxyprosta-$\Delta^{8,12}$10,13-*trans*-trienoic acid		$C_{18}H_{29}NO_4$ (323)	From ethyl 5-(6-oxycarbonylhexyl)-oxazole-4-carboxylate[45]	
35	8-Aza-9-thia-15α-hydroxyprosta-5-*cis*-13-*trans*-dienoic acid 9,9-dioxide		$C_{18}H_{31}NO_5$ (373)	From 3-hydroxymethyl isothiazolidine sulfone[46]	Increased cAMP concentration 14-fold over control in mouse ovary assay at 1.0 µg/mℓ (8-fold for PGE_2 at 0.01 µg/mℓ)[46]
36	(+) 8-Aza-9-keto-10-thia-12S,15S-hydroxyprosta-13-*trans*-enoic acid		$C_{18}H_{31}NO_4S$ (357)	From S (+) methyl thiazolidin-2-one-4-carboxylate[47]	Bronchial dilator in anesthetized dogs at 50 µg/kg[47]
37	8-Aza-9-keto-11-thia-15α- and 15β-hydroxyprostanoic acid		$C_{18}H_{33}NO_4$ (359)	Thermal addition of mercaptoacetic acid to imine derived from 4-acetoxydecanal and methyl 7-aminoheptanoate, followed by hydrolysis[12]	Increased cAMP 8-fold, in mouse ovary assay, at 25 µg/mℓ: (PGE_2 gave 60-fold increase at this conc.)[12]
38	9-Aza-11-thia-15-hydroxyprosta-$\Delta^{8,12}$; 9,13-*trans*-trienoic acid		$C_{18}H_{29}NO_3S$ (339)	From methyl 4-(6-methoxycarbonylhexyl)-thiazole-5-carboxylate[48]	

| 39 | 9-Thia-11-aza-15-hydroxyprosta-$\Delta^{8,12};10,13$-*trans*-trienoic acid | | $C_{18}H_{29}NO_3S$ (339) | From methyl 5-(6-methoxycarbonylhexyl)-thiazole-4-carboxylate[49] | At 1 mg/kg (p.o.) of the ($\pm$) compound in conscious dogs, renal blood flow was increased from 1st through 5th hr to a maximum of 70% at 5th hr; a modest increase in lower aortic blood flow occurred during 2nd and 3rd hr but returned to control values at hr 4 and 5; blood pressure was decreased modestly with an associated tachycardia; single injections of 25 and 50 ng/kg into renal artery of anesthetized dogs reduced renal vascular resistance to the same extent as infusion of PGE_2 at 60 ng/kg/min[50,51] |
| 40 | 2,3,4-Trinor-1,5-inter-*p*-phenylene-9-thia-11-keto-12-aza-15-hydroxy-15,15-pentamethylene-ω-pentanorprostanoic acid | | $C_{21}H_{29}NO_4S$ (391) | Reaction of methyl thioglycolate with Schiff's base derived from 1-(2-aminoethyl)-cyclohexanol and 4-(4-ethoxycarbonylphenyl)-butyraldehyde followed by hydrolysis[50]
Resolved via ($-$)-camphanic acid salt[50] | In open-chested anesthetized dogs, i.v. bolus doses of 5, 20, and 100 µg/kg decreased total peripheral resistance from 74 to 59, 37, and 24 mm/L/min, respectively; cardiac output was increased by the 100 µg dose, but HR, left ventricular contractility and left ventricular end diastolic pressure were unchanged; this is a profile similar to PGI_2, i.e., potent vasodilitation with no direct effect on cardiac function;[58] in healthy humans, i.v. infusion (0.5—4 µg/kg/min for 10 min) lowered diastolic BP and increased HR but did not increase effective renal plasma flow; the higher doses caused nausea and flushing of the skin
In mildly hypertensive patients, oral administration (7—280 µg) had variable effects on BP and HR, but symptoms related to vasodilation were observed (flushing, headache, nasal congestion, and postural symptoms)[59] |

Table 6 (continued)
HETEROCYCLIC PG ANALOGS

No.	Systematic name	Structure	Mol. formula (mol. wt.)	Synthesis	Biological actions
41	8,12-Diaza-9,11-di-keto-10-thia-15-hy-droxy-15-cyclohexyl-ω-pentanorprostanoic acid		$C_{18}H_{30}N_2O_5S$ (386)	From 1,2,4-thiadiazoli-dine-3,5-dione[32]	Inhibited aggregation of HPRP induced by ADP; $0.01 \times PGE_1$[32]
42	10-Oxa-12-aza-11-keto-15α-hydroxy-15-cy-clohexyl-ω-pentanor-prostanoic acid		$C_{19}H_{33}NO_5$ (344)	Tetramethylguanidine induced conjugate ad-dition of 4-(6-meth-oxycarbonylhexyl)-ox-azolidin-2-one to cy-clohexylvinyl ketone, followed by sodium borohydride reduction tlc separation and saponificaiton[60]	Inhibited ADP-induced aggregation of HPRP ($0.93 \times PGE_1$); potent, but short acting (ca. 0.5 hr) inhibitor of ex vivo ADP induced platelet aggregation on i.v. administration to guinea pigs[60]
43	10,12-Diaza-11-keto-15α-hydroxy-15 cy-clohexyl-ω-pentanor-prostanoic acid		$C_{19}H_{34}N_2O_4$ (354)	From methyl 8-azido-9-hydroxynonanoate[61]	Potent inhibitor of ADP induced aggrega-tion of HPRP ($2.2 \times PGE_1$) in vitro but inactive in guinea pigs at doses up to 2 mg/kg p.o.[61]

REFERENCES

1. **Bolliger, G. and Muchowski, J. M.**, Synthesis of 11-desoxy-8-azaprostaglandin E₁, *Tetrahedron Lett.*, 2931—2934, 1975.
2. **Bruin, J. W. de Koning, H., and Huisman, H. O.**, Synthesis of 8-azaprostaglandin E₁ and E₂, *Tetrahedron Lett.*, 4599—4602, 1975.
3. **Zoretic, P. A., Branchaud, P., and Sinha, N. D.**, Synthesis of 11-deoxy-8-azaprostaglandin E₁, *J. Org. Chem.*, 42, 3201—3203, 1977.
4. **Saijo, S., Wada, M., Noguchi, K., Muraki, M., Ishida, A., and Himizu, J.**, Heterocyclic prostaglandins. IV. Synthesis of 8-aza-11-deoxyprostaglandin E₁ and its related compounds, *Yakugaki Zasshi*, 100, 389—395, 1980.
5. **Saijo, S., Wada, M., Himizu, J., and Ishida, A.**, Heterocyclic prostaglandins. V. Synthesis of (12R, 15S)-(−)-11-deoxy-8-azaprostaglandin E₁ and related compounds, *Chem. Pharm. Bull.*, 28, 1449—1458, 1980.
6. **Rooks, W. H. and Tomolonis, A. J.**, Unpublished data.
7. **Guzman, A. and Muchowski, J. M.**, Unpublished data.
8. **Zoretic, P. A., Sinha, N. D., and Branchaud, B.**, Synthesis of 11-desoxy-8-azaprostaglandin E₂, *Synth. Commun.*, 7, 299—303, 1977.
9. **Zoretic, P. A., Jardin, J., and Angus, R.**, Synthesis of 11-deoxy-5,6-didehydro-13,14-dihydro-8-azaprostaglandin E₁, *J. Heterocyclic Chem.*, 17, 1623—1624, 1980.
10. **Bolliger, G. and Muchowski, J. M.**, Unpublished data.
11. **Zoretic, P. A. and Chiang, J.**, Synthesis of 11-deoxy-13,14-dihydro-8-azaprostaglandin E₁, *J. Org. Chem.*, 42, 2103—2105, 1977.
12. **Smith, R. L., Lee, T., Gould, N. P., Dragoe, E. J., Orien, H. G., and Kuehl, F.A.**, Prostaglandin isosteres. I. (8-Aza-, 8,10-diaza-, and 8-aza-11-thia)-9-oxoprostanoic acids and their derivatives, *J. Med. Chem.*, 20, 1292—1299, 1977. These authors did not separate the 15-eipmers and the biological activity reported is for the 1:1 mixture of 15-epimers.
13. **Zoretic, P. A. and Soja, P.**, Synthesis of 9-oxo-15-hydroxy-8-azaprost-13-ene, *J. Heterocyclic Chem.*, 14, 1267—1269, 1977.
14. **Wang, C. J.**, Azaprostanoids. I. Synthesis of (rac)-8-aza-11-deoxy-15-deoxy-16-hydroxy-16-methylprostaglandins, *Tetrahedron Lett.*, 23, 1067—1070, 1982.
15. **Zoretic, P. A., Xinha, N. D., Shiah, F., Maestrone, T., and Branchaud, B.**, Synthesis of 8-aza-15-hydroxy-7-oxo-12S-13E-prostenoic acids, *J. Heterocyclic Chem.*, 15, 1025—1026, 1978.
16. **Pailer, M. and Schläger, I.**, Synthese von 2-[3-oxo-(1E)-octenyl]-1-pyrrolheptansäure, *Mh. Chem.*, 109, 313—317, 1978.
17. **Barco, A., Bennetti, S., Pollini, G. P., Baraldi, P. G., Guarneri, M., Simoni, D., Vincentini, C. B., Borasio, P. G., and Capuzzo, A.**, Synthesis and prostaglandin-like activity of 2-(trans-3-hydroxy-1-octenyl)-3-indoleheptanoic acid, *J. Med. Chem.*, 21, 988—990, 1978.
18. **Rozing, G. P., de Koning, H., and Huisman, H. O.**, Synthesis of 9-azaprostaglandin analogs, *Heterocycles*, 5, 325—330, 1976.
19. **King, R. W.**, Synthesis of 10-azaprostaglandins, *Tetrahedron Lett.*, 22, 2837—2840, 1981.
20. **Kuhlein, K., Linkies, A., and Reuschling, D.**, Synthese von 10-aza-dihydro-A-Prostaglandin, *Tetrahedron Lett.*, 4463—4466, 1976.
21. **Zoretic, P. A. and Barcelos, F.**, Synthesis of 1-azaprostaglandin E₁, 529—532, 1979; **Zoretic, P. A., Barcelos, F., Jardin, J., and Bhakta, C.**, Synthetic approaches to 10-azaprostaglandins, *J. Org. Chem.*, 45, 810—814, 1980.
22. **Reuschling, D., Mitzlaff, M., and Kuhlein, K.**, Synthese von 10-aza-Prostaglandinen, *Tetrahedron Lett.*, 4467—4468, 1976.
23. **Barco, A., Benetti, S., Pollini, G. P., Baraldi, P. G., Guarneri, M., Gandolfi, C., Ceserani, R., and Longiare, D.**, Azaprostaglandin analogues. Synthesis and biological properties of 11-azaprostaglandin derivatives, *J. Med. Chem.*, 24, 625—628, 1981.
24. **Scribner, R. M.**, Azaprostanoids. I. Synthesis of (rac)-11-deoxy-12-azaprostanoids, *Tetrahedron Lett.*, 3853—3856, 1976.
25. **Scribner, R. M.**, (±-11-Oxo-10,11,13,14-tetrahydro-12-azaprostaglandin A₁ methyl ester, *Prostaglandins*, 13, 677—679, 1977.
26. **Higgs, G. A., Armstrong, J. M., and Reed, P. M.**, The synthesis and biological activities of some 12-azaprostaglandin analogues, *Prostaglandins*, 16, 773—787, 1978.
27. **Lapierre Armand, J. C. and Pandit, U. K.**, A simple approach to 12-azaprostaglandin congeners, *Rec. Trav. Chim.*, 99, 87—91, 1980.
28. **Depres, J.-P., Greene, A. E., and Crabbé, P.**, β-Lactam prostaglandins, *Tetrahedron Lett.*, 2191—2194, 1978.

29. **Saijo, S., Wada, M., Himizu, J., and Ishida, A.,** Heterocyclic prostaglandins. VI. Synthesis of 11-deoxy-8,10-diazaprostaglandin E$_1$ and its 10-methyl derivative, *Chem. Pharm. Bull.*, 28, 1459—1467, 1980.

30. **Caldwell, A. C., Harris, C. J., Stepney, R., and Whittaker, N.,** Heterocyclic prostaglandin analogues. II. Hydantoins and other imidazole analogues, *J. Chem. Soc. Perkin Trans. I*, 495—505, 1980.

31. **Pailer, M. and Gutwillinger, H.,** Synthese von rac. 2-[3-Hydroxy-1(E)-octenyl]-1-imidazolheptansäure, *Mh. Chem.*, 108, 1059—1066, 1977.

32. **Barraclough, P., Caldwell, A. G., Harris, C. J., and Whittaker, N.,** Heterocyclic prostaglandin analogues. IV. Piperazine-2,5-diones, pyrazolidone-3,5-diones, 1,2,4-triazolidinediones, 1,3,4-oxazolidine-diones, and 1,3,4-thiazolidinediones, *J. Chem. Soc. Perkin Trans. I*, 2096—2105, 1981.

33. **Pailler, M. and Gutwillinger, H.,** Synthese der rac. 1-[3-Hydroxy-1-(E)-octenyl]-2-imidazol-hetpansäure, *Mh. Chem.*, 108, 653—664, 1977.

34. **Brockwell, M. A., Caldwell, A. G., Whittaker, N., and Begley, M. J.,** Heterocyclic prostaglandin analogues. III. The relationship of configuration to biological activity for some hydantoin prostaglandin analogues, *J. Chem. Soc. Perkin Trans. I*, 706—711, 1981.

35. **Whittle, B. J. R., Moncada, S., Mullane, K., and Vane, J. R.,** Platelet and cardiovascular activity of the hydantoin BW245C, a potent prostaglandin analogue, *Prostaglandins*, 25, 205—223, 1983.

36. **Town, M.-H., Casals-Stenzel, J., and Schilliinger, E.,** Pharmacological and cardiovascular properties of a hydantoin derivative, BW 245C, with high affinity and selectivity for PGD$_2$ receptors, *Prostaglandins*, 25, 13—27, 1983.

37. **Orchard, M. A., Ritter, J. M., Shepherd, G. L., and Lewis, P. J.,** Cardiovascular and platelet effects in man of BW 245C, a stable mimic of epoprostenol (PGE$_2$), *Br. J. Clin. Pharmacol.*, *15, 509—511, 1983.*

38. **Ishida, A., Saijo, S., Noguchi, K., Wada, M., Takaita, O., and Himizu, J.,** Synthesis of 1-oxa-11-deoxyprostaglandin E$_2$, *Chem. Pharm. Bull.*, 27, 625—632, 1979.

39. **Carroll, F. I., Hauser, F. M., Huffman, R. C., and Coleman, M.C.,** Synthesis and biological evaluation of 10-oxa-11-deoxyprostaglandin E$_1$ and 10-Nor-9,11-secoprostaglandin F$_2$ and their derivatives, *J. Med. Chem.*, 21, 321—325, 1979.

40. **Ishida, A., Saijo, S., and Himizu, J.,** Heterocyclic prostaglandins. III. Synthesis of 10-oxa-11-deoxy-prostaglandin E$_2$, *Chem. Pharm. Bull.*, 28, 783—788, 1980.

41. **Ansell, M. F., Caton, M. P. L., and North, P. C.,** The synthesis of 3,4-disubstituted furan prostanoids, *Tetrahedron Lett.*, 22, 1727—1728, 1981.

42. **Klinch, M., Taliani, L., and Buendia, J.,** Synthese d'un isostere soufre de la prostaldandine A$_2$, *Tetrahedron Lett.*, 22, 4387—4390, 1979.

43. **Jones, D. N., Kogan, T. P., and Newton, R. F.,** Synthesis of some thietanoprostanoids, *J. Chem. Soc. Perkin Trans. I*, 1333—1342, 1982.

44. **Kubodera, N., Nagano, H., Takagi, M., and Matsunaga, I.,** Synthesis of (±)-8-aza-11-deoxy-10-oxaprostaglandin E$_1$, *Heterocycles*, 19, 1285—1290, 1982.

45. **Ambrus, G., Barta, I., Horvath, G., Soti, N., and Sohar, P.,** Heterocyclic analogues of prostaglandins, Oxazole derivatives, *Acta Chim. Acad. Sci. Hung.*, 99, 421—432, 1979.

46. **Jones, J. H., Bicking, J. B., and Cragoe, E. J.,** Synthesis of the sultam analog of 11-deoxy PGE$_2$, *Prostaglandins*, 17, 223—226, 1979.

47. **Kubodera, N., Nagano, H., Takagi, M., and Matsunaga, I.,** Synthesis of both enantiomers of 8-aza-11-deoxy-10-thiaprostaglandin E$_1$, *Heterocycles*, 18, 259—263, 1982.

48. **Ambrus, G., Barta, I., Horvath, G., Mehesfalvi, Z., and Sohar, P.,** Heterocyclic analogues of prostaglandins. Thiazoles. I. *Heterocycles*, 97, 413—428, 1978.

49. **Barta, I., Ambrus, G., Horvath, G., Soti, M., and Sohar, P.,** Heterocyclic analogues of prostaglandins. Thiazoles. II., *Heterocycles*, 98, 463—477, 1978.

50. **Bicking, J. B., Bock, M. G., Cragoe, E. J., Dipardo, R. M., Gould, N. P., Holz, W. J., Lee, T. H., Robb, C. M., Smith, R. C., Springer, J. P., and Blaine, E. H.,** Prostaglandin isosteres. II. Chain modified thiazolidinone prostaglandin analogues as renal vasodilators, *J. Med. Chem.*, 26, 342—348, 1983.

51. **Blaine, E. H., Russo, H. F., Schorn, T. W., and Snyder, C.,** An orally active prostaglandin analog with renal vasodilatory activity in the dog, *J. Pharmacol. Exp. Ther.*, 222, 152—158, 1982.

52. **Baraldi, P. G., Simoni, D., Casolari, A., and Manfredini, S.,** Azaprostaglandins. Synthesis and antiulcer activity of 11-deoxy-8-azaprostaglandin analogues, *Il Farmaco Ed. Sci.*, 38, 498—507, 1983.

53. For a review of heteroprostanoids see **Orth, D. and Radunz, H. E.,** Synthesis and activity of hetero-prostanoids, *Top. Curr. Chem.*, 72, 520097, 1977.

54. **Zoretic, P. A., Bhakta, C., and Jardin, J.,** Synthesis of (E)-7-[[2-[4-(m-Trifluoromethylphenoxy)-3α and 3β-Hydroxy-1-butenyl]-5-oxo-1-pyrrolidinyl]]heptanoic acids, *J. Heterocyclic Chem.*, 20, 465—466, 1983.

55. **Corey, E. J. and Pyne, S. G.,** Synthesis of a new series of potent inhibitors of thromboxane A$_2$ biosynthesis, *Tetrahedron Lett.*, 24, 3291—3294, 1983.

56. **Bermudez, J., Cassidy, F., and Thompson, M.,** The synthesis and bronchodilator activity of 5,6-E-, 5,6-Z-, and 5-Thia-8,10,12-triazaprostaglandin analogues, *Eur. J. Med. Chem.,* 18, 545—550, 1983.

57. **Al-Sinawi, L. A-H., Mekki, Q. A., Hassan, S., Hedges, A., Burke, C., Moody, S. G., and O'Grady, J.,** Effect of a hydantoin prostaglandin analog, BW245C, during oral dosing in man, *Prostaglandins,* 29, 99—111, 1985.

58. **Siegl, P. K. S. and Wenger, H. C.,** Cardiovascular responses to an isosterically modified prostaglandin analog in the anesthetized dog, *Eur. J. Pharmacol.,* 113, 305—311, 1985.

59. **Ritter, J. M., Ludgin, J. R., Scharschmidt, L. A., Smith, R. D., and Dunn, M. J.,** Effects of a stable prostaglandin analogue L-646,122 in healthy and hypertensive men, *Eur. J. Clin. Pharmacol.,* 23, 685—688, 1985.

60. **Carpio, H., Guzman, A., De La Torre, J. A., Mendoza, S., Bruno, J. J., Chang, L-F., Yang, D., and Muchowski, J. M.,** ($\pm$)-9-Desoxi-10-oxa-12-aza-13,14-dihidroprostaglandina D_1 y sus derivados. Una nueva clase de inhibidores potentes de la aggregacion de plaquetas, *Rev. Soc. Quim. Mex.,* 29, 35—36, 1985.

61. **Cervantes, A., Guzman, A., Bruno, J. J., Chang, L-F., and Muchowski, J. M.,** ($\pm$)-9-Deoxy-10,12-diaza-13,14-dihydro-16-cyclohexyl-ω-pentanor prostaglandin D_1 and the 2,3-dehydro derivative thereof. New, powerful inhibitors of ADP induced platelet aggregation, *Rev. Soc. Quim. Mex.,* in press.

Table 7
MISCELLANEOUS PGs

No.	Systematic name	Structure	Mol. formula (mol. wt.)	Synthesis	Biological actions
1	9β,15α-Dihydroxy-11-ketoprosta-5-cis-13-trans-dienoic acid		$C_{20}H_{32}O_5$ (352)	Oxidation of 9β-trimethylsilyloxy-11β-hydroxy-15-THP PGF$_2$ derivative[1]	Inhibition of ADP-induced aggregation of HPRP; IC$_{50}$ = 0.1 ng/mℓ (32 × PGD$_2$); rat BP, 1 × PGD$_2$ (hypotensive, 0.1—0.3 × PGE$_1$)[1]
2	11-Keto-15α-hydroxy-prosta-5-cis-13-trans-dienoic acid		$C_{20}H_{32}O_4$ (336)	LAH reduction of 9-mesyloxy PGF$_{2α}$ 15-THP methyl ester followed by oxidation at C-1 and C-11 and deprotection[1]	Inhibition of ADP-induced aggregation of HPRP; IC$_{50}$ = 1 ng/mℓ; rat BP; 1 × PGD$_2$ (hypotensive)[1]
3	11-Keto-15α-hydroxy-prosta-5-cis-9,13-trans-trienoic acid		$C_{20}H_{30}O_4$ (334)	Oxidation of 15-THP-1,9-olide, followed by removal of THP group and silica gel chromatography[1]	Inhibition of ADP-induced aggregation of HPRP; IC$_{50}$ 3.2 ng/mℓ (1 × PGD$_2$) Rat BP; 1 × PGD$_2$ (hypotensive)[1] Potent inhibitor of L 1210 leukemia cells[2]
4	9α,11α-Dihydroxy-11-keto-15β-methyl-prosta-5-cis-13-trans-dienoic acid		$C_{21}H_{34}O_5$ (366)	Jones[19] oxidation of 15-methyl PGF$_{2α}$[1]	Inhibition of ADP-induced aggregation of HPRP; IC$_{50}$ = 320 ng/mℓ Rat BP; 0.1—0.3 × PGF$_{2α}$ (pressor) Hamster antifertility: 100% at 10 μg[1]
5	Methyl 11-keto-16α, and 16β-hydroxy-prosta-5-cis-14-trans-dienoate		$C_{21}H_{34}O_4$ (350)	Elaboration of product derived from Michael addition of ethyl cyanoacetate to 2-allylcyclopent-2-enone[3]	
6	9α,15α-Dihydroxy-11-keto-17-phenyl-ω-trinorprosta-5-cis-13-trans-dienoic acid		$C_{23}H_{30}O_5$ (386)	Jones oxidation and deprotection of 9,15-bis-THP derivative[1]	Rat BP; pressor (0.1—0.3 × PGF$_{2α}$) Hamster antifertility; 100% at 10 μg[1]

No.	Name	Structure	Formula (MW)	Method	Activity
7	9α,15α-Dihydroxy-11-ketoprosta-5-cis-13-trans-17-cis-trienoic acid		$C_{20}H_{30}O_5$ (350)	Selective Jones oxidation of $PGF_{2\alpha}$-15-THP;[1] also by a total synthesis of the Corey[20] type[4]	Inhibition of ADP-induced aggregation of HPRP; $IC_{50} = 3.2$ ng/mℓ Rat BP; $1 \times PGD_2$ (hypotensive)[1]
8	Methyl 9-keto-15α-hydroxyprosta-10,13-trans-dien-4-ynoate		$C_{21}H_{30}O_4$ (346)	Trifluoracetic acid hydrolysis of 11,15-bis-THP PGE derivative[5]	
9	Methyl 9-keto-15α-hydroxyprosta-10,13-trans-dien-5-ynoate		$C_{21}H_{30}O_4$ (346)	Acidic dehydration of PGE derivative[6]	Less active than PGA_2 in gerbil colon and hamster antifertility assays[6]
10	5,6-Bis-nor-4,7-inter-o-phenylene-9-keto-15α-hydroxyprosta-10,13-trans-dienoic acid		$C_{24}O_{32}O_4$ (384)	Acidic dehydration of PGE derivative[7]	Inactive on isolated smooth muscle assays and no effect on BP of SH rats up to 10 mg/kg[7]
11	Methyl 9-keto-11-methyl-15α-hydroxyprosta-5-cis-10,13-trans-trienoate		$C_{22}H_{34}O_4$ (362)	Acidic dehydration of PGE derivative[8]	
12	9-Keto-15α-hydroxy-16,16-dimethyl-18-oxaprosta-5-cis-10,13-trans-trienoic acid (HR-466)		$C_{21}H_{32}O_5$ (364)	Hydrochloric acid-induced dehydration of PGE derivative[9]	Decreased mean arterial BP in anesthetized dogs by >20% at 10 μg/kg (i.v.), and systemic BP by 15—25% (>5 hr duration) in conscious renal hypertensive dogs (p.o.)
13	9-Keto-15α-hydroxy-18-cyano-ω-bis-norprosta-5-cis-10,13-trans-trienoic acid		$C_{19}H_{25}NO_4$ (331)	Hydrochloric acid-induced dehydration of PGE derivative[11]	Contracted isolated rat fundus, $0.02 \times PGF_2$;[9,10]

Table 7 (continued)
MISCELLANEOUS PGs

No.	Systematic name	Structure	Mol. formula (mol. wt.)	Synthesis	Biological actions
14	9-Keto-11α-homo-15α-hydroxyprosta-10,13-*trans*-dienoic acid		$C_{21}H_{34}O_4$ (350)	Cuprate addition of protected ω-chain to cyclohexenone functionalized with α-chain	Less effective than PGE_1 as a hypotensive in rats (i.v.)[12]
15	Methyl 10-keto-15α-hydroproxyprosta-5-*cis*-13-*trans*-dienoate		$C_{21}H_{34}O_4$ (350)	Unsaturation at C-10 introduced by phenylsulfenation followed by oxidative elimination[12] From 5-norbornen-1-ol[13]	
16	15α-Hydroxyprosta-13-*trans*-enoic acid		$C_{20}H_{36}O_3$ (324)		Potent noncompetitive inhibitor of swine lung 15-hydroxy PG dehydrogenase; K_i = 7.5 μM[14]
17	13α,14α- and 13β,14β-Methano-15-ketoprostanoic acid		$C_{21}H_{38}O_3$ (338)		Potent noncompetitive inhibitor of swine lung 15-hydroxy PG dehydrogenase; K_i = 0.14 μM Completely inhibited metabolism of PGE_1 by guinea pig lung homogenates (400 mM) Co-infusion (250 μg/kg/min) with PGE_1 (0.5 μg/kg/min) in rats with elevated serum free fatty acid levels (stimulated by epinephrine) enhanced anti-lipolytic effect of PGE_1, returning free fatty acids to basal levels[14]

No.	Name	Structure	Formula (MW)	Synthesis	Properties
18	13-Azaprostanoic acid		$C_{19}H_{37}NO_2$ (311)	Reductive amination of 2-(6-methoxycarbonylhexyl) cyclopentanone, chromatographic separation of isomers and hydrolysis[15]	Direct antagonist of the human platelet thromboxane/endoperoxide receptor[16,17] Deaggregatory properties are potentiated by PGI_2[18]
19	2-(6-Carboxy-hexyl)cyclopentanone hexylhydrazone		$C_{18}H_{34}NO_2O_2$ (310)	Reaction of 2-(6-carboxyhexyl)cyclopentanone with *n*-hexylhydrazine[21]	Potent cyclooxygenase inhibitor Complete time-dependent inhibition of AA-induced aggregation of HPRP at 0.1 μM No effect on aggregation induced by U46619 or ADP up to 100 μM Aggregation induced by PGH_2 inhibited 89% at 100 μM[21]
20	9-(4-Phenylbenzyloxy)-11-keto-12-(morpholin-1-yl)-ω-octanor-prosta-4-cis-enoic acid (AH23848)		$C_{29}H_{35}NO_5$ (477)		Potent, specific thromboxane receptor blocking agent with oral activity and prolonged action; antagonized the effects of U-46619 on human platelets, human pulmonary artery, rat aorta, and dog saphenous vein with pA_2 values of 7.75—8.25; in HPRP, inhibited aggregation induced by collagen (0.3 μM) as well as PGG_2/PGH_2, TXA_2, and U-46619; in the anesthetized guinea pig (i.v.), inhibited the bronchoconstrictor action of i.v. collagen; in the anesthetized dog (i.v.), inhibited the vasoconstrictor responses of U-46619; in the conscious dog, oral administration (1 mg/kg) produced a marked inhibition of collagen induced platelet aggregation in whole blood ex vivo which was still evident at 11 hr; in humans, oral administration inhibited platelet aggregation induced by U-46619, ex vivo, for at least 8 hr after doses of ≥ 0.5 mg/kg[22]

REFERENCES

1. **Bundy, G. L., Morton, D. R., Peterson, D. C., Nishizawa, E. E., and Miller, W. L.,** Synthesis and platelet aggregation inhibiting activity of prostaglandin D analogues, *J. Med. Chem.,* 26, 790—799, 1983.

2. **Fukushima, M., Kato, T., Ota, K., Arai, Y., Noriyama, S., and Hayashi, O.,** 9-Deoxy-Δ^9-prostaglandin D$_2$, a prostaglandin D$_2$ derivative with a potent antineoplastic and weak smooth muscle-contracting activities, *Biochem. Biophys. Res. Commun.,* 109, 626—633, 1982.

3. **Traverso, G. and Pirillo, D.,** Synthesis of (±)-8,12-*trans*-Δ^5en-*cis*-11-keto-14,15-*trans*-en-16-R,S-hydroxyprostanoic acid methyl ester, *Il Farmaco Ed. Sci.,* 31, 438—441, 1976.

4. **Konishi, Y., Wakatsuka, H., and Hayashi, M.,** Synthesis of prostaglandin D$_3$, *Chem. Lett.,* 377—378, 1980.

5. **Lin, C. H. and Stein, S. J.,** The synthesis of 4,5-acetylenic prostaglandins, *Synth. Commun.,* 6, 503—508, 1976.

6. **Lin, C. H., Stein, S. J., and Pike, J. E.,** The synthesis of 5,6-acetylenic prostaglandins, *Prostaglandins,* 11, 377—380, 1976.

7. **Buckler, R. T., Ward, F. E., Hartzler, H. E., and Kurchacova, E.,** The synthesis and biological activity of benzo[5,6]prostaglandins A$_2$, and F$_{2\alpha}$, *Eur. J. Med. Chem.,* 12, 463—465, 1977.

8. **Lin, C. H.,** Synthesis of 11-methyl prostaglandins, *Chem. Ind. (London),* 994—995, 1976.

9. **Beck, G., Bartmann, W., Lerch, U., Teufel, H., and Schölkens, B.,** Antihypertensive activity of 16,16-dimethyl-oxa-alkyl-prostaglandins, of the PGA$_2$, PGE$_2$ and *trans*-$\Delta^{2,11}$-deoxy-PGE$_2$ series: structure-activity relationships, *Prostaglandins,* 20, 153—169, 1980.

10. HR-466 and HR-601, *Drugs Future,* 6, 620—622, 1981.

11. **Disselnkötter, H., Lieb, F., Oediger, H., and Wendisch, D.,** Synthese von Prostaglandin-Analoga, *Liebigs Ann. Chem.,* 150—166, 1982.

12. **Floyd, M. B. and Weiss, M. J.,** Prostaglandins and congeners. XXI. Synthesis of some cyclohexyl analogues (11a-homoprostaglandins), *J. Org. Chem.,* 44, 71—75, 1979.

13. **Arndt, H. C. and Ranjani, C.,** Preparation of methyl 10-oxo-15S-hydroxyprosta-5E, 13Z-dienaote, *Tetrahedron Lett.,* 2365—2368, 1982.

14. **Yamazaki, M., Ohuchi, K., Sasaki, M., and Sakai, K.,** Inhibition of 15-hydroxyprostaglandin dehydrogenase by 9,11-deoxyprostaglandins in vitro and in vivo, *Mol. Pharmacol.,* 19, 456-462, 1981.

15. **Venton, D. L., Enke, S. E., and LeBreton, G. C.,** Azaprostanoic acid derivatives. Inhibitors of arachidonic acid induced platelet aggregation, *J. Med. Chem.,* 22, 824—830, 1979.

16. **LeBreton, G. C., Venton, D. L., Enke, S. E., and Halushka, P. V.,** 13-Azaprostanoic acid: a specific antagonist of the human blood platelet thromboxane/endoperoxide receptor, *Proc. Natl. Acad. Sci. U.S.A.,* 76, 4097—4101, 1979.

17. **Cook, J. A., Wise, W. C., and Halushka, P. V.,** Elevated thromboxane levels in the rat during endotoxic shock, *J. Clin. Invest.,* 65, 227—230, 1980.

18. **Parise, L. V., Venton, D. L., and LeBreton, G. C.,** Prostacyclin potentiates 13-azaprostanoic acid-induced platelet deaggregation, *Thrombosis Res.,* 28, 721—730, 1982.

19. **Bowden, K., Heilbron, I. M., Jones, E. R. H., and Weedon, B. C. L.,** Researches on acetylenic compounds. I. The preparation of acetylenic ketones by oxidation of acetylenic carbinols and glycols, *J. Chem. Soc.,* 39—45, 1946.

20. A very good account of the Corey and other PG syntheses is given in the books cited in Reference 3 of the textual section of this review.

21. **Ghali, N. I., Venton, D. L., Hung, S. C., and Le Breton, G. C.,** 2-(6-Carboxyhexyl)cyclopentanone hexylhydrazone. A potent and time dependent inhibitor of platelet aggregation, *J. Med. Chem.,* 26, 1056—1060, 1983.

22. **Brittain, R. T., Boutal, L., Carter, M. C., Coleman, R. A., Collington, E. W., Geisow, H. P., Hallett, P., Hornby, E. J., McCabe, P. J., Skidmore, I. F., Thomas, M., and Wallis, C. J.,** AH23848: A thromboxane receptor-blocking drug that can clarify the pathophysiologic role of thromboxane A$_2$, *Circulation,* 72, 1208—1218, 1985.

Table 8
SECO PGs

No.	Systematic name	Structure	Mol. formula (mol. wt.)	Synthesis	Biological actions
1	(5Z,10E)-(8RS,12S)-8-Acetyl-12-hydroxy-5,10-heptadecadioenoic acid		$C_{19}H_{32}O_4$ (324)	From methyl vinyl ketone[1]	Induced platelet aggregation (56% at 100 µg/mℓ) with a potency approx. that of AA; aggregation was inhibited by PGI_2 but not by indomethacin[1]
2	8-Acetyl-12-hydroxyheptadecanoic acid		$C_{19}H_{36}O_4$ (328)	Sequential alkylation of *t*-butylacetoacetate with ethyl 7-bromoheptanoate and 1-chloro-4-acetoxynonane followed by alkaline hydrolysis[1]	Elevated mouse ovary cAMP level 11-fold at 10 µg/mℓ (PGE_1 caused a 54-fold increase at 1 µg/mℓ); increased renal blood flow by 22% in anesthetized dogs by i.v. infusion (0.5 mg/kg/min) Oral administration to guinea pigs caused ex vivo inhibition of collagen induced platelet aggregation ($ED_{50} = 6.5$ mg/kg)[2]
3	4-(4-Acetyl-8-hydroxytridecyl)benzoic acid		$C_{22}H_{34}O_4$ (362)	Sequential alkylatin of *t*-butylacetoacetate with ethyl 4-(3-bromopropyl) benzoate and 1-chloro-4-acetoxynonane followed by acid catalyzed decarboxylation and saponification[3]	Decreased renal vascular resistance by 47% in anesthetized dogs at 0.5 mg/kg (i.v., bolus); in conscious dogs, oral dosing (5 mg/kg) significantly increased renal blood flow (max. 75%) at 2 and 3 hr; the effects on HR and BP were small and nonsignificant[3]
4	4-(4-Acetyl-8-hydroxy-9-propoxynonyl)benzoic acid		$C_{21}H_{32}O_5$ (364)	Sequential alkylation of *t*-butylacetoacetate with ethyl 4-(3-bromopropyl)benzoate and 5-bromo-1-pentene followed by acid-induced decarboxylation, epoxidation, epoxide cleavage with sodium propoxide, and saponification[3]	Decreased renal vascular resistance by 43% in anesthetized dogs at 0.5 mg/kg (i.v.)[3]

Table 8 (continued)
SECO PGs

No.	Systematic name	Structure	Mol. formula (mol. wt.)	Synthesis	Biological actions
5	4-[4-Acetyl-7-(1-hydroxycyclohexyl)heptyl]benzoic acid		$C_{22}H_{32}O_4$ (360)	Sequential alkylation of ethyl acetoacetate with ethyl 4-(3-bromopropyl) benzoate and 1-acetoxy-1-(3-bromo-1-propynyl)cyclohexane, followed by saponification, decarboxylation, and catalytic reduction[3]	Decreased renal vascular resistance by 42% in anesthetized dogs at 0.5 mg/kg (i.v.)[3]
6	8-Methylsulfonyl-12-hydroxyheptadecanoic acid		$C_{18}H_{36}O_5S$ (364)	Alkylation of dimethyl 2-methylsulfonylazelate with 1-iodo-4-acetoxynonane followed by decarbomethoxylation and saponification[4]	Elevated mouse ovary cAMP level 14-fold at 1 μg/mℓ as compared to 54-fold for PGE_1 at this concentration[4]
7	(E)-6-Methoxy carbonylhexyl-7-hydroxy-5-dodecenoate		$C_{20}H_{36}O_5$ (356)	From sodium 5-hydroxypentanoate[5]	In pylorus ligated rats 2 doses (s.c.) of 1.5 mg/kg (at 0 and 2 hr) of compound was 0.25 times as effective as PGE_2 in decreasing volume of gastric acid secretion and 0.17 times as effective in decreasing total acid output[5]
8	(E)-5-Hydroxy-3-decenyl-8-methoxy-carbonyloctanoate		$C_{20}H_{36}O_5$ (356)	From 8-methoxycarbonyloctanoyl chloride[6]	In pylorus ligated rats, 3 mg/kg (s.c.) decreased total acid output by 34%
9	(E)Methyl 7-(N(-methyl-7-hydroxy-5-dodecamido)heptanoate		$C_{21}H_{39}NO_4$ (369)	From N-methyl-5-hydroxyvaleramide[6]	In pylorus ligated rats, 3 mg/kg (s.c.) decreased total acid output by 32%[6]

No.	Name	Formula (MW)	Synthesis	Activity
10	7-[1-(4-Hydroxynonyl)-ureido]heptanoic acid	$C_{17}H_{34}N_2O_4$ (330)	Alkylation of ethyl 7-cyanamido-heptanoate with 1-chloro-4-acetoxynonane followed by reaction with alkali[7]	Elevated mouse ovary cAMP 23-fold at 10 $\mu g/m\ell$ (PGE$_1$ caused a 54-fold increase at 1 $\mu g/m\ell$); increased renal blood flow by 46% in anesthetized dog upon i.v. infusion (0.25 mg/kg/min)[7]
11	7-[N-(4-(S)-Hydroxy-nonyl]methane-sulfonamido]heptanoic acid	$C_{17}H_{35}NO_5S$ (365)	Alkylation of ethyl 7-methanesul-fonamidoheptanoate with 1-bromo-4-(R)-acetoxy-2-nonyne followed by hydrogenation and hydrolysis[8]	Elevated mouse ovary cAMP 54-fold at 10 $\mu g/m\ell$ (PGE$_1$ caused 54-fold increase at 1 $\mu g/m\ell$); oral administration to guinea pigs caused ex vivo inhibition of collagen-induced platelet aggregation with ED$_{50}$ = 5 mg/kg[x]
12	7-(N-[4-Hydroxy-5-(4-fluorophenoxy)-(E)2-pentenyl] methansulfon-amide]-(Z)-5-heptenoic acid	$C_{19}H_{26}FNO_4S$ (383)	From ethyl 7-methanesulfonami-dohept-5-ynoate[9]	Elevated mouse ovary cAMP by 2-fold at 10 $\mu g/m\ell$/(PGE$_1$ gave a 54-fold increase at 1 $\mu g/m\ell$): inhibited (ex vivo) collagen induced platelet aggregation in guinea pigs on oral administration with ED$_{50}$ = 0.01 mg/kg (PGE$_1$ had ED$_{50}$ = 0.02 mg/kg, i.p.)[9]
13	2-Methyl-3(4-hydroxy-nonyl)-5-(6-carboxybu-tyl) furan	$C_{19}H_{32}O_4$ (324)	Sequential alkylation of *t*-butyla-cetoacetate with 1-iodo-4-acetoxynonane and methyl 7-bromoheptynoate followed by acid catalyzed cyclization and saponification[10]	Low level of activity in the inhibition of ADP-induced aggregation of HPRP and as a diuretic on oral administration to salt-loaded rats[10]
14	2-(3-Ethoxycarbonyl)pro-pylthio)-4-(5-ethoxy-4,4-dimethyl-3α-hy-droxy-(E)1-pentenyl)-1-pyrroline	$C_{19}H_{33}NO_4S$ (371)	Conversion of methyl pyroglutamate to thione, followed by sodium borohydride reduction to 4-hydroxymethyl-2-pyrrolidine thione and sequential attachment of α- and ω-chains by standard techniques[11]	Hypotensive and gastric secretion inhibiting properties[11]

Table 8 (continued)
SECO PGs

No.	Systematic name	Structure	Mol. formula (mol. wt.)	Synthesis	Biological actions
15	10-Nor-9,11-seco-9,11-peroxido-15α-hydroxy-prosta-5-*cis*-13-*trans*-dienoic acid		$C_{19}H_{32}O_5$ (340)	From PGA$_2$[12]	Induced aggregation of HPRP; 0.2 × PGH$_2$; this aggregation is not blocked by 9α,11α-imi-noepoxyprosta-5(Z),13(E)-dienoic acid, a thromboxane synthesis inhibitor (compound 8 Table 1), but it is induced by 9α,11α-epoxyinimoprosta-5(Z),13(E)-dienic acid, (compound 9, Table 1), a receptor-level TX antagonist; induced contractions of rat aortic strip; ED$_{50}$ = 20 ng/mℓ[12]
16	10-Nor-9,11-seco-9,11-epidithia-15α-hydroxy-prosta-5-*cis*-13-*trans*-dienoic acid		$C_{19}H_{32}O_3S_2$ (372)	From PGA$_2$[12]	Activity analogous to the oxy-gen analog above except for potency on rat aortic strip; ED$_{50}$ = 5 ng/mℓ[12]
17	5-(3α-Hydroxy-*trans*-1-octenyl)-2-(3-ethoxycar-bonylpropylthio)-1-methylimidazoline		$C_{18}H_{32}N_2O_3S$ (356)	From 1-methyl-3-benzyloxycar-bonyl-5-methoxycarbonylimida-zolidin-2-one[13]	Less active than PGI$_2$ methyl ester[13]

REFERENCES

1. **Tanaka, T., Bannai, K., Toru, T., Oba, T., Okamura, N., Watanabe, K., and Kurozumi, S.,** Synthesis of new secoprostaglandins as inducers of platelet aggregation, *Chem. Pharm. Bull.,* 30, 51—62, 1982.

2. **Bicking, J. B., Robb, C. M., Smith, R. L., Cragoe, E. J., Kuehl, F. A., and Mandel, L. R.,** 11,12-Secoprostaglandins, I. Acylhydroxyalkanoic acids and related compounds, *J. Med. Chem.,* 20, 35—43, 1977.

3. **Bicking, J. B., Robb, C. M., Cragoe, E. J., Blaine, E. H., Watson, L. S., and Dunlay, M. C.,** 11,12-Secoprostaglandins. VI. Interphenylene analogues of acylhydroxyalkanoic acids and related compounds as renal vasodilators, *J. Med. Chem.,* 26, 335—341, 1983.

4. **Smith, R. L., Bicking, J. B., Gould, N. P., Lee, T., Robb, C. M., Kuehl, F. A., Mandel, L. R., Cragoe, E. J.,** 11,12-Secoprostaglandins. III. 8-Alkylthio (sulfinyl and sulfonyl)-12-hydroxyalkanoic acids and related compounds, *J. Med. Chem.,* 20, 540—547, 1977.

5. **Zoretic, P. A., Soja, P., and Shiah, T.,** 8,12-Secroprostaglandins. 10-aza and 8-oxa analogs, *Prostaglandins,* 16, 555—561, 1978.

6. **Zoretic, P. A., Soja, P., and Shiah, T.,** Synthesis and gastric antisecretory Properties of an 8-aza and a 10-oxa-8,12-secroprostaglandin, *J. Med. Chem.,* 21, 1330—1332, 1978.

7. **Jones, J. H., Holz, W. J., Bicking, J. B., Cragoe, E. J., Mandel, L. R., and Kuehl, F. A.,** 11,12-secroprostaglandins. II. *N*-acyl-*N*-alkyl-7-aminoheptanoic acids, *J. Med. Chem.,* 20, 44—48, 1977.

8. **Jones, J. H., Holz, W. J., Bicking, J. B., Cragoe, E. J., Mandel, L. R., and Kuehl, F. A.,** 11,12-Secroprostaglandins. IV. 7-(*N*-Alkylmethanesulforamido)heptanoic acids, *J. Med. chem.,* 20, 1299—1304, 1977.

9. **Bicking, J. B., Jones, j. H., holz, W. J., Robb, C. M., Kuehl, F. A., Minsker, D. H., and Cragoe, E. J.,** 11,12-Secroprostaglandins. V. 8-Acetyl- or 8-(1-hydroxyethyl)-12-hydroxy-13-aryloxytridecanoic acids and sulfonamide isoteres as inhibitors of platelet aggregation, *J. Med. Chem.,* 21, 1011—1018, 1978.

10. **Saunders, J., Tipney, D. C., and Robins, P.,** The synthesis of furan-based secoprostacyclins, *Tetrahedron Lett.,* 23, 4147—4150, 1982.

11. **Bartmann, W., Beck, G., Knolle, J., and Rupp, R. H.,** Synthese von (±)-(E)-2-(1-thia-4-äthoxycarbonylbutyl)-4-(3-hydroxy-1-octenyl)-Δ¹-pyrrolin und Seiner Analogen, *Tetrahedron Lett.,* 23, 2947—2950, 1982.

12. **Lin, C., Alexander, C. L., Chidester, C. L., Gorman, R. R., and Johnson R. A.,** 10-Nor-9,11-secroprostaglandins. Synthesis, structure and biology of endoperoxide analogues, *J. Am. Chem. Soc.,* 104, 1621—1628, 1982.

13. **Bartmann, W., Beck, G., Lau, H. H., and Wess, G.,** Synthesis of ethyl (±)-(E)-5-(4- and 5-(3-hydroxy-1-octenyl)-1-methyl-2-imidazolin-2-yl-5-thiapentanoates, *Tetrahedron Lett.,* 25, 733—736, 1984.

PROPERTIES OF SYNTHETIC PROSTANOIDS (TO CIRCA 1976)*

Anthony L. Willis and K. John Stone

PROSTAGLANDIN ANALOGS

Prostaglandins (PGs) exert their biological effect by interacting with "receptors" which show a selectivity for compounds with a particular geometry. Thus, stereoselective chemical syntheses of PGs have been devised which introduce a particular configuration at each of the asymmetric centers. The most successful of these syntheses, which combines a high degree of stereocontrol with high yield, is the Corey cyclopentyl-lactone route.[1-4] In many cases these syntheses have been adapted to generate PG analogs, although sometimes the design of specific analogs has dictated the development of a new synthesis.

Synthetic PG analogs have been prepared with the intention of introducing more specific and potent biological properties than those of the natural PGs. Of particular importance has been the development of analogs resistant to those enzymes which metabolize and inactivate PGs (see contributions by Smith et al. and Jackson-Roberts). A similar strategy has resulted in the development of PG antagonists. Examples of the above types of analogs are shown in Table 1 which is intended to be a representative, but not a comprehensive, list of those compounds which have shown interesting biological properties.

Possibly the simplest modification which can be made to the PG molecule is esterification of the terminal carboxylic acid function. This modification selectively decreases the smooth muscle and vasodepressor properties of PGE_2 and $PGF_{2\alpha}$.[5]

The introduction of alkyl and halogen atoms close to the site of attack by PG 15-hydroxydehydrogenase blocks the metabolism of the biologically important 15-hydroxyl function. Thus, analogs with this modification are generally more potent and longer lasting in vivo than the parent compounds. Some of these compounds are orally active. The only analogs which do not possess a functional group designed to block PG 15-hydroxydehydrogenase but which have the important property of being orally active are ICI 74205 (dihomo-$PGF_{2\alpha}$) and 3-oxa-phenylene PGE_1.

The introduction of oxygen or a *trans*-double bond to the carboxyl end of the PG molecule prevents metabolism by β-oxidative enzymes. However, such analogs have not been reported to be longer acting than the parent compound except where metabolism by PG 15-hydroxydehydrogenase is also blocked, e.g., ONO 747 which, like 3-oxa-phenylene PGE_1 and 2-*trans*-PGE_1, is a potent mimic of PGE_1 and blocks platelet aggregation. Some 15-methylated analogs of PGE_2 which are distinguished by the absence of an 11α-hydroxyl function have the opposite effect and induce platelet aggregation, unlike PGE_2 which usually modulates platelet aggregation induced by other agents. A shift of the double bond in $PGF_{2\alpha}$ from the Δ^5 to the Δ^4 position decreases the rate of metabolism in rats by β-oxidation and by 15-hydroxyprostaglandin dehydrogenase.[87]

The introduction of a methyl group at C-8 of PGC_2 must prevent its isomerization by PGC isomerase to a PGB_2 derivative. As PGC_2 is a potent hypotensive agent and PGB_2 is not, this modification might be expected to impart prolonged hypotensive properties. Unfortunately, reports of the biological properties of 8-methyl PGC_2[6] have not included such data. The synthesis of stable derivatives of PG endoperoxides[7-9] has resulted in compounds more potent than the parent natural products. Their ability to induce platelet aggregation suggests that they, like Wy 17186, might find a potential use in preventing postsurgical bleeding.[10]

---

* Reprinted from Willis, A. L. and Stone, K. J., in *Handbook of Biochemistry and Molecular Biology*, Fasman, G. D., Ed., Vol. 2, 3rd ed., CRC Press, Cleveland, 1976.

Table 1
PROSTAGLANDIN ANALOGS

No.	Trivial name	Systematic name	Structure	Molecular formula	Molecular weight	Synthesis	Biological actions	Comments
1	Ethylene-PGH_1	15-Hydroxy-9α,11α(ethylene)-prost-13-enoic acid		$C_{21}H_{36}O_3$	348		Specifically inhibits PGE_1 synthesis in sheep vesicular gland extracts, and increases the synthesis of $PGF_{1α}$, C_{17}- and C_{20}-monohydroxy fatty acids[7]	Stable analog of PGH_1
2	9-Methano-PGH_2	15-Hydroxy-9α,11α(methanoepoxy)prosta-5,13-dienoic acid		$C_{21}H_{34}O_4$	350	From PGE_2[8]	Potent bronchoconstrictor in laboratory animals; 3.7 times more potent than PGG_2 as an inducer of human platelet aggregation; 6.2 times more potent than PGH_2 in eliciting contractions of isolated rabbit aorta; less active than PGH_2 in initiating platelet release reaction; induced release of (^{14}C)-serotonin is inhibited by indomethacin suggesting a requirement for endogenous endoperoxide synthesis[18]	Stable analog of PGH_2; other stable PGH_2 analogs (U44069 and U46619) enhance ADP-induced aggregation of platelets from man and rabbit; the platelet shape change and reversible aggregation induced by 9-methano-PGH_2 are not associated with the release reaction and are not inhibited by indomethacin[19]
3	11-Methano-PGH_2	15-Hydroxy-9α,11α(epoxymethano)prosta-5,13-dienoic acid		$C_{21}H_{34}O_4$	350	From PGA_2[8]	Potent bronchoconstrictor in laboratory animals[18]	Stable analog of PGH_2
4	Azo-endoperoxide	9,11-Azo-15-hydroxyprosta-5,13-dienoic acid		$C_{20}H_{33}O_3N_2$	348	—[9]	Potent mimic of PGG_2 and PGH_2 with reference to platelet aggregation and the release of serotonin from human platelet-rich plasma; 8 times more potent than PGG_2 in causing aggregation, 6 times more potent and more prolonged effect than PGG_2 in stimulating serotonin release; more active than PGH_2 (7 times) and PGE_2 (1450 times) in contracting isolated rabbit aorta strip[9]	Stable form of PGH_2
5	11,15-epi-PGE_1	dl-11β,15β-Dihydroxy-9-ketoprost-13-enoic acid		$C_{20}H_{34}O_5$	354	—[20]	Twice as active as dl-PGE_1 on rat uterus smooth muscle; less active than PGE_1 as a vasodepressor in rats[20]	

No.	Name	Structure	Formula	MW	Synthesis/Source	Biological activity	Remarks
6	PGF$_3\beta$ 9β,11α,15α-trihydroxyprosta-5,13-dienoic acid		$C_{20}H_{34}O_5$	354	Borohydride reduction of PGE$_2$	Reverses bronchoconstrictor effect of PGF$_2\alpha$ in cats and guinea pigs; bronchodilator effect is 1/100 as potent as PGE$_2$, or PGF$_2$, i.e., equipotent with isoprenaline[21]	Substrate for PG-15-hydroxydehydrogenase
7	4-*cis*-PGE$_1$ 11,15-Dihydroxy-9-ketoprosta-4,13-dienoic acid		$C_{20}H_{32}O_5$	352	Biosynthesized from 4,8,11,14-eicosatetraenoic acid[32]	Equipotent with PGE in producing contractions of isolated guinea pig ileum; poor inhibitor of platelet aggregation but at low concentrations is more potent than PGE$_2$ in producing enhancement[22]	
8	5-Dehydro-PGE$_2$ ester 11,15-Dihydroxy-9-ketoprost-13-ene-5-ynoic acid methyl ester		$C_{22}H_{32}O_5$	364	From parent prostaglandin[23]	As potent as PGE in lowering rat blood pressure; less active than PGE$_2$ in producing contractions of isolated gerbil colon[24]	
9	5-Dehydro-PGF$_2\alpha$ ester 9,11,15-Trihydroxyprosta-13-ene-5-ynoic acid methyl ester		$C_{21}H_{34}O_5$	366	From parent prostaglandin[23]	Less potent than PGF$_2\alpha$ on gerbil colon[24]	
10	5-Dehydro-PGA$_2$ ester 15-Hydroxy-9-ketoprosta-10,13-diene-5-ynoic acid methyl ester		$C_{21}H_{30}O_4$	346	From parent prostaglandin[23]	Less potent than PGA$_2$ on gerbil colon[24]	
11	K 8548 11,15-Dihydroxy-9-ketoprost-5-ene-13-ynoic acid		$C_{20}H_{30}O_5$	350		2—3 times as potent as PGE$_2$ at inhibiting gastric secretion and the formation of stress-induced gastric ulcers in rats; half as active as PGE$_2$ on guinea pig ileum and rat uterus[25]	16(S)-methyl analogue is 43-fold more potent than PGE$_2$ at inhibiting gastric secretion when given orally, and 70-fold when given subcutaneously 16(R)-methyl analogue is 106-fold more active orally and 43-fold when given subcutaneously[25]
12	13-Dehydro-PGF$_2\alpha$ 9,11,15-Trihydroxyprost-5-ene-13-ynoic acid		$C_{20}H_{32}O_5$	352	—[26]	Stimulates cAMP synthesis in mouse ovary; terminates pregnancy in hamsters[26,55]	Inhibits PG-15-hydroxydehydrogenase

Table 1 (continued)
PROSTAGLANDIN ANALOGS

No.	Trivial name	Systematic name	Structure	Molecular formula	Molecular weight	Synthesis	Biological actions	Comments
13	13-Dehydro-PGF$_{2\beta}$	9β,11β,15α-Trihydroxy-8-ent-12-ent-prost-5-ene-13-ynoic acid		C$_{20}$H$_{32}$O$_5$	352	26	Stimulates cAMP synthesis in mouse ovary; terminates pregnancy in hamsters[26,55]	Inhibits PG-15-hydroxydehydrogenase
14	14-Chloro-PGF$_{2\alpha}$	14-Chloro-9,11,15-trihydroxyprosta-5,13-dienoic acid		C$_{20}$H$_{33}$O$_5$Cl	388		Potent antifertility agent[27]	Not a substrate for PG-15-hydroxy-dehydrogenase
15	15(S)-Methyl-PGE$_1$	15(S)-15-Methyl-11,15-dihydroxy-9-keto-prost-13-enoic acid		C$_{21}$H$_{36}$O$_5$	368	Oxidation of 15-methyl-PGF$_{1\alpha}$[28,29]	Methyl ester is equipotent, with PGE$_1$; -Me ester in producing contractions of isolated gerbil colon; 15-Me-PGE$_1$, -Me ester reduces blood pressure of rat, but dose-response curve not parallel with that for PGE$_1$, methyl ester; is about eightfold more potent than PGE$_1$, -Me ester as an antifertility agent in hamsters (through luteolytic action)[58]	Not a substrate for PG-15-hydroxy-dehydrogenase[24]
16	15(S)-Methyl-PGE$_2$	15(S)-15-Methyl-11,15-dihydroxy-9-keto-prosta-5,13-dienoic acid		C$_{21}$H$_{34}$O$_5$	366	Oxidation of 15-methyl-PGF$_{2\alpha}$[24,30]	Methyl ester is equi-active with parent compound in vitro in producing contractions of gerbil colon;[24] reduces blood pressure in rat; about 50-fold more potent than parent compound as an antifertility agent in hamsters;[24] 3—10 times more active than PGE$_2$ in vivo on pregnant uterus in monkeys and several hundred times more potent in similar tests in humans; given intravenously, intramuscularly, or intra-amniotically is about fivefold more potent than parent compound as an abortifacient; mechanism probably involves uterine smooth muscle contraction;[31] orally it induces nausea and vomiting and is equipotent with PGE$_2$ -methyl ester in producing diarrhea;[31] inhibits penicillin-induced epilepsy in cats;[32] inhibits histamine-induced gastric secretion	Rationale for using methyl ester of 15-methyl PGE$_2$ is based upon the expected reduction in side effects; thus Me esters of PGE$_2$ or PGF$_{2\alpha}$ have only one third to half potency of parent PG in reducing blood pressure in man or baboon and about 3—6 times the smooth muscle stimulatory activity (bird colon and guinea pig ileum); methyl esters are equi-active with parent

No.	Name	Structure	Formula	M.W.	Preparation	Biological activity	Remarks
						(acid, pepsin, and volume) in dogs when given parenterally, intrajejunally, or orally; parent compound is inactive orally;[33] not a substrate for PG-15-hydroxydehydrogenase; 100—400 times as active as PGE₂ in terminating second trimester pregnancy with at least threefold longer action[28]	PGs in producing diarrhea in man (oral route) but are equi-fold active or fivefold more potent in producing contractions of human uterus in vivo
17	15(R)-Methyl-PGE₂; 15(R)-15-Methyl-11,15-dihydroxy-9-keto-prosta-5,13-dienoic acid		$C_{21}H_{34}O_5$	366	Oxidation of 15-methyl-PGF₂α[28,30] separated from 15(S)-analog (which is more polar) by TLC[28]	Given by intramuscular, intravenous, intra-vaginal, or intra-amniotic routes, 15(R)-15 methyl PGE₂, methyl ester is 10—40-fold more potent than parent PGE₂; methyl ester in inducing abortion (through uterine contractions) in women;[31] however, it is about 10-fold less potent than 15(S) analog and produces nausea and vomiting in some cases;[31] methyl ester (up to 200 μg) given orally in man produces a marked and prolonged reduction in basal or pentagastrin-stimulated gastric acid secretion[34] but is ineffective when infused intra-venously; these actions contrast with those of PGE₂, whose inhibitory effects on gastric secretion are not seen after oral administration; side effects (such as nausea) seen with orally administered 15(S) analog are minimized[33,34]	Not a substrate for PG-15-hydroxy-dehydrogenase; orally active
18	15(S)-Methyl-PGF₁α; 15(S)-15-Methyl-9,11,15-trihydroxyprost-13-enoic acid		$C_{21}H_{38}O_5$	370	Methylation of PGF₁α using dichlorodicyano-benzoquinone and methyl-Mg-bromide[28]	Methyl ester contracts gerbil colon[28] with 75% potency of parent PGF₁α; methyl ester, 12-fold more potent than PGF₁α; methyl ester as an antifertility agent in hamsters[28]	Not a substrate for PG-15-hydroxy-dehydrogenase
19	15(S)-Methyl-PGF₂α; 15(S)-15-Methyl-9,11,15-trihydroxyprosta-5,13-dienoic acid		$C_{21}H_{36}O_5$	368	Methylation of PGF₂α using dichlorodicyano-benzoquinone and methyl-Mg-bromide[28]	Shortens luteal phase in women and decreases progesterone levels; methyl ester contracts gerbil colon, equipotent with parent compound;[28] potent abortifacient in hamsters, monkeys, and man; more potent than parent PGF₂α;[3,4-36] 10 times as active as PGF₂α in inducing therapeutic abortion in second trimester patients[35] with at least a threefold longer action;[28] raises pulmonary blood pressure in dogs[37]	Not a substrate for PG-15-hydroxy-dehydrogenase

Table 1 (continued)
PROSTAGLANDIN ANALOGS

No.	Trivial name	Systematic name	Structure	Molecular formula	Molecular weight	Synthesis	Biological actions	Comments
20	15(R)-Methyl-PGF$_{2α}$	15(R)-15-Methyl-9α,11α,15-trihydroxyprosta-5,13-dienoic acid	[chemical structure]	C$_{21}$H$_{36}$O$_5$	368		Methyl ester is 2.5 times more potent than PGF$_{2α}$ methyl ester on rat blood pressure and gerbil colon; 16-fold more potent as an abortifacient in hamsters[28]	Not a substrate for PG-15-hydroxy-dehydrogenase
21	15(S)-Methyl-PGF$_{2β}$	15(S)-15-Methyl-9β,11α,15α-trihydroxy-prosta-5,13-dienoic acid	[chemical structure]	C$_{21}$H$_{36}$O$_5$	368	As for 15-methyl PGF$_{2α}$ but using PGF$_{2β}$ as starting compound	Methyl ester is 40% less active than PGF$_{2β}$ methyl ester in producing contractions of isolated gerbil colon;[28] equipotent with PGF$_{2β}$ methyl ester as an anti-fertility agent in hamsters[28]	
22	15(S)-Methyl-PGA$_1$	15(S)-15-Methyl-15-hydroxy-9-ketoprosta-10,13-dienoic acid	[chemical structure]	C$_{21}$H$_{34}$O$_4$	350	Reaction of 15-methyl-PGE$_1$ with dicyclo-hexylcarbodiimide and CuCl$_2$[28]	Methyl ester has depressor action on rat blood pressure[28]	
23	15(S)-Methyl-PGA$_2$	15(S)-15-Methyl-15-hydroxy-9-ketoprosta-5,10,13-trienoic acid	[chemical structure]	C$_{21}$H$_{32}$O$_4$	348	Reaction of 15-methyl-PGE$_2$ with dicyclo-hexylcarbodiimide and CuCl$_2$[28]	Low doses have a depressor effect on rat blood pressure; high doses have a pressor effect[28]	
24	K-10134	16(S)-16-Methyl-11,15-dihydroxy-9-keto-prost-5-enoic acid	[chemical structure]	C$_{21}$H$_{36}$O$_5$	368		Intravenous administration at 0.1–1 μg/kg inhibits gastric secretion induced by various stimuli in conscious dogs and cats; 30–90 min effect duration[28]	
25	ONO 464	16(R)-16-Methyl-11,15-dihydroxy-9-keto-prost-5-enoic acid methyl ester	[chemical structure]	C$_{22}$H$_{38}$O$_5$	382		10–100-fold depressor activity of PGE$_2$ in dogs, 5–10-fold depressor activity of PGE$_2$ in spontaneously hypertensive rats (dosed orally); hypotensive effect due to peripheral vasodilatation; more potent (10-fold) than PGE$_2$ in stimulating dog smooth muscle; like PGE$_2$ it is not diuretic in rats[39]	

No.	Common name	Systematic name	Formula	MW	Preparation	Biological activity	Additional notes
26	16,16-dimethyl-PGE₂	16,16-Dimethyl-11,15-dihydroxy-9-keto-prosta-5,13-dienoic acid	$C_{22}H_{36}O_5$	380	Oxidation of 16,16-dimethyl PGF₂α[40]	Methyl ester more potent than parent PGE₂; methyl ester or 15(R)-methyl PGE₂ at inhibiting gastric secretion[24] in man;[41] antifertility agent in hamsters;[24] contracts gerbil colon;[37] given orally to dogs inhibits gastric secretion by a local effect on oxyntic glands;[40,42] has a stimulating effect on uterus of pregnant women[43]	Incidence of nausea, vomiting, and diarrhea limits use of the Me-ester for induction of abortion;[43] is not a substrate for PG-15-hydroxydehydrogenase
27	16,16-Dimethyl-PGF₂α	16,16-Dimethyl-9,11,15-trihydroxy-prosta-5,13-dienoic acid	$C_{22}H_{38}O_5$	382	Modified Corey synthesis[40,44]	Abortifacient;[40] contracts gerbil colon[37]	
28	16,16-Dimethyl-PGA₂	16,16-Dimethyl-15-hydroxy-9-ketoprosta-5,10,13-trienoic acid	$C_{22}H_{34}O_4$	362	Reaction of 16,16-dimethyl PGE₂ with acid		
29	16,16-Dimethyl-PGB₂	16,16-Dimethyl-15-hydroxy-9-ketoprosta-5,8(12),13-trienoic acid	$C_{22}H_{34}O_4$	362	Reaction of 16,16-dimethyl PGE₂ with base		
30	16-Fluoro-PGF₂α	16-Fluoro-9,11,15-trihydroxyprosta-5,13-dienoic acid	$C_{20}H_{31}O_5F$	370		78% potency of PGF₂α in gerbil colon smooth muscle assay[45]	
31	16,16-Difluoro-PGE₂	16,16-Difluoro-11,15-dihydroxy-9-keto-prosta-5,13-dienoic acid	$C_{20}H_{30}O_5F_2$	388	—[45]	Potent antifertility agent[45]	Not a substrate for PG-15-hydroxy-dehydrogenase[45]
32	16-Methyl-16-hydroxy-PGE₂	(±)15-Deoxy-11,16-dihydroxy-9-keto-15-methyl-prosta-5,13-dienoic methyl ester	$C_{22}H_{36}O_5$	380	—[46]	Potent gastric antisecretory agent in dogs[46]	

Table 1 (continued)
PROSTAGLANDIN ANALOGS

No.	Trivial name	Systematic name	Structure	Molecular formula	Molecular weight	Synthesis	Biological actions	Comments
33	ONO-747	17(ξ)-Ethyl-11,15-dihydroxy-9-keto-prosta-2-trans,13-trans-dienoic acid	(chemical structure)	$C_{22}H_{36}O_5$	380		ONO-747 is 15.8, 12.0, and 2.1 times more potent than PGE$_1$ in producing inhibition of ADP-induced platelet aggregation with human platelets, rat platelets, and with rabbit platelets, respectively; infused intravenously at 2 μg/kg/min. ONO-747 completely inhibited thrombus formation in mesenteric blood vessels of rabbits, whereas urokinase or aspirin were ineffective; when infused intravenously. aggregation in PRP of rabbits was inhibited by ONO-747 at 3 μg/kg/min and only 0.2 μg/kg/min when infused intra-arterially.[47]	β-Oxidation blocked; is a substrate for PG-15-hydroxydehydrogenase
34	2-*trans*-PGE$_1$	11,15-Dihydroxy-9-ketoprosta-2-*trans*,13-*trans*-dienoic acid	(chemical structure)	$C_{20}H_{32}O_5$	352	Biosynthesized from 2-*trans*-8,11,14-*cis*-eicosatetraenoic acid;[48] synthesis[22,49]	2.5 times more potent than PGE$_1$, as an inhibitor of ADP-induced platelet aggregation;[22] less active than PGE$_1$ on guinea pig ileum[48]	
35	5-*trans*-PGE$_2$	11α,15α-Dihydroxy-9-ketoprosta-5-*trans*-13-*trans*-dienoic acid	(chemical structure)	$C_{20}H_{32}O_5$	352		Has weak ability to produce contractions of isolated guinea pig ileum, or inhibition of platelet aggregation[91]	Formed chemically from 5-*trans*-PGA$_2$ or biosynthetically from 5-*trans*-arachidonic acid[90,91]
36	PGE$_2$-tetrazole	2-Decarboxyl-2(tetrazol-5-yl)-11,15-dihydroxy-9-ketoprosta-5,13-dienoic acid	(chemical structure)	$C_{23}H_{32}O_3N_4$	376		Equipotent with PGE$_2$ in contracting gerbil colon and reducing rat blood pressure[50]	
37	PGF$_{2α}$-tetrazole	2-Decarboxy-2-(tetrazol-5-yl)-9,11,15-trihydroxyprosta-5,13-dienoic acid	(chemical structure)	$C_{23}H_{34}O_3N_4$	378		Diminished gerbil colon activity; corresponding PGF$_{1α}$ analogue has a slight depressor activity on rat blood pressure rather than the expected pressor response[50]	pKa of 5-methyl tetrazole is 5.6, i.e., very similar to that of carboxylic acid

Comments

No.	Name	Structure	Formula	MW	Synthesis	Biological activity	Comments
38	dl-3-Oxa-PGF₁	dl-3-Oxa-11,15-dihydroxy-9-ketoprost-13-enoic acid		$C_{19}H_{32}O_6$	356	—[29]	Ethyl ester and ethyl ester of parent PGE₁ have equipotent activity in contracting gerbil colon; PGE₁, however, is very much more active than dl-oxa-PGE ethyl ester[29]	β-Oxidation blocked
39	dl-3-Oxa-PGF₁,α	dl-3-Oxa-9α-11,15-trihydroxyprost-13-enoic acid		$C_{19}H_{34}O_6$	358	Reduction of dl-3-oxa-PGE₁ with borohydride produces a mixture of α and β isomers[29]	Contracts smooth muscle[29]	β-Oxidation blocked
40	3-Oxa-PGF₂α	9,11,15-Trihydroxy-3-oxaprosta-5,13-dienoic acid		$C_{19}H_{32}O_6$	356			Is not metabolized by β-oxidation
41	dl-3-Oxa-PGF₁,β	dl-3-Oxa-9β,11,15-trihydroxyprost-13-enoic acid		$C_{19}H_{34}O_6$	358	Reduction of dl-3-oxa-PGE₁ with borohydride produces a mixture of α and β isomers[29]	Contracts smooth muscle[29]	β-Oxidation blocked
42	3-Oxa-phenylene-PGE₁	4,5,6-Trinor-3,7-inter-m-phenylene-3-oxa-11,15-dihydroxy-9-ketoprost-13-enoic acid		$C_{22}H_{30}O_6$	390	—[51]	More potent than PGE₁ as an inhibitor of platelet aggregation in vitro; orally active on platelet aggregation in rats at 20 mg/kg; less potent than PGE₁ in producing contractions of smooth muscle and depressor effects on blood pressure[52]	
43	7-Oxa-PGE₁	7-Oxa-11,15-dihydroxy-9-ketoprost-13-enoic acid		$C_{19}H_{32}O_6$	356	—[53]	Produces contractions of gerbil colon[53]	Effects of 7-oxa derivatives on smooth muscle are dependent upon degree of hydroxylation of these substances, i.e., transition from pure agonism for substances with oxygen substituents on C-9 and C-11 to pure antagonism for substances devoid of hydroxyl or keto groups[53]

Table 1 (continued)
PROSTAGLANDIN ANALOGS

No.	Trivial name	Systematic name	Structure	Molecular formula	Molecular weight	Synthesis	Biological actions	Comments
44	7-Oxa-PGF$_{1\alpha}$	7-Oxa-9,11,15-trihydroxyprost-13-enoic acid		$C_{19}H_{34}O_6$	358	44,53	Stimulates cAMP formation in mouse ovary at 100 µg/ml; is a substrate for PG-15-hydroxydehydrogenase;[54] causes contractions of gerbil colon[53]	8-isomer inhibits PGF$_{1\alpha}$ effect on gerbil colon[54]
45	Homo-PGE$_2$	11,15-Dihydroxy-9-keto-2b-methyl-prosta-5,13-dienoic acid		$C_{21}H_{34}O_5$	366	Biosynthesized from homo-eicosatetraenoic acid (Δ5,11,14, 21:4)[83]	More active than PGE$_2$ on platelet aggregation	Homo-PGE$_1$ is more potent than PGE$_1$ as an inhibitor of platelet aggregation.[83]
46	ICI 74205 dihomo-PGF$_{2\alpha}$	9,11,15-Trihydroxy-20,20b-dihomoprosta-5,13-dienoic acid		$C_{22}H_{38}O_5$	382		20 times as active as PGF$_{2\alpha}$ when given p.o. in terminating pregnancy in hamsters[55]	Orally active
47	17-Oxadihomo-PGF$_{2\alpha}$	17-Oxa-9,11,15-trihydroxy-20-ethyl-prosta-5,13-dienoic acid		$C_{21}H_{36}O_6$	384	Modified Corey synthesis[44,56]	Leuteolytic but is not a potent stimulator of uterine smooth muscle[56]	
48	17-Phenyl-PGE$_2$	17-Phenyl-11,15-dihydroxy-9-keto-(ω-trinor)-prosta-5,13-dienoic acid		$C_{23}H_{30}O_5$	386	Modified Corey synthesis[44,57]	Contracts gerbil colon and rat uterus;[58] has transient depressor activity followed by more prolonged pressor action in rats (PGE$_2$ is depressor only); antifertility agent in hamster; stimulates uterine contractility in pregnant monkeys; abortifacient;[59] reduces gastric secretion in dogs (equipotent with PGE$_2$)[58]	Not an efficient substrate for prostaglandin-15-hydroxy-dehydrogenase[58]
49	17-Phenyl-PGF$_{2\alpha}$	17-Phenyl-9,11,15-trihydroxy-4(ω-trinor)-prosta-5,13-dienoic acid		$C_{23}H_{32}O_5$	388	Modified Corey synthesis[44,57]	More potent than PGF$_{2\alpha}$ in synchronizing estrus in cows; 3 times more potent than PGF$_{2\alpha}$ in contracting rat uterus,[58] but equiactive on gerbil colon; 5 times more potent than PGF$_{2\alpha}$ in its pressor action in rats; 90 times more potent	Not an efficient substrate for prostaglandin-15-hydroxy-dehydrogenase[58]

No.	Name	Systematic name	Structure	Formula	MW	Synthesis	Biological activity	Notes
							than PGF₂ₐ in hamster antifertility test, probably works by luteolytic action; stimulates uterine contractility in pregnant monkeys; in high doses is an abortifacient[59]	
50	4-cis-17-Phenyl-PGF₂ₐ	17-Phenyl-9,11,15-trihydroxy-(ω-trinor)-prosta-4,13-dienoic acid		$C_{23}H_{32}O_5$	388		200 times more active than PGF₂ₐ as an antifertility agent in hamster[27]	Δ⁴ double bond unusual
51	16-Phenoxy-PGF₂ₐ	18-Phenyl-17-oxa-9,11,15-trihydroxy-(ω-trinor)-prosta-5,13-dienoic acid		$C_{22}H_{30}O_6$	390	Modified Corey synthesis[44,46]	10 times abortifacient activity of PGF₂ₐ in rat[46]	
52	ICI 79939	16-Fluorophenoxy-9,11,15-trihydroxy-(ω-tetranor)-prosta-5,13-dienoic acid		$C_{22}H_{29}O_6F$	408		200 times as potent as PGF₂ₐ in terminating pregnancy in hamsters; 10 times as potent as PGF₂ₐ on guinea pig uterus and gerbil colon; side effects parallel potency as a smooth muscle stimulant; terminates life of corpus luteum in the pregnant pig and hence controls the onset of parturition[61]	Used to synchronize estrus in horses, cattle,[16] and sheep[15]
53	ICI 81008	16-Trifluoromethylphenoxy-9,11,15-trihydroxy-(ω-tetranor)-prosta-5,13-dienoic acid		$C_{23}H_{29}O_6F_3$	458		100 times as potent as PGF₂ₐ in terminating pregnancy in hamsters; 1/50 as potent as PGF₂ₐ on guinea pig uterus and gerbil colon[61]	Used to synchronize estrus in horses and cattle[16]
54	16-Me, 17-Cyclopentyl-PGA₂	17-Cyclopentyl-16-methyl-15-hydroxy-9-keto-(ω-trinor)-prosta-5,10,13-trienoic acid		$C_{23}H_{34}O_4$	374			Good competitive inhibitor of PG-15-hydroxydehydrogenase[27]
55	16-Me, 17-Cyclohexyl-PGA₂	17-Cyclohexyl-16-methyl-15-hydroxy-9-keto-(ω-trinor)-prosta-5,10,13-trienoic acid		$C_{24}H_{36}O_4$	388			Good competitive inhibitor of PG-15-hydroxydehydrogenase[27]

Table 1 (continued)
PROSTAGLANDIN ANALOGS

No.	Trivial name	Systematic name	Structure	Molecular formula	Molecular weight	Synthesis	Biological actions	Comments
56		12-Hydroxycyclohexen-9-keto-(ω-octanor)-13,14,15-prost-5-enoic acid		$C_{18}H_{26}O_4$	306		Specific antagonist of PGF$_{2\alpha}$ on guinea pig ileum[11]	
57		12-Hydroxycyclohexen-9-keto-(ω-octanor)-13,14,15 prost-5-enoic acid		$C_{18}H_{26}O_4$	306		Specific antagonist of PGF$_{2\alpha}$ on guinea pig ileum[11]	
58	8-Me-PGC$_2$	8-Methyl-15-hydroxy-9-ketoprosta-5,11,13-trienoic acid		$C_{21}H_{32}O_4$	348		1/30 activity of PGE$_2$ on guinea pig uterus, methyl ester inhibits gastric secretion with 1/60 of PGE$_2$ activity; effects on blood pressure not given[6]	Stable form of PGC$_2$
59	PG-Thiosemicarbazone	11,15-Dihydroxy-9-thiosemicarbazone-prosta-5,13-dienoic acid		$C_{21}H_{35}O_4N_3S$	425	PGE$_2$ thiosemicarbazide[62]	Bronchodilator[62]	
60	11-Desoxy-PGE$_1$ AY-23578	15-Hydroxy-9-ketoprost-13-enoic acid		$C_{20}H_{34}O_4$	338	From hexamethyl-phosphorus triamide cuprate[63-65]	Bronchodilator, protects guinea pigs against histamine-induced convulsions, equipotent with PGE$_2$ in inhibiting bronchoconstriction induced by histamine; reduces or prevents neostigmine, PGF$_{2\alpha}$ or carbachol-induced increases in pulmonary resistance in the cat;[66] is bronchodilator in guinea pigs, when given intravenously or by aerosol; decreases blood pressure when given intravenously but not when administered by aerosol[67]	In vitro, 15-methyl analogue is equipotent with parent compound but in vivo, 15-methyl compound is 32–78 times as active as parent compound, because it is not a substrate for PG-15-hydroxydehydrogenase[67]

No.	Name	Chemical name	Structure	Formula	MW	Notes	Activity	Effect
61	Wy-17186	15(S)-15-Methyl-15-hydroxy-9-ketoprosta-5,13-dienoic acid		$C_{21}H_{34}O_4$	350		Induces platelet aggregation in vitro; shortens clotting time in rats and reduces bleeding time from a wound surface;[10] does not induce platelet aggregation in washed human platelets, unless ADP or fibrinogen is present[8,9]	Prostaglandin endoperoxide-like effects of potential use in preventing postsurgical bleeding
62	Wy-17185	15(S)-15-Methyl-9,15-dihydroxyprosta-5,13-dienoic acid		$C_{21}H_{36}O_4$	352		Induces platelet aggregation in vitro;[10] shortens clotting time in rats	Prostaglandin endoperoxide-like effect
63	Wy-16991	11,15-Dimethyl-9,15-dihydroxyprosta-5,13-dienoic acid		$C_{22}H_{38}O_4$	366		Induces platelet aggregation in vitro[9]	Prostaglandin endoperoxide-like effect
64	11-Me-PGE$_2$	11-Methyl-11β,15-dihydroxy-9-keto-prosta-5,13-dienoic acid		$C_{21}H_{34}O_5$	366	From PGA$_2$ methyl ester[6,8]		
65	11-Me-PGF$_{2\beta}$	11-Methyl-9β,11β,15-trihydroxyprosta-5,13-dienoic acid		$C_{21}H_{36}O_5$	368			
66	11,15-diMe-PGF$_{2\beta}$	11α,15-Dimethyl-9β,11β,-15-trihydroxy-prosta-5,13-dienoic acid		$C_{22}H_{38}O_5$	382	From the methyl ester of 11-methyl-PGF$_2$,β using Me-Mg-Br[6,8]		
67	11-Deoxyprostynoic acid	13-Hydroxy-9-ketoprost-14-ynoic acid		$C_{20}H_{32}O_4$	336		Antagonizes PGE$_2$ induced formation of cAMP in rat anterior pituitary[6,9]	
68	5-Oxaprostenoic acid	5-Oxaprost-13-enoic acid		$C_{19}H_{34}O_3$	310		Inhibits prostaglandin synthetase in vitro; inhibits effect of PGF$_2$ on gerbil colon at 2.5—10 $\mu g/ml^{-3}$	

Table 1 (continued)
PROSTAGLANDIN ANALOGS

No.	Trivial name	Systematic name	Structure	Molecular formula	Molecular weight	Synthesis	Biological actions	Comments
69	7-Oxaprostynoic acid	7-Oxaprost-13-ynoic acid		$C_{20}H_{31}O_3$	320		Prostaglandin antagonist which has a weak dose-related affinity for the adipocyte PG_i-receptor;[68] potentiation of the lipolytic activity of epinephrine[70] may result from interaction with a site other than the PGE-receptor on rat adipocyte plasma membranes[68] has a marked inhibitory effect on pure muscle adenylate kinase, but activates Na^+/K^+ ATPase from human erythrocytes and platelets;[71] inhibits PG-synthetase;[53] competitively inhibits the elevation of cAMP by PGE_1, or PGE_2 in isolated mouse ovary (Ki, 6×10^{-5} M);[72] blocks ability of PGE_2 to produce contractions of isolated gerbil colon,[53] guinea pig, and rabbit ileum;[73] stimulates rat fundus and longitudinal muscle from human stomach;[74] antagonizes the stimulatory effect of prostaglandins on cAMP function in various tissues[75-78] and antagonizes the prostaglandin induced release of some hormones;[78,79] suggested that it acts at the catalytic site of adenyl cyclase;[80] does *not* block increase in protein synthesis caused by PGE_1 in isolated rat ovary but acts like PGE_2 and luteinizing hormone in increasing glycolysis; unlike PGE_2 it does not stimulate protein synthesis;[81] promotes lysis of human erythrocytes and dissolution of erythrocyte membranes[84]	Has been widely employed in the study of the relationship between prostaglandins and cyclic nucleotides
70	Oxystearic acid	9,10-Oxystearic acid		$C_{18}H_{34}O_3$	298	From stearic acid derivative[82]	Inhibits PGE_2-induced diarrhea; inhibitor of prostaglandin synthetase[82]	
71	Sulphinyldioxahomoprostanoate	10,12-Dioxa-11-sulphinyl-ent-α-homo-prostanoic acid methyl ester		$C_{19}H_{36}O_5S$	376	From stearic acid derivative[82]	100 times as active as PGE_1 in tracheal chain bioassay[82]	

No.	Name	Structure	Formula	MW	Source	Activity
72	Ketodioxahomo-prostanoate	10,12-Dioxa-11-keto-ent-α-homo-prostanoic acid methyl ester	COOCH₃ $C_{20}H_{36}O_5$	356	From stearic acid derivative[82]	100 times as active as PGE₂ in tracheal chain bioassay[82]
73	Sulphinyldioxahomo-prostanoate	10,12-Dioxa-11-sulphinyl-ent-α-homo-prostanoic acid methyl ester	COOCH₃ $C_{19}H_{36}O_5S$	376	From stearic acid derivative[82]	Inhibits PGE₂ induced diarrhea[82]
74	Dioxahomopros-tanoic acid	10,12-Dioxa-ent-α-homoprostanoic acid	COOH $C_{19}H_{36}O_4$	328	From stearic acid derivative[82]	Inhibits PGE₂ induced diarrhea[82]
75	Dithiathionehomo-prostanoic acid	10,12-Dithia-11-thione-ent-α-homo-prostanoic acid	COOH $C_{19}H_{34}O_2S_3$	390	From stearic acid derivative[84]	Inhibitor of prostaglandin synthetase[82]

The development of PG analogs which block PG receptors is potentially as important as the development of PG mimetics. The PGE antagonist, 7-oxa-13-prostynoic acid, has properties unlike that of polyploretin phosphate and has been employed widely in the study of the relationship between the PGs and cyclic nucleotides. Specific antagonists of $PGF_{2\alpha}$ in which the β-chain is replaced by a hydroxycyclohexene function have also been reported.[11] These antagonists have not yet found a clinical application.

Some PG analogs have been used in clinical studies and are marketed as therapeutic agents. Karim and others[12,13] pioneered the use of PG analogs in abortion and the induction of labor. The demonstration that intra-amniotic or intravaginal application reduces the effective dose and as a consequence, the frequency and severity of side effects was an important advance.[5] The induction of menstruation in women has led to the development of a "morning-after" contraceptive pill. Synchronization of estrus in cattle and sheep by PG analogs[14-16] is one of the most significant commercial developments of PG research. Such synchronization results in increased rates of conception[17] and enables the breeding program to be chronologically planned.

The diversity of PG action is an indication of the many areas in which analogs might find a future application as therapeutic agents. Antihypertensive therapy which might use analogs of PGA or PGC, antithrombotic therapy using such PGE_1-analogs as ONO 747, 3-oxa-phenylene PGE_1, and thromboxane antagonists, protected PGE analogs which selectively bronchodilate to relieve asthma or inhibit gastric secretion and reverse peptic ulceration are obvious areas of interest.

ACKNOWLEDGMENT

We acknowledge Mr. P. Marples and Mr. J. Denton for assistance with the artwork and bibliography. We are especially grateful to Mrs. E. Botfield for help with the initial compilation of data.

REFERENCES

1. **Corey, E. J., Weinshenker, N. M., Shaaf, T. K., and Huber, W.**, Stereo-controlled synthesis of prostaglandins $F_{2\alpha}$ and E_2 (dl), *J. Am. Chem. Soc.*, 91, 5675—5677, 1969.
2. **Corey, E. J., Schaaf, T. K., Huber, W., Koelliker, U., and Weinshenker, N. M.**, Total synthesis of prostaglandins $F_{2\alpha}$ and E_2 as the naturally occurring forms, *J. Am. Chem. Soc.*, 92, 397—398, 1970.
3. **Corey, E. J., Albonico, S. M., Koelliker, U., Schaaf, T. K., and Varma, R. K.**, New reagents for stereoselective carbonyl reduction. An improved synthetic route to the primary prostaglandins, *J. Am. Chem. Soc.*, 93, 1491—1493, 1971.
4. **Corey, E. J., Shirahama, H., Yamamoto, H., Terashima, S., Venkateswarlu, A., and Shaaf, T. K.**, Stereospecific total synthesis of prostaglandins E_3 and $F_{3\alpha}$, *J. Am. Chem. Soc.*, 93, 1490—1491, 1971.
5. **Karim, S. M. M., Sharma, S. D., Filshie, G. M., Salmon, J. A., and Ganesan, P. A.**, Termination of pregnancy with prostaglandin analogs, *Adv. Biosci.*, 9, 811—830, 1973.
6. **Corey, E. J. and Sachdev, H. S.**, A simple synthesis of 8-methylprostaglandin C_2, *J. Am. Chem. Soc.*, 95, 8483—8484, 1973.
7. **Wlodawer, P., Samuelsson, B., Albonico, S. M., and Corey, E. J.**, Selective inhibition of prostaglandin synthetase by a bicyclo (2.2.1) heptene derivative, *J. Am. Chem. Soc.*, 93, 2815—2816, 1971.
8. **Bundy, G. L.**, The synthesis of prostaglandin endoperoxide analogs, *Tetrahedron Lett.*, 24, 1957—1960, 1975.
9. **Corey, E. J., Nicolaou, K. C., Machida, Y., Malmsten, C. L., and Samuelsson, B.**, Synthesis and biological properties of a 9,11-azo-prostanoid: highly active biochemical mimic of prostaglandin endoperoxides, *Proc. Natl. Acad. Sci. U.S.A.*, 72, 3355—3358, 1975.

10. **Fenichel, R. L., Stokes, D. D., and Alburn, H. E.,** Prostaglandins as haemostatis agents, *Nature (London),* 253,537—538, 1975.

11. **Arndt, H. C., Biddlecom, W. G., Kluender, H. C., Perozzotti, G. P., and Woessner, W. D.,** The synthesis of prostaglandin analogs containing hydroxycyclohexenyl rings, *Prostaglandins,* 9, 521—525, 1975.

12. **Karim, S. M. M., Trussell, R. R., Patel, R. C., and Hillier, K.,** Response of pregnant human uterus to prostaglandin $F_{2\alpha}$-induction of labour, *Br. Med. J.,* 4, 621—623, 1968.

13. **Karim, S. M. M.,** Physiological role of prostaglandins in the control of parturition and menstruation, *J. Reprod. Fert. (Suppl.),* 16, 105—119, 1972.

14. **Cooper, M. J. and Jackson, P. S.,** The application of cloprostenol (compound ICI 80,996) in the control of the bovine oestrous cycle, *Adv. Prostaglandin and Thromboxane Res.,* 2, 922—923, 1976.

15. **Furr, B. J. A.,** Induction of oestrus in the ewe with two synthetic analogues of prostaglandin $F_{2\alpha}$ cloprostenol (ICI 80, 996) and fluprostenol (ICI 81,008), *Adv. Prostaglandin Thromboxane Res.,* 2, 926, 1976.

16. **Allen, W. R. and Rowson, L. E. A.,** Control of the mare's oestrous cycle by prostaglandins, *J. Reprod. Fertil.,* 33, 539—543, 1973.

17. **Deletang, F.,** Control of oestrous cycles in cattle with a prostaglandin $F_{2\alpha}$ analogue (ICI 80,996): synchronization of oestrus and fertility following artificial insemination, *Adv. Prostaglandin Thromboxane Res.,* 2, 924, 1976.

18. **Malmsten, C.,** Some biological effects of prostaglandin endoperoxide analogs, *Life Sci.,* 18, 169—176, 1976.

19. **Smith, J. B., Ingerman, C. M., and Silver, M. J.,** Effects of arachidonic acid and some of its metabolites on platelets, in *Prostaglandins in Hematology,* Silver, M. J., Bryan Smith, J., and Kocsis, J. J., Eds., Spectrum, Jamaica, New York, 1977, 277—292.

20. **Corey, E. J., Vlattas, I., Andersen, N. H., and Harding, K.,** A new total synthesis of prostaglandins of the E_1 and F_1 series including 11-epiprostaglandins, *J. Am. Chem. Soc.,* 90, 3247—3248, 1968.

21. **Rosenthale, M. E.,Dervinis, A., Kassarich, J., Blumenthal, A., and Gluckman, M. I.,** Bronchodilating properties of the prostaglandin $F_{2\beta}$ in the guinea pig and cat, *Prostaglandins,* 3, 767—772, 1973.

22. **Van Dorp, D.,** Recent developments in the biosynthesis and the analyses of prostaglandins, *Ann. N.Y. Acad.Sci.,* 180, 181—199, 1971.

23. **Lin, C. H., Stein, S. J., and Pike, J. E.,** The synthesis of 5,6-acetylenic prostaglandins, *Prostaglandins,* 11, 377—380, 1976.

24. **Weeks, J. R., Ducharme, D. W., Magee, W. E., and Miller, W. L.,** The biological activity of the (15S)-15-methyl analogs of prostaglandins E_2 and $F_{2\alpha}$, *J. Pharmacol. Exp. Ther.,* 186, 67—74, 1973.

25. **Usardi, M. M., Franceschini, J., and Mizzotti, B.,** Structure-activity relationship of some acetylenic-PGE_2 derivatives, *Adv. Prostaglandin Thromboxane Res.,* 2, 948, 1976.

26. **Fried, J. and Lin, C. H.,** Synthesis and biological effects of 13-dehydro derivatives of natural prostaglandin $F_{2\alpha}$ and E_2 and their 15-epienantiomers, *J. Med. Chem.,* 16, 429—430, 1973.

27. **Horton, E. W.,** Prostaglandins: advances by analogy, *New Sci.,* 69, 9—12, 1976.

28. **Bundy, G. L., Yankee, E. W., Weeks, J. R., and Miller, W. L.,** The synthesis and biological activity of a series of 15-methyl prostaglandins, *Adv. Biosci.,* 9, 125—133, 1973.

29. **Bundy, G., Lincoln, F., Nelson, N., Pike, J., and Schneider, W.,** Novel prostaglandin synthesis, *Ann. N. Y. Acad. Sci.,* 180, 76—90, 1971.

30. **Yankee, E. W. and Bundy, G. L.,** (15S)-15-Methylprostaglandins, *J. Am. Chem. Soc.,* 94, 3651—3652, 1972.

31. **Karim, S. M. M., Sharma, S. D., Filshie, G. M., Salmon, J. A., and Ganesan, P. A.,** Termination of pregnancy with prostaglandin analogs, *Adv. Biosci.,* 9, 811—830, 1973.

32. **Quesney, L. F., Gloor, P., Wolfe, L. S., and Jozsef, S.,** Effect of $PGF_{2\alpha}$ and 15(S)-15-methyl-PGE_2 methyl ester on feline generalized penicillin epilepsy, *Adv. Prostaglandin Thromboxane Res.,* 1, 387—390, 1976.

33. **Robért, A. and Magerlein, B. J.,** 15-Methyl PGE_2 and 16,16-dimethyl PGE_2: potent inhibitors of gastric secretion, *Adv. Biosci.,* 9, 247—253, 1973.

34. **Karim, S. M. M., Carter, D. C., Bhana, D., and Ganesan, P. A.,** Effect of orally and intravenously administered prostaglandin 15(R)15-methyl E_2 on gastric secretion in man, *Adv. Biosci.,* 9, 255—264, 1973.

35. **Bydgeman, M., Beguin, F., Toppozada, M., Wiqvist, N., and Bergström, S.,** Intrauterine administration of 15(S)-15-methyl-prostaglandin $F_{2\alpha}$ for induction of abortion, *Lancet,* 1, 1336—1337, 1972.

36. **Wiqvist, N., Beguin, F., Bygdeman, M., and Toppozada, M.,** 15(S)-15-Methyl-prostaglandin $F_{2\alpha}$: myometrial response and abortifacient efficacy, *Adv. Biosci.,* 9, 831—842, 1973.

37. **Weir, E. K., Reeves, J. T., Droesmueller, W., and Grover, R. F.,** 15-Methylation augments the cardiovascular effects of prostaglandin $F_{2\alpha}$, *Prostaglandins,* 9, 369—376, 1975.

38. **Impicciatore, M., Bertaccini, G., and Usardi, M. M.,** Effect of a new synthetic prostaglandin on acid gastric secretion in different laboratory animals, *Adv. Prostaglandin Thromboxane Res.,* 2, 945, 1976.

39. **Kawasaki, A., Ishii, K., Ishii, M., and Tatsumi, M.,** Hypotensive and other pharmacological activities of prostaglandin analogue ONO-464, *Adv. Prostaglandin Thromboxane Res.*, 2, 916—917, 1976.

40. **Magerlein, B. J., Ducharme, D. W., Magee, W. E., Miller, W. L., Robért, A., Weeks, J. R.,** Synthesis and biological properties of 16-alkylprostaglandins, *Prostaglandins*, 4, 143—145, 1973.

41. **Karim, S. M. M., Carter, D. C., Bhana, D., and Ganesan, P. A.,** The effect of orally and intravenously administered prostaglandin 16,16 dimethyl E_2 methyl ester on human gastric acid secretion, *Prostaglandins*, 4, 71—83, 1973.

42. **Andersson, S. and Nylander, B.,** Local inhibitory action of 16,16-dimethyl PGE_2 on gastric acid secretion in the dog, *Adv. Prostaglandin Thromboxane Res.*, 2, 943, 1976.

43. **Karim, S. M. M., Sivasamboo, R., and Ratnam, S. S.,** Abortifacient action of orally administered 16,16-dimethyl prostaglandin E_2 and its methyl ester, *Prostaglandins*, 6, 349—354, 1974.

44. **Corey, E. J., Weinshenker, N. M., Shaaf, T. K., and Huber, W.,** Stereo-controlled synthesis of prostaglandins $F_{2\alpha}$ and E_2 (dL), *J. Am. Chem. Soc.*, 91, 5675—5677, 1969.

45. **Magerlein, B. J. and Miller, W. L.,** 16-Fluoroprostaglandins, *Prostaglandins*, 9, 527—530, 1975.

46. **Bruhn, M., Brown, C. H., Collins, P. W., and Palmer, J. R.,** Synthesis and properties of 16-hydroxy analogs of PGE_2, *Tetrahedron Lett.*, 24, 235—238, 1976.

47. **Ojima, M. and Fujita, K.,** Inhibition of platelet aggregation and white thrombus formation by prostaglandin analogue ONO-747, *Adv. Prostaglandin Thromboxane Res.*, 2, 781—785, 1976.

48. **Beerthuis, R. K., Nugteren, D. H., Pabon, H. J. J., Steenhoek, A., and Van Dorp, D. A.,** Synthesis of a series of polyunsaturated fatty acids. Their potencies as essential fatty acids and as precursors of prostaglandins, *Recl. Trav. Chim. Pays. Bas.*, 90, 943—960, 1971.

49. **Wakatsuka, H., Kori, S., and Hayashi, M.,** Letter to the editor. The synthesis of delta 2-prostaglandin, *Prostaglandins*, 8, 341—344, 1974.

50. **Nelson, N. A., Jackson, R. W., and Au, A. T.,** Synthesis and biological activity of 2-decarboxy-2-(tetrazol-5-yl) prostaglandins, *Prostaglandins*, 10, 303—306, 1975.

51. **Nelson, N. A., Jackson, R. W., Au, A. T., Wynalda, D. J., and Nishizawa, E. E.,** Synthesis of DL-4,5,6-trinor-3,7-inter-m-phenylene-3-OXA-prostaglandins including one which inhibits platelet aggregation, *Prostaglandins*, 10, 795—806, 1975.

52. **Nishizawa, E. E.,** Prostaglandin analogs in the treatment of thrombosis, in *Prostaglandins in Hematology*, Silver, M. J., Smith, J. B., and Kocsis, J. J., Eds., Spectrum, Jamaica, New York, 1977, 321—330.

53. **Fried, J., Lin, C., Mehra, M., Kao, W., and Dalven, P.,** Synthesis and biological activity of prostaglandins and prostaglandin antagonists, *Ann. N. Y. Acad. Sci.*, 180, 38—63, 1971.

54. **Fried, J. Mehra, M. M., and Kao, W. L.,** Synthesis of (+)- and (−)-7-oxaprostaglandin $F_{1\alpha}$ and their 15-epimers, *J. Am. Chem. Soc.*, 93, 5594—5595, 1971.

55. **Labhsetwar, A. P.,** New antifertility agent — an orally active prostaglandin, ICI 74,205, *Nature (London)*, 238, 400—401, 1972.

56. **Bowler, J., Crossley, N. S., and Dowell, R. I.,** The synthesis and biological activity of alkyloxy prostaglandin analogues, *Prostaglandins*, 9, 391—396, 1975.

57. **Magerlein, B. J., Bundy, G. L., Lincoln, F. H., and Youngdale, G. A.,** Synthesis of 17-phenyl-18,19,20-trinorprostaglandins. II. PG_2 series, *Prostaglandins*, 9, 5—8, 1975.

58. **Miller, W. L., Weeks, J. R., Lauderdale, J. W., and Kirton, K. T.,** Biological activities of 17-phenyl-18,19,20-trinorprostaglandins, *Prostaglandins*, 9, 9—18, 1975.

59. **Wiquist, N., Martin, J. N., Bygdeman, M., and Green, K.,** Prostaglandin analogues and uterotonic potency: a comparative study of seven compounds, *Prostaglandins*, 9, 255—269, 1975.

60. **Binder, D., Bowler, J. Brown, E. D., Crossley, N. S., Hutton, J. Senior, M., Slater, L., Wilkinson, P., and Wright, N. C. A.,** 16-Aryloxyprostaglandins: a new class of potent luteolytic agent, *Prostaglandins*, 6, 87—90, 1974.

61. **Dukes, M., Russell, W., and Walpole, A. L.,** Potent leuteolytic agents related to prostaglandin $F_{2\alpha}$, *Nature (London)*, 250, 330—331, 1974.

62. **Lapidus, M., Grant, N. H., Rosenthale, M. E., and Alburn, H. E.,** Prostaglandin Thiosemicarbazones, 1-ester and 1-carbinol Derivatives as Bronchodilators, U.S. Patent No. US-3504020.

63. **Arndt, H. C., Biddlecom, W. G., Peruzzotti, G. P., and Woessner, W. D.,** Prostaglandin analog synthesis using hexamethylphosphorous triamidealkenylcuprate (I), *Prostaglandins*, 7, 387—391, 1974.

64. **Bagli, J. and Bogri, T.,** Prostaglandins. V. Utility of the NEF reaction synthesis of prostanoic acids. A total synthesis (+ −)-11-deoxy-PGE_1, -PGE_2, and their C-15 epimers, *Tetrahedron Lett.*, 36, 3815—3817, 1972.

65. **Pike, J. E., Kupiecki, F. P., and Weeks, J. R.,** Biological activity of the prostaglandins and related analogs, in Bergström, S. and Samuelsson, B., Eds., *Prostaglandins. Proceedings of the Second Nobel Symposium, Stockholm, 1966*, Wiley Interscience, New York, 1967, 161—171.

66. **Greenberg, R. and Smorong, K.,** The bronchodilator activity of an 11-deoxyprostaglandin (AY-23,578), *Can. J. Physiol. Pharmacol.*, 53, 799—809, 1975.

67. **Greenberg, R.,** A comparison of the bronchodilator activity of an 11-deoxy prostaglandin (AY-23,578) with its 15-methyl analogue (Ay-24,559), *Adv. Prostaglandin Thromboxane Res.*, 2, 961, 1976.
68. **Guzman, A., Vera, M., and Crabbé, P.,** Synthesis of isomeric 11-hydroxy 11-methyl prostaglandins, *Prostaglandins*, 8, 85—91, 1974.
69. **Lippmann, W.,** Inhibition of prostaglandin E_2-induced cyclic AMP accumulation in the rat anterior pituitary by 11-deoxyprostaglandin E analogs (9-ketoprostynoic acids), *Prostaglandins*, 10, 479—491, 1975.
70. **Illiano, G. and Cuatrecasas, P.,** Modulation of adenylate cyclase activity in liver and fat cell membranes by insulin, *Science*, 175, 906—908, 1972.
71. **Johnson, M. and Ramwell, P. W.,** Prostaglandin modification of membrane-bound enzyme activity, *Adv. Biosci.*, 9, 205—212, 1973.
72. **Kuehl, F. A., Jr., Cirillo, V. J., Ham, E. A., and Humes, J. L.,** The regulatory role of the prostaglandins on the cyclic 3′,5′-AMP system, *Adv. Biosci.*, 9, 155—172, 1973.
73. **Fried, J., Santhanakrishnan, T. S., Himizu, J., Lin, C. H., Ford, S. H., Rubin, B., and Grigas, E. O.,** Prostaglandin antagonists: synthesis and smooth muscle activity, *Nature (London)*, 223, 208—210, 1969.
74. **Bennett, A. and Posner, J.,** Studies on prostaglandin antagonists, *Br. J. Pharmacol.*, 42, 584—594, 1971.
75. **Burke, G. and Sato, S.,** Effects of long-acting thyroid stimulator and prostaglandin antagonists on adenyl cyclase activity in isolated bovine thyroid cells, *Life Sci.*, 10, 969—981, 1971.
76. **Kowalski, K., Sato, S., and Burke, G.,** Thyrotropin- and prostaglandin E_2-responsive adenyl cyclase in thyroid plasma membranes, *Prostaglandins*, 2, 441—452, 1972.
77. **Sato, S., Szabo, M., Kowalski, K., and Burke, G.,** Role of prostaglandin in thyrotropin action on thyroid, *Endocrinology*, 90, 343—356, 1972.
78. **Ratner, A., Wilson, M. C., and Peake, G. T.,** Antagonism of prostaglandin-promoted pituitary cyclic AMP accumulation and growth hormone secretion in vitro by 7-oxa-13-prostnoic acid, *Prostaglandins*, 3, 413—419, 1973.
79. **Vale, W., Rivier, C., and Guillemin, R.,** A prostaglandin receptor in the mechanisms involved in the secretion of anterior pituitary hormones, *Fed. Proc.*, 30, 363ABS, 1971.
80. **Wenke, M., Cernohorsky, M., Cepelik, J., and Hynie, S.,** Prostaglandin-antagonists OPA and PPP non-specifically antagonizing stimulatory effects of different drugs on adenylate cyclase, *Adv. Prostaglandin Thromboxane Res.*, 2, 848, 1976.
81. **Ahren, K. and Perklev, T.,** Effects of PGE_1 and 7-oxa-13-prostynoic acid on the isolated prepubertal rat ovary, *Adv. Biosci.*, 9, 717—721, 1973.
82. **Bender, A. D., Berkoff, C. E., Groves, W. G., Sofranko, L. M., Wellman, G. R., Liu, J. H., Besosch, P. P., and Horodniak, J. W.,** Synthesis and biological properties of some novel heterocyclic homoprostanoids, *J. Med. Chem.*, 18, 1094—1098, 1975.
83. **Kloeze, J.,** Relationship between chemical structure and platelet-aggregation activity of prostaglandins, *Biochim. Biophys. Acta*, 187, 285—292, 1969.
84. **Swislocki, N. I., Tierney, J., and Ritterstein, S.,** Disruption of human erythrocytes with 7-oxa-13-prostynoic acid, *Prostaglandins*, 7, 401—410, 1974.
85. **Gordon, J. L. and MacIntyre, D. E.,** Stimulation of platelets by bis-enoic prostaglandins, *Br. J. Pharmacol.*, 58, 298P—299P, 1976.
86. **Gorman, R. R. and Miller, O. V.,** Specific prostaglandin E_1 and A_1 binding sites in rat adipocyte plasma membranes, *Biochim. Biophys. Acta*, 323, 560—572, 1973.
87. **Green, K., Samuelsson, B., and Magerlein, B. J.,** Decreased rate of metabolism induced by a shift of the double bond in prostaglandin $F_{2\alpha}$ from the delta-5 to the delta-4 position, *Eur. J. Biochem.*, 62, 527—537, 1976.
88. **Kuehl, F. A., Jr. and Humes, J. L.,** Direct evidence for a prostaglandin receptor and its application to prostaglandin measurements, *Proc. Natl. Acad. Sci. U.S.A.*, 69, 480—484, 1972.
89. **Smith, J. B., Ingerman, C. M., and Silver, M. J.,** Platelet prostaglandin production and its implications, *Adv. Prostaglandin Thromboxane Res.*, 2, 747—753, 1976.
90. **Bundy, G. L., Daniels, E. G., Lincoln, F. H., and Pike, J. E.,** Isolation of a new naturally occurring prostaglandin, 5-trans-PGA_2, synthesis of 5-trans-PGE_2 and 5-trans-$PGF_{2\alpha}$, *J. Am. Chem. Soc.*, 94, 2124, 1972.
91. **Van Dorp, D.,** Recent developments in the biosynthesis and the analysis of prostaglandins, *Ann. N. Y. Acad. Sci.*, 180, 181—199, 1972.

SYNTHESIS OF THE LEUKOTRIENES AND OTHER LIPOXYGENASE-DERIVED PRODUCTS

Joseph G. Atkinson and Joshua Rokach

INTRODUCTION

Following extensive work on the cyclooxygenase pathways of arachidonic acid (AA) metabolism,[1] which give rise to the prostaglandins, Hamberg and Samuelsson[2] then discovered the first example of a lipoxygenase-derived product, 12-HPETE,[3] in mammalian systems. Prior to this discovery, lipoxygenases had been found only in plant tissues, and the finding of Hamberg and Samuelsson was destined to open up a whole new and exciting area of AA metabolism[4] which is still under vigorous investigation. Probably the single most exciting development in this area was the discovery in 1979 by Murphy et al.[5] that the long-known[6] but very elusive "slow-reacting substance of anaphylaxis" (SRS-A)[7] was in fact a product of the mammalian lipoxygenase pathway of AA metabolism.

Because of their great biological importance, and the difficulty in isolating them in quantity from natural sources, a considerable chemical effort has been carried out, following Samuelsson's structure elucidation, to synthesize these new compounds arising from the lipoxygenase metabolism of AA. The major synthetic efforts have been carried out in the laboratories of Corey at Harvard University and in the laboratories of Merck Frosst Canada in Montreal, with a number of significant contributions from many other laboratories.

Two general remarks concerning the synthetic efforts in this area are worth making. One is that because of the great similarity in physical and chemical properties of impurities in many of the intermediates and final products, every research group in this area has found high pressure liquid chromatography (HPLC) to be virtually indispensable for the handling and purification of these compounds. Second, since none of the natural products in question has yet been obtained crystalline, no X-ray structural verification has been possible. Consequently, the classical structural verification by total synthesis and comparison of the natural with the synthetic materials has been far from an academic exercise, in contrast with the synthesis of most natural products over the past 15 years.

In this chapter, the synthetic effort to date (summer 1982) in this new area of important biochemical transmitters will be reviewed. Also reviewed is the known chemistry of the natural products, although apart from degradative work done during the original structure proof, relatively little chemistry of the compounds themselves has been carried out. For the most part, the chemistry has been left to speak for itself through extensive use of flow charts, and the discussion is generally limited to those aspects which are of particular novelty or interest.

In order to see at a glance the compounds which have been the object of synthetic efforts, the reader is referred to Charts 1 and 2. The most important and extensive series of products discovered to date are those derived from the lipoxygenase-initiated reaction at position 5 of AA, as illustrated in Chart 1. In Chart 2 are indicated the principal products derived from the lipoxygenase reaction at the remaining susceptible sites of AA. These charts do not indicate all the known lipoxygenase-derived products, but are limited to the major compounds, especially those which have been synthesized. It must be emphasized with respect to Chart 1 that the biochemical origin of the sulfones of LTC_4, LTD_4, and LTE_4 has not been elucidated and they are indicated as arising from the corresponding sulfides only in order to highlight their structural relationships.

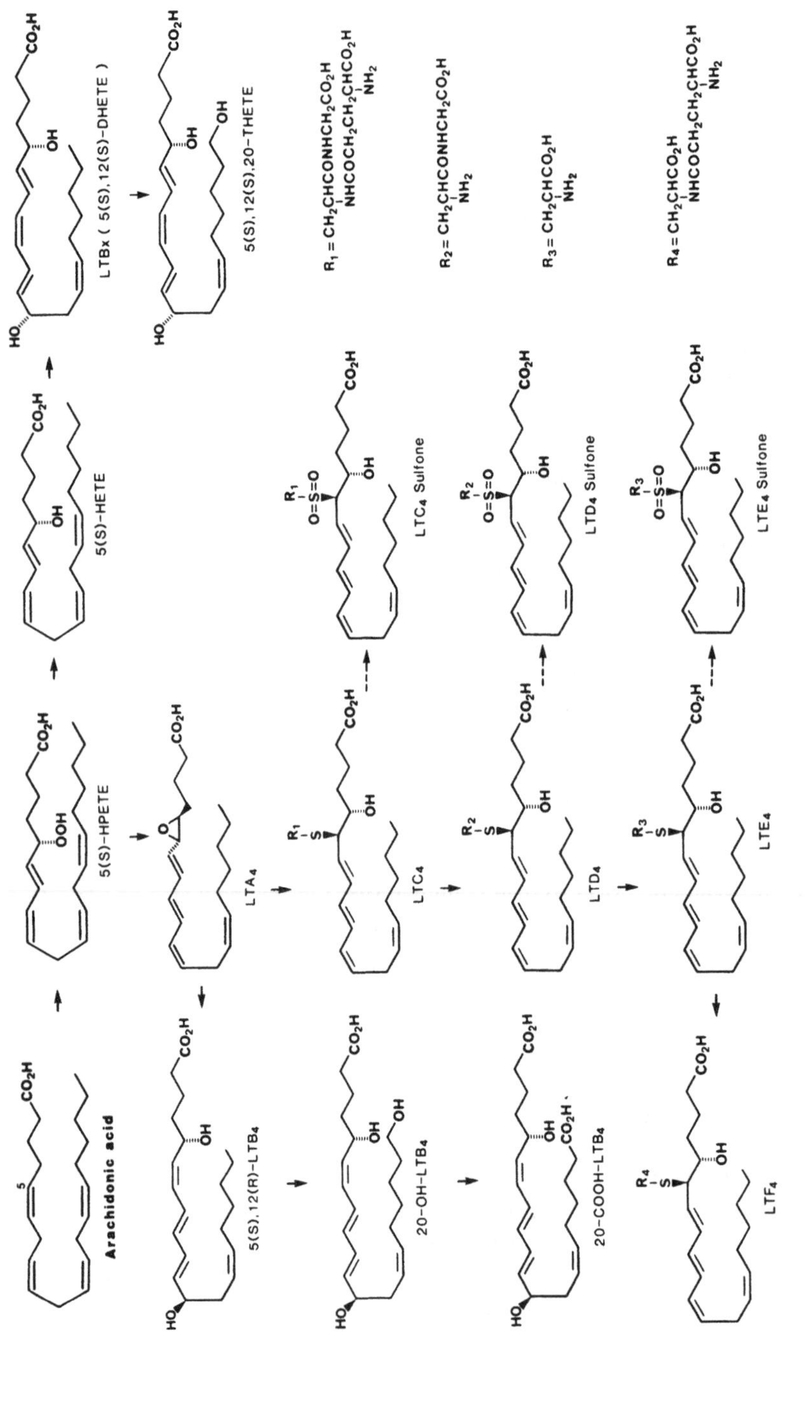

CHART 1. AA metabolism: principal products of the 5-lipoxygenase pathway.

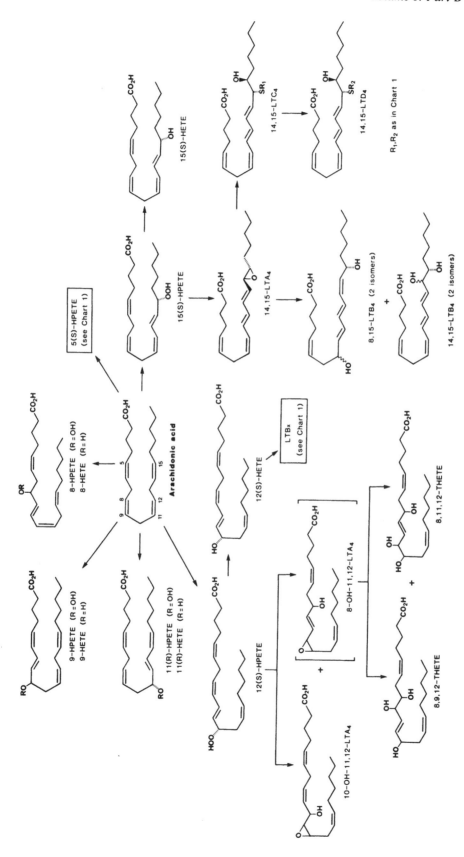

CHART 2. AA metabolism: principal products of the lipoxygenase pathways.

The chapter is divided by following the major structural classifications of the products of the lipoxygenase pathway, as indicated in the following:

SYNTHESIS OF THE MONOHYDROXY AND MONOHYDROPEROXY EICOSATETRAENOIC ACIDS (HETEs AND HPETEs)

The HPETEs and HETEs are the primary products of the action of lipoxygenase on AA, and from them, by further biochemical transformations, are derived all the remaining important biochemical transmitters shown in Charts 1 and 2. As pointed out in a recent review on arachidonate lipoxygenase,[8] all of the six theoretically possible HPETEs (or the derived HETEs) have now been found in mammalian systems. Of these, only the 8- and 9-HETE isomers have not been the object of synthetic efforts to date.

Synthesis of the 5-HETEs and 5-HPETEs

Synthesis of (±)-5-HETE and its Hydroperoxide, (±)-5-HPETE)

Two syntheses of (±)-5-HETE methyl ester have been reported starting from arachidonic acid **1** (Scheme 1) using electrophilic reagents to form δ-lactones, thus introducing the desired oxygen function directly into the 5-position. In the first synthesis[9] (path A), iodolactonization gave the lactone **2**, which, after elimination of HI gave the lactone **3** of (±)-5-HETE. Methanolysis then produced the methyl ester **4**, which was hydrolyzed to **5**, (±)-5-HETE. Mesylation of **4** followed by solvolysis with H$_2$O$_2$ and hydrolysis gave **6**, (±)-5-HPETE, which was reduced to **5** with NaBH$_4$. The second synthesis[10] (path B) employed selenolactonization to give the desired lactone **7**. After conversion to the methyl ester **8**, the latter was oxidized to its selenoxide which underwent elimination at room temperature to yield **4**, the methyl ester of (±)-5-HETE, in 48% yield along with 12% of the 8,9-E-isomer of **4**. The selenoxide elimination was studied extensively and it was found that strongly basic conditions were necessary to avoid extensive isomerization of the 8,9-Z-double bond to the E isomer.

A third synthesis of racemic 5-HETE has been reported recently by Rokach and co-workers[11] as outlined in Scheme 2. Its basis stems from several observations made during the course of their original synthesis of LTC$_4$ (to be discussed later). A key synthetic finding was that addition of diazocarbonyl compounds to furan to yield cyclopropafurans (e.g., **10**) is very efficiently catalyzed by rhodium (II) acetate, compared to the photochemical (low yield) or thermal (vigorous conditions) additions. A second realization was that the resulting cyclopropafurans undergo retrocyclic ring-opening under mild conditions to directly deliver conjugated E,Z-diene units (e.g., **11**). Since the E,Z-diene unit occurs in virtually all of the lipoxygenase-derived products (see Charts 1 and 2), this methodology provides a powerful tool in approaches to the synthesis of many of these compounds.

One key aspect of the sequence in Scheme 2 was the use of CeCl$_3$ to effect selective carbonyl reductions as pioneered by Gemal and Luche.[12] The CeCl$_3$ serves two purposes in the reduction of **11** to **12**, most importantly, that of allowing selective reduction of the ketone group over the aldehyde. It also exerts a buffering action so that the ester group in **12** and in the alcohol precursor of **15** is not hydrolyzed by water in the solvent. Furthermore,

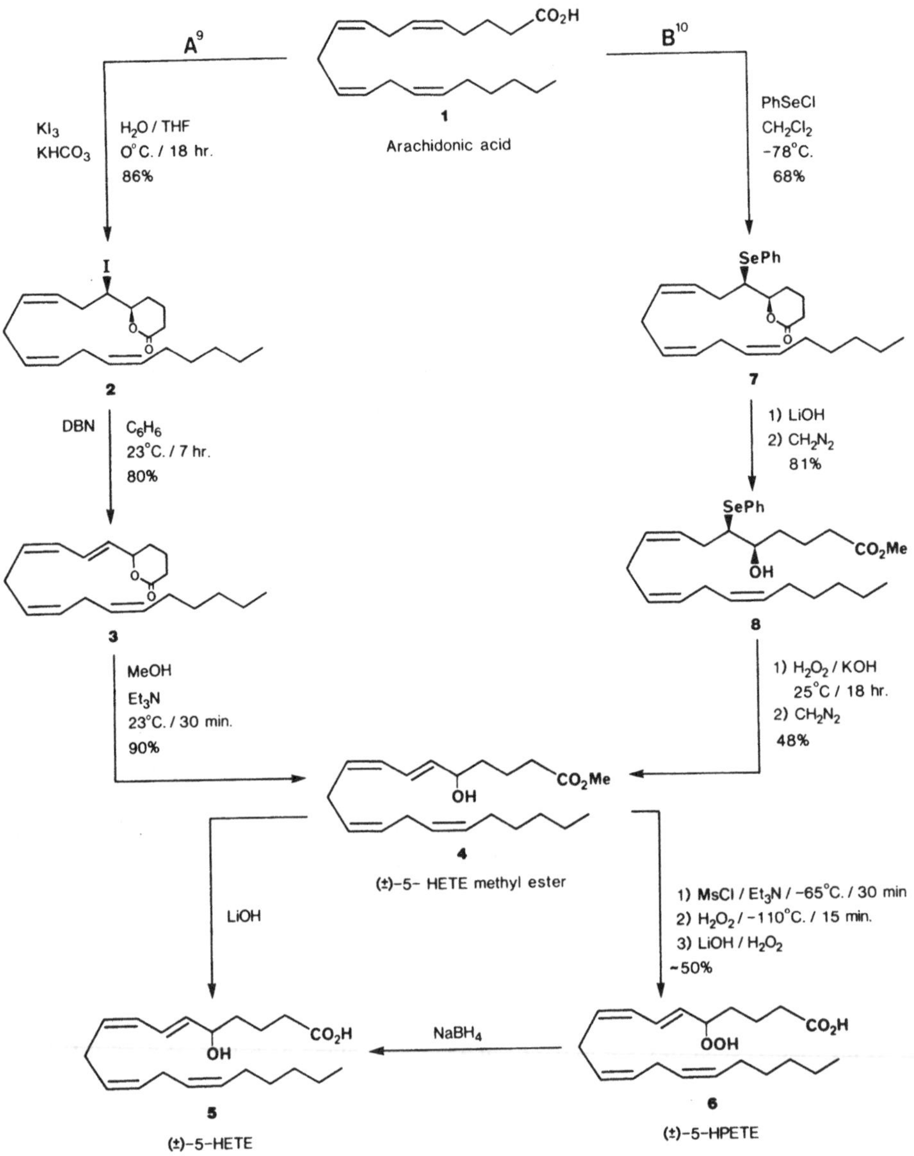

SCHEME 1. Synthesis of (±)-5-HETE and (±)-5-HPETE.

it probably also serves to suppress 1,4-reduction in these conjugated systems. Another point of some interest was the surprising stability of the hemiacetal **12**. Evaporation with benzene, benzene-acetic acid, chloroform, acetone, or water only slowly removed the methanol, but, unexpectedly, two or three evaporations with 10% water in acetone yielded the free aldehyde **13** in good yield. The preparation of the bromide **15** was first carried out using CBr$_4$-triphenylphosphine, but the yield was rather low (30 to 40%). It was then found that by using diphos in place of triphenylphosphine resulted in a much cleaner and higher yielding (75%) reaction. Whether the use of diphos in such reactions is of some general utility awaits further study.

Synthesis of 5(R)- and 5(S)-HETE and 5(R)- and 5(S)-HPETE

The naturally occurring isomer of 5-HETE has the (S) configuration at C$_5$. Corey and Hashimoto[13] have reported the chemical resolution of (±)-5-HETE methyl ester (**4**) as indicated in Scheme 3. After investigating a number of more well-known reagents without

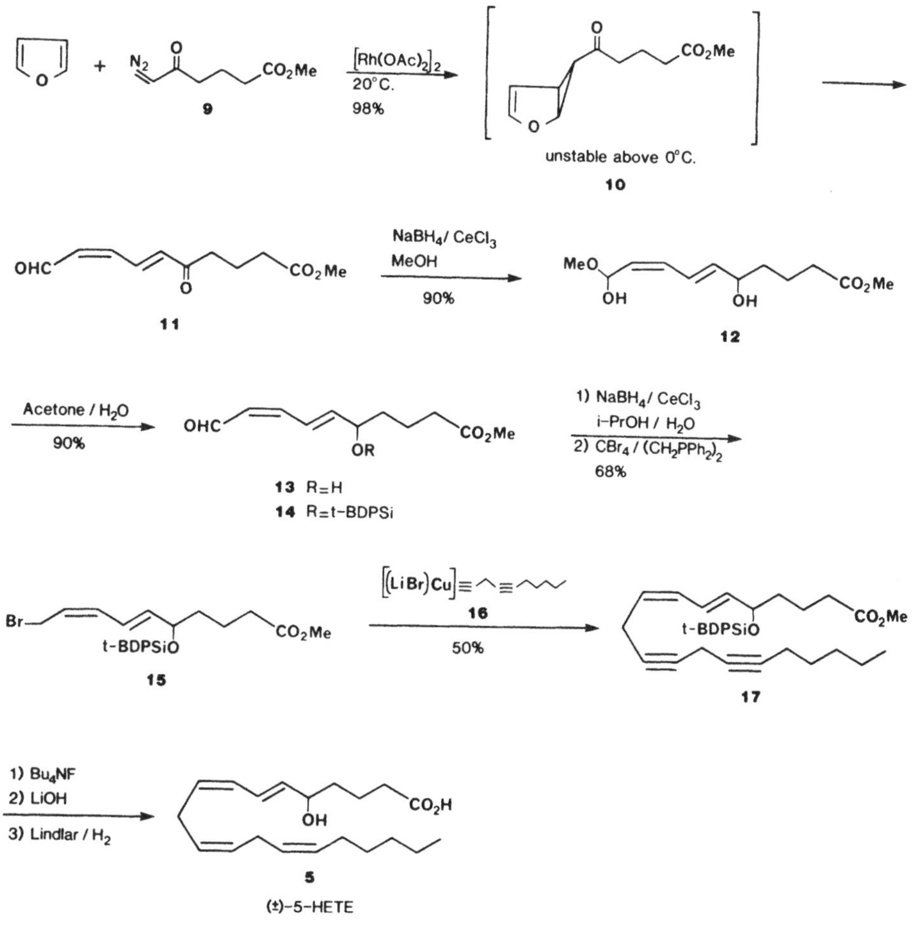

SCHEME 2. Synthesis of (±)-5-HETE.[11]

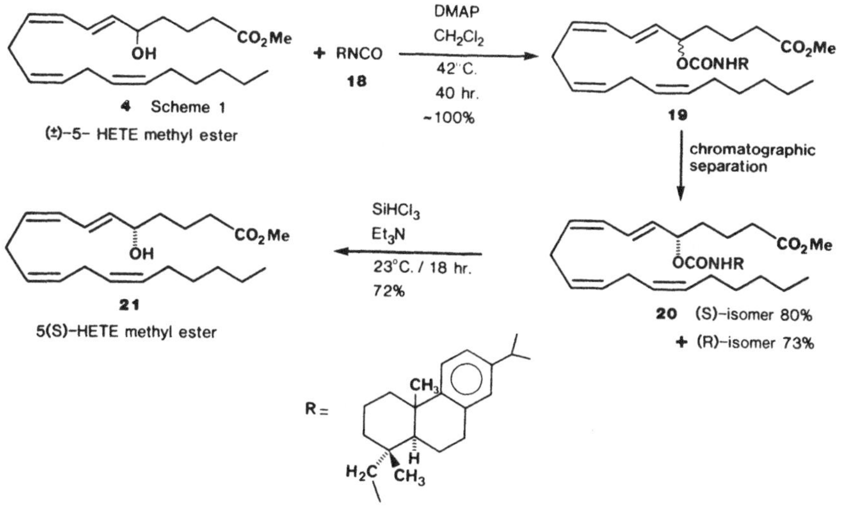

SCHEME 3. Resolution of 5-HETE methyl ester.[13]

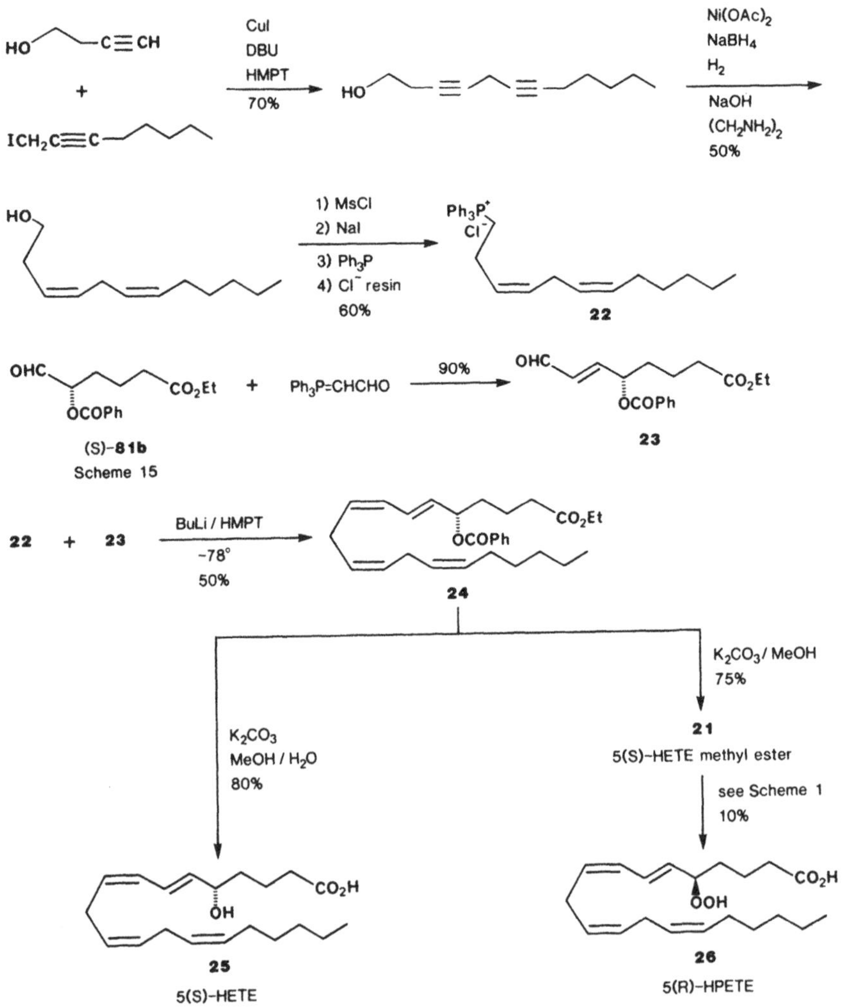

SCHEME 4. Synthesis of 5(S)-HETE and 5(R)-HPETE.[11]

success, it was found that the diastereomeric carbamates **19**, derived from the isocyanate **18** of dehydroabietylamine could be separated by column chromatography. A point of some note is the mild cleavage of the carbamate (**20 → 21**) using trichlorosilane-triethylamine,[14] a reaction which would otherwise be very difficult to carry out on such sensitive molecules.

More recently, Rokach and co-workers[11] have reported the first stereospecific total synthesis of both the natural 5(S) and unnatural 5(R) isomers of the 5-HETEs and the 5-HPETEs. The availability of the unnatural isomers of the HETEs is of considerable importance since, by conversion of the hydroxyl to a suitable leaving group and displacement by H_2O_2, the natural HPETEs can be obtained. Since the latter are the primary products of the lipoxygenase reaction, it is very useful to have them available in order to study this critical biochemical step. The key to these syntheses, outlined in Schemes 4 and 5, was the use of the two enantiomeric aldehydes, (S)-**81b** and (R)-**81b**, containing the asymmetric center which becomes C_5 in the final products. The detailed synthesis of these key intermediates, both obtained from 2-deoxy-D-ribose, will be presented fully in a later section.

Synthesis of the 11-, 12-, and 15-HETEs and HPETEs
Synthesis of (±)-11-,12-, and 15-HETE and (±)-15-HPETE

The syntheses of racemic 11-, 12-, and 15-HETE and 15-HPETE have been reported by

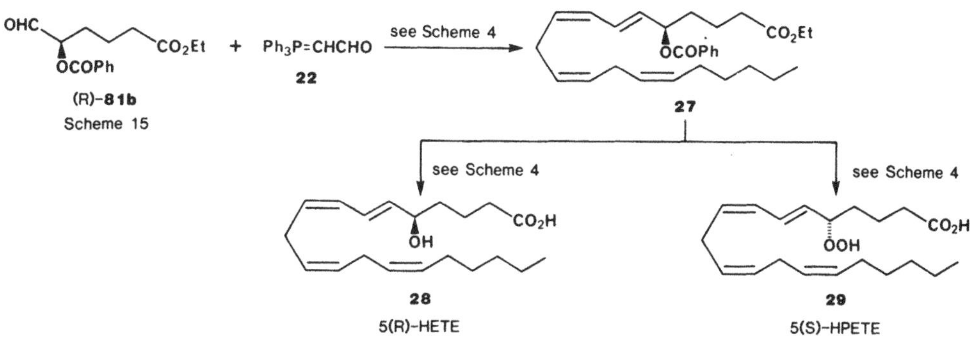

SCHEME 5. Synthesis of 5(R)-HETE and 5(S)-HPETE.[11]

Corey et al.[15] and are all based on the earlier remarkable finding[16] that the peroxide of AA (**30**, Scheme 6) undergoes spontaneous intramolecular epoxidation to give exclusively 14,15-epoxyarachidonic acid, **31**. As shown in Scheme 6, all four title compounds were then obtained from epoxide **31** by appropriate manipulations. The conversion of an allylic or a homoallylic epoxide to a diene alcohol by the magnesium amide of isopropylcyclohexylamine (MICA) was of considerable utility and was used to obtain (±)-15-HETE (**33**) directly from **31** and to obtain a mixture of (±)-11-HETE (**37**) and (±)-12-HETE (**38**) from epoxide **36**, and was used in an alternative synthesis of (±)-12-HETE to obtaine the precursor **39**. The stereospecific removal of the elements of hypobromonous acid from **35** "presented formidable obstacles" and was finally effected by reductive elimination of the bromo triflate with hexamethylphosphophorous triamide, which generated the 14,15-Z double bond in **36** with ≃95% stereospecificity. The conversion of (±)-15-HETE (**33**) to (±)-15-HPETE (**34**) was carried out by a sequence analogous to that used to prepare (±)-5-HPETE (**6**, Scheme 1).

Synthesis of (±)-11-HETE

Just and Luthe[17] have recently reported the synthesis of the racemic methyl ester of 11-HETE, **43**, starting from a derivative of glycidol (Scheme 7). Since optically active glycidol is available, they point out that their synthesis could also be applied to obtain the natural R isomer. They found that the ketal blocking group in **40** could be removed with pyridinium *p*-toluenesulfonate (PPTS) at 0°C without removing the acetal blocking group. Then, in order to obtain **42**, the acetal group was removed from **41** by treatment with the same reagent at 40°C, thus demonstrating a delicate but useful differentiation between these two types of alcohol protecting groups.

Synthesis of 11(R)- and 11(S)-HETE

A stereospecific total synthesis of 11(R)-HETE, carried out by Corey and Kang,[18] is outlined in Scheme 8. The optically active starting material is the acetonide **44** of D-glyceraldehyde. Of note is the condensation of lithium acetylide with epoxide **47**, which takes place predominantly at the nonallylic position, a phenomenon which depends on the presence of HMPT as a cosolvent. Another reaction of considerable potential interest is the nucleophilic displacement of the allenic bromide **51** by the reagent **50** (see **51**, arrows). Although allenic bromides are not a widely used species, the reaction, if general, would provide a novel synthesis of skipped diynes. The shorter and more obvious synthesis of **52** by coupling the acetylenic anion of **54** with **47** was frustrated by the finding that the CH_2 group between the two acetylenic bonds was of comparable acidity to the terminal acetylenic hydrogen. The unexpected acidity of a similarly located CH_2 group has also been found in the laboratories of Rokach and co-workers[11] during work with the anion of hydrocarbon **55**, from which a significant amount of allenic by-products were obtained.

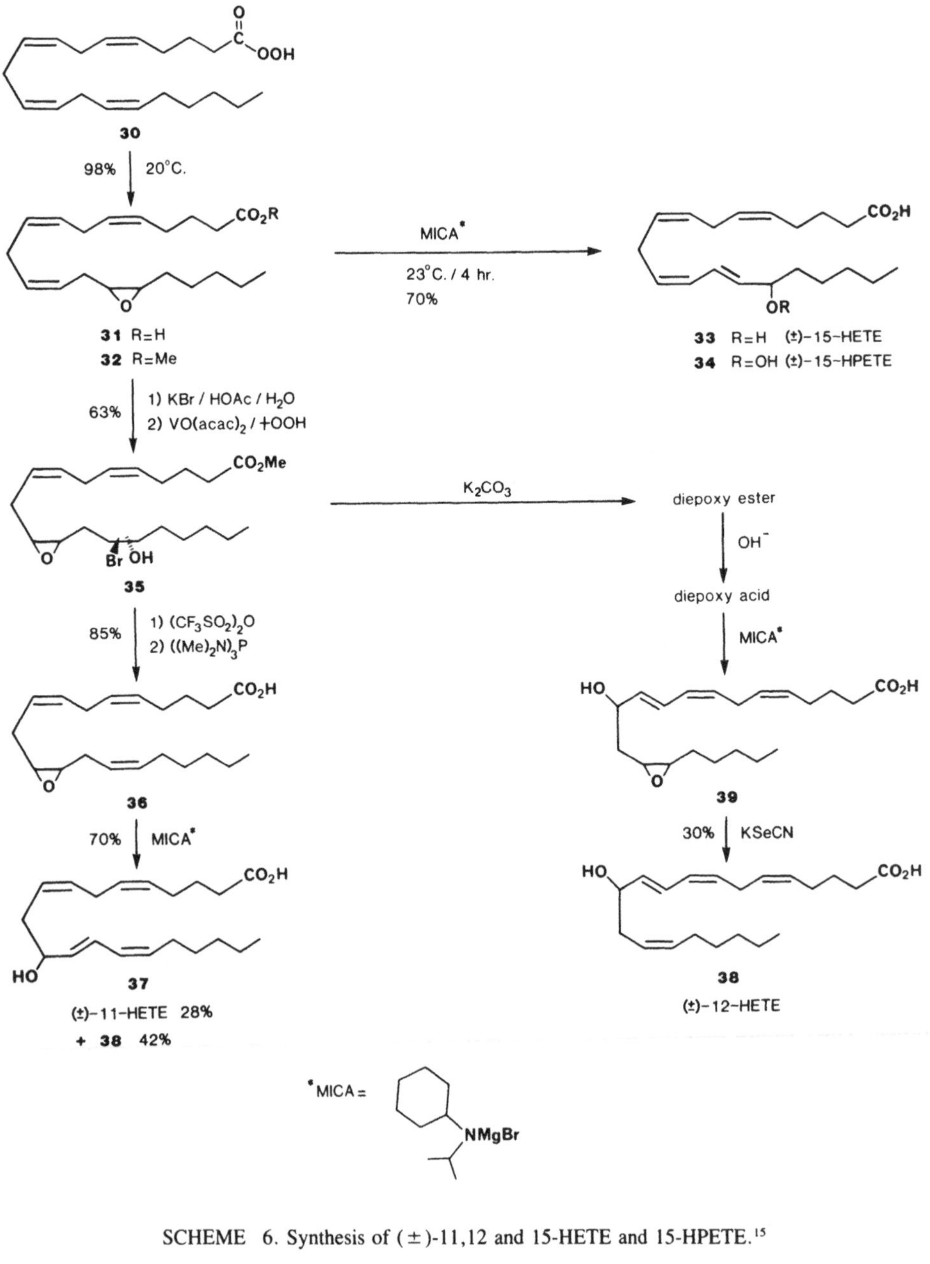

SCHEME   6. Synthesis of (±)-11,12 and 15-HETE and 15-HPETE.[15]

Just and his co-workers,[19] making good use of their experience with the diastereomeric oxazolidines obtained by the condensation of 1-ephedrine with α-acetoxyaldehydes, have accomplished a synthesis of both 11(R)- and 11(S)-HETE (Scheme 9). The overall synthetic approach was the same as that used in their synthesis of (±)-11-HETE[17] (Scheme 7) with two significant improvements on their published chemistry. By using acetic anhydride instead of *t*-butyldimethylsilyl chloride as an alcohol protecting group they were able to obtain the α-acetoxyaldehyde **57** directly instead of converting the silylated analog (**42**, Scheme 7) to **57**.[20] This was important because the silylated oxazolidines corresponding to **58** and **59** were not easily separable.[20] A second improvement was the oxidation of the primary alcohol precursor to **57** using dimethylsulfoxide-oxalyl chloride at low temperature in place of pyridinium chlorochromate (PCC), which raised the yield from 45 to 85%.[20] The diaster-

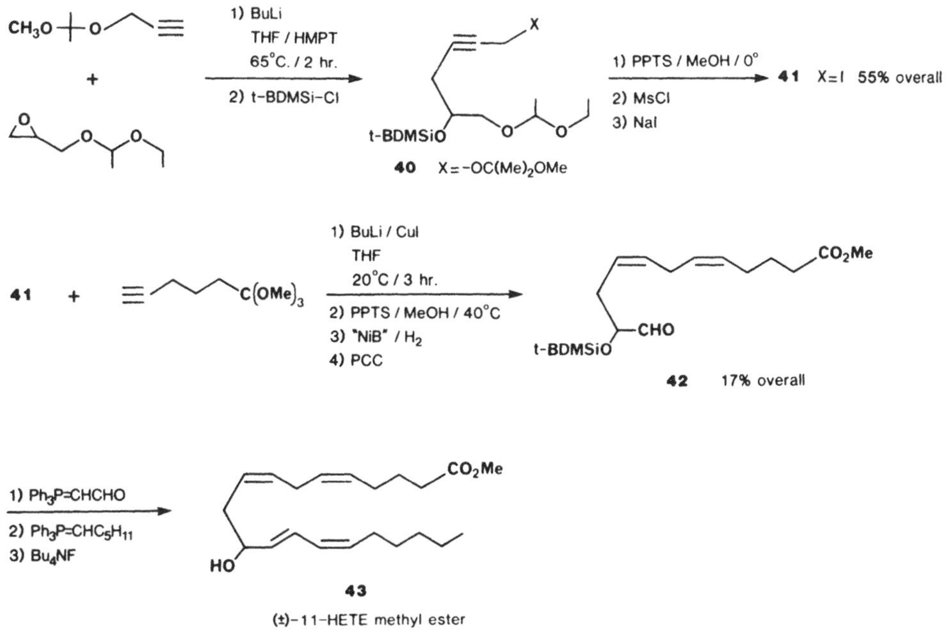

SCHEME 7. Synthesis of (±)-11-HETE.[17]

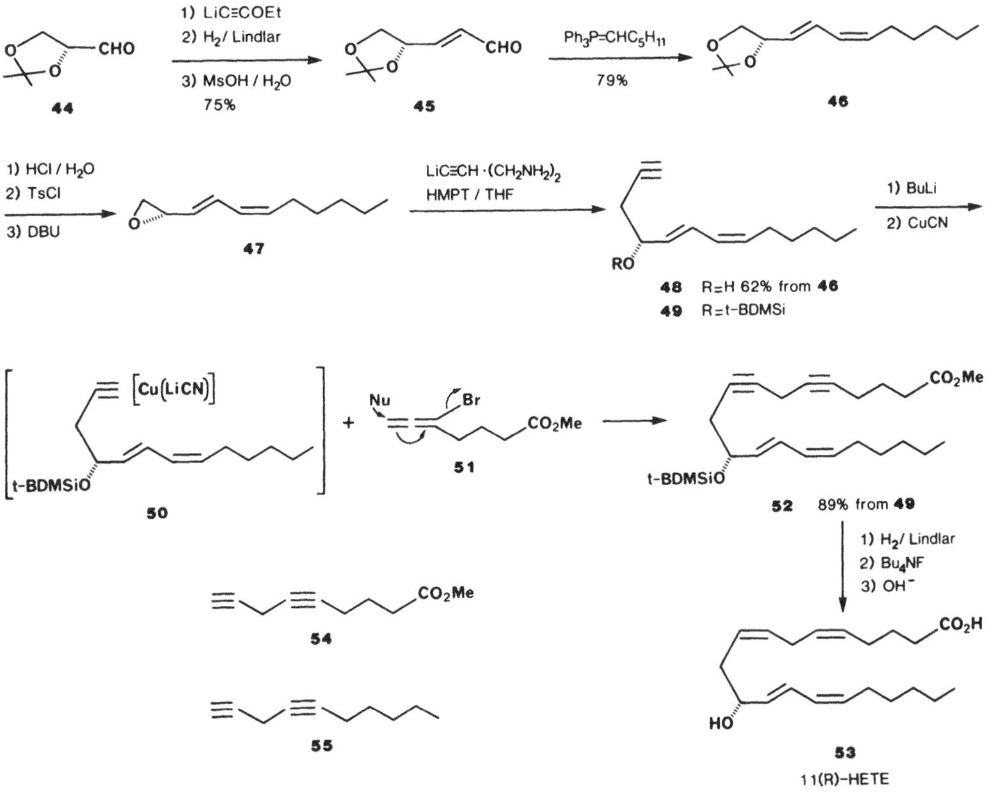

SCHEME 8. Synthesis of 11(R)-HETE.[18]

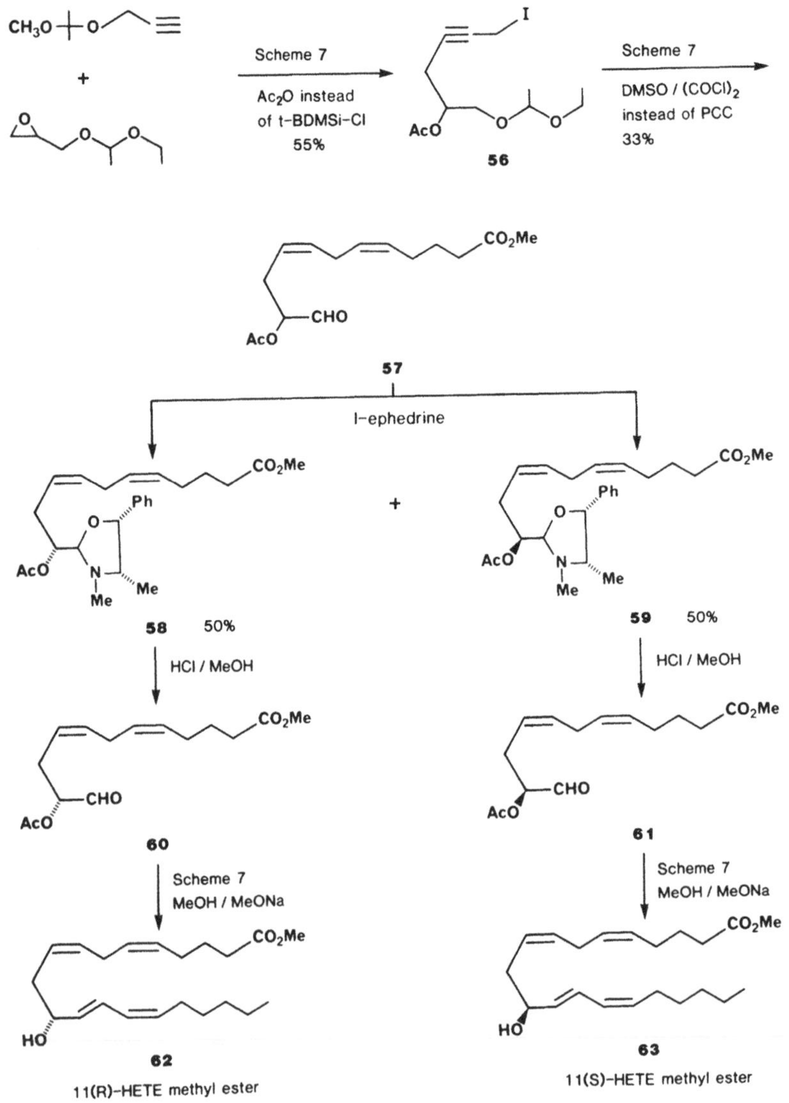

SCHEME 9. Synthesis of 11(R)- and 11(S)-HETE.[19]

eomeric oxazolidines **58** and **59** were then separable by flash chromatography and easily hydrolyzed to the desired resolved aldehydes **60** and **61**. The synthesis of the methyl esters of 11(R)-HETE (**62**) and 11(S)-HETE (**63**) was then completed essentially as outlined in Scheme 7.

Synthesis of 12(S)-HETE

As pointed out in the Introduction, 12(S)-HETE was the first product shown to arise from a lipoxygenase reaction of AA in mammalian systems.[2] Chronologically it was also the first lipoxygenase product to be synthesized. As outlined in Scheme 10, Corey et al.[21] carried out a stereospecific total synthesis from acetonide alcohol **64**, which was readily prepared from diethyl (S)-malate. The unsaturated aldehyde **70** was found to be stereochemically very unstable and had to be used immediately in the Wittig reaction with **66**. The Wittig reaction was a particularly critical step and the desired E-stereochemistry was obtained by taking advantage of earlier work from Corey's laboratory which had demonstrated that β-oxido

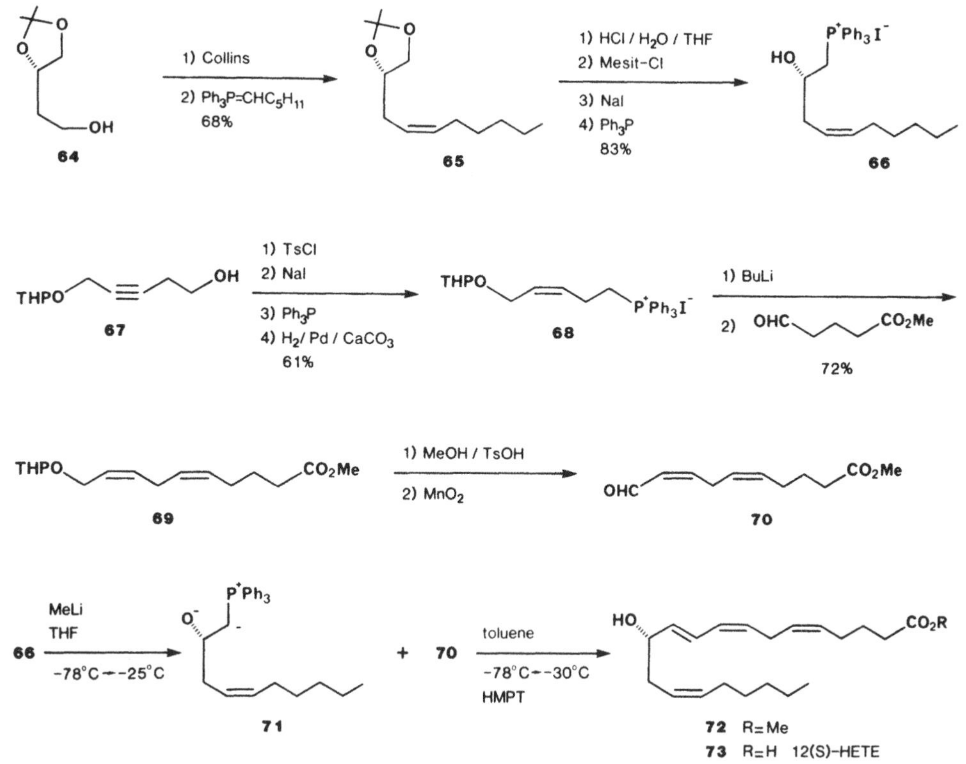

SCHEME 10. Synthesis of 12(S)-HETE.[21]

ylides such as **71** gave predominantly E-olefins, rather than Z-olefins as is usually observed with unstabilized phosphonium ylides.

Biochemical Synthesis of 15(S)-HETE and 15(S)-HPETE

No stereospecific synthesis of 15-HETE has yet been reported, in large part due to the fact that good biochemical syntheses make this last of the HETEs readily available. Although not strictly within the scope of the present review, it is worthwhile pointing out the excellent preparation optimized by Baldwin and his co-workers.[22] Using a commercially available soybean lipoxygenase, they describe conditions for preparing 500 mg of pure 15(S)-HETE (**75**, Scheme 11) in about 1 working day. Somewhat surprisingly, they found that by adding $NaBH_4$ directly to the reaction mixture, the yield was improved and the reaction was somewhat faster. Conditions are described for isolating either 15(S)-HETE itself (**75**) or its methyl ester (**76**).

Random Chemical Oxygenation of AA

When it became clear that the lipoxygenase pathway in mammalian systems was the source of previously unknown, but very important biochemical transmitters, the need was immediately felt to have the mono-HETEs and HPETEs readily available for comparison purposes. Consequently, the direct oxidation of AA under free radical conditions and with singlet oxygen was investigated as a possible direct route to these compounds. Both processes are of course nonstereospecific and the compounds were all obtained in their racemic form.

Free Radical Oxygenation of AA

Two laboratories have independently reported details of the free radical oxygenation of AA. Under free radical conditions the doubly allylic methylene groups at C_7, C_{10}, and C_{13}

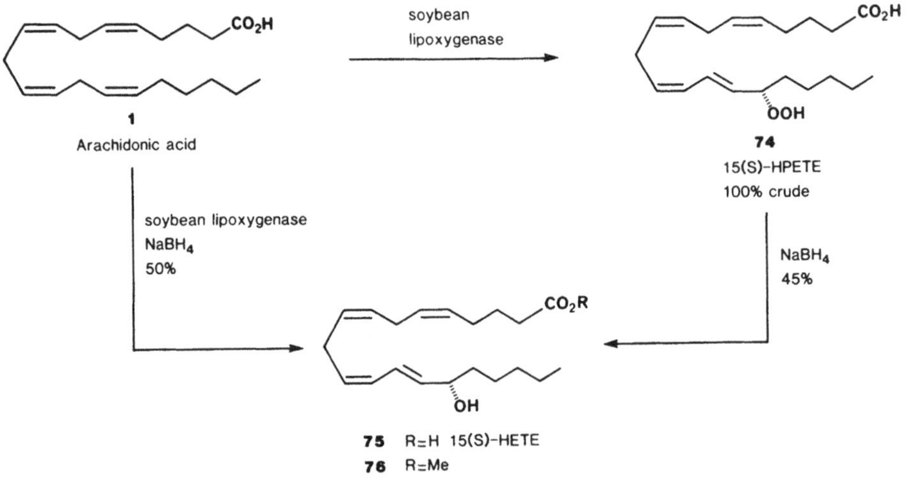

SCHEME 11. Biochemical synthesis of 15(S)-HETE and 15(S)-HPETE.[22]

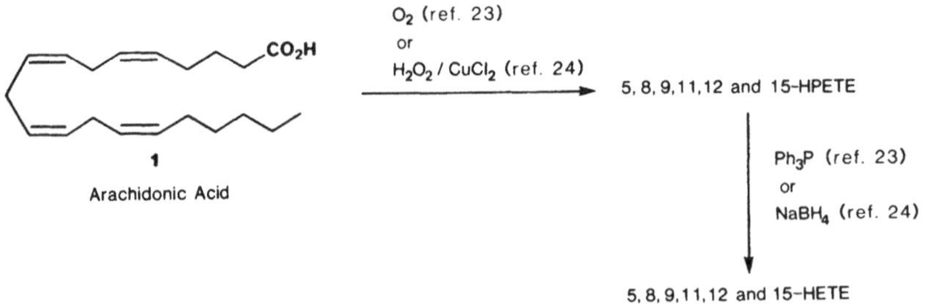

SCHEME 12. Free radical oxygenation of AA.

are much more susceptible to hydrogen abstraction that the monallylic methylenes at C_4 and C_{16}. As a result, only the six naturally occurring HPETEs are formed under these conditions. Porter et al.[23] exposed neat arachidonic **1** (Scheme 12) or its methyl ester to 0_2 at room temperature for 48 hr and subsequently isolated the HPETEs produced by HPLC. The HETEs were obtained by reduction of the peroxides with triphenylphosphine. Boeynams et al.[24] generated the mixture of HPETEs using $H_2O_2/CuCl_2$ in methanol and directly reduced them to the mixture of HETEs for HPLC analysis. Both groups proved the structure of the individual HETEs by mass spectral analysis of the trimethylsilylated methyl esters. These methods are limited in that only milligram quantities can be prepared easily, but they do provide a facile one-step preparation of all the naturally occurring HETEs and HPETEs which are useful for comparison purposes.

Singlet Oxygen Oxygenation of AA

Porter's group[25] also studied the products of singlet oxygen oxygenation of AA, **1** (Scheme 13). Singlet oxygen is nonselective in its site of attack on double bonds and as a result eight mono-HPETEs were obtained from AA. Both the HPETEs and the HETEs, obtained by reduction, were analyzed as their methyl esters. It is worth noting that a virtually identical study was carried out on the ethyl ester of AA 13 years earlier by workers in the Unilever Laboratories[26] (Scheme 13). These workers did not attempt to isolate the individual HPETEs or HETEs, but in a monumental piece of degradative work combined with gas chromatography (GC) and mass spectrometry (MS), came to conclusions identical with those of Porter and associates[25] with regard to the position of oxygenation in the eight HPETE ethyl esters obtained.

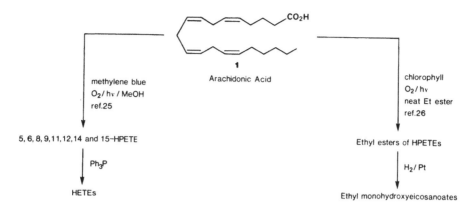

SCHEME 13. Singlet oxygen oxygenation of AA.

SYNTHESIS OF THE DIHYDROXY AND MORE HIGHLY OXYGENATED EICOSATETRAENOIC ACIDS (LTB₄, ETC.)

Of the more highly oxygenated products from AA, the compound which has attracted the greatest attention both biologically and synthetically is leukotriene B₄ (LTB₄, Chart 1). A number of other polyoxygenated derivatives are known in addition to those shown in Charts 1 and 2, but less is known about their biological roles and none of them have been the object of synthetic efforts.

Strategies of Synthetic Approaches to LTB₄

$$R = CH_3 \quad LTB_4$$
$$R = CH_2OH \quad 20\text{-}OH\text{-}LTB_4$$
$$R = CO_2H \quad 20\text{-}CO_2H\text{-}LTB_4$$

In approaching an analysis of the strategies of approach to molecules such as LTB₄, one can discern several key factors to be addressed in their synthesis. These general strategies will be discussed here, with the details being presented in the following sections. All syntheses reported to date have been convergent in nature, i.e., key pieces of the molecules have been built up independently and combined in two or three steps to form the full 20 carbon structure.

Chirality at C₅

The source of chirality at C₅ has had a common solution in all of the syntheses reported to date. Starting from 2-deoxy-D-ribose, it has been used to prepare the common synthon (S)-5-benzoyloxy-6-oxohexanoic acid as its methyl or ethyl ester (S-**81**, Schemes 14 and 15). Through this synthon, which represents carbons 1 to 6 in the target structure, C₃ of the parent sugar eventually becomes the chiral center at C₅ in LTB₄ and related compounds.

2-deoxy-D-ribose — — — → (S)-**81** ⇒ LTB₄ / 20-OH-LTB₄ / 20-CO₂H-LTB₄ / LTBx

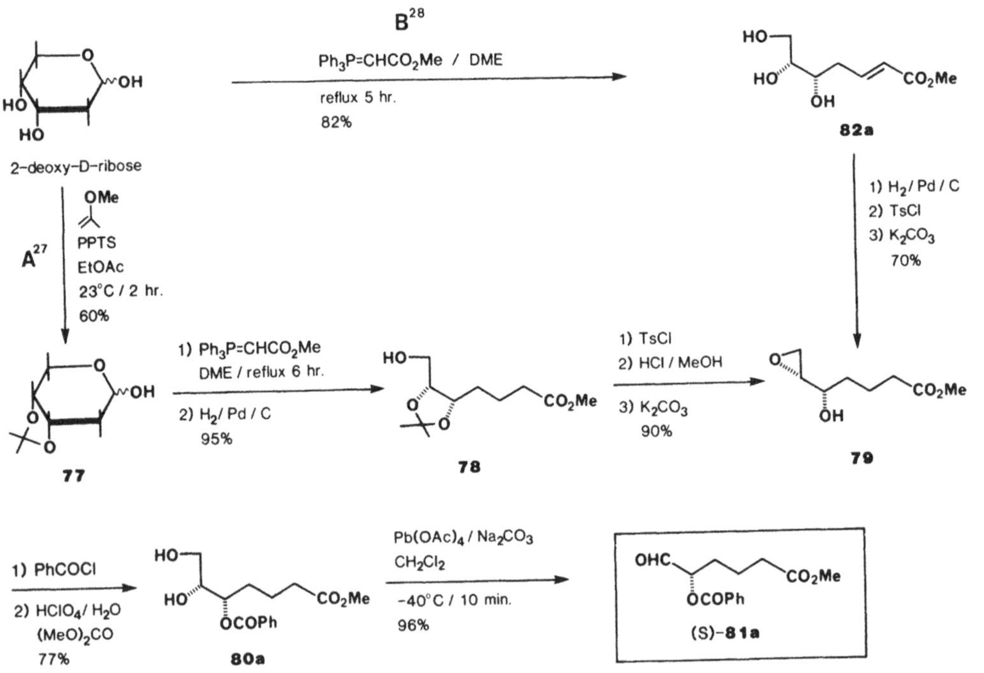

SCHEME 14. Synthesis of methyl (S)-5-benzoyloxy-6-oxohexanoate.[27,28]

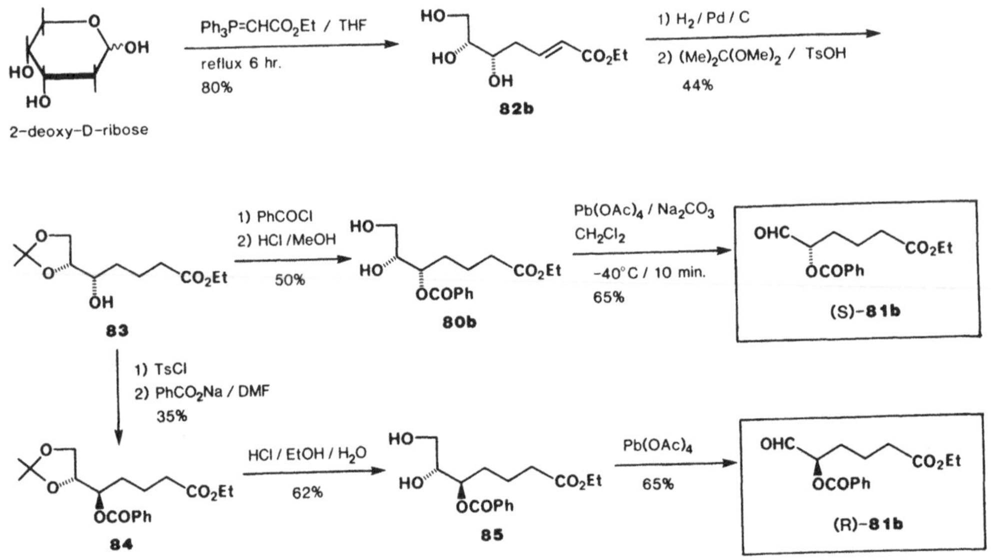

SCHEME 15. Synthesis of ethyl (S)- and (R)-5-benzoyloxy-6-oxohexanoate.[11,29]

Chirality at C_{12}

1.   Acyclic chiral precursors. This approach was employed by Corey in his first synthesis of LTB$_4$ in which C_3 of D-mannose became C_{12} via the hemiacetal **90** (Scheme 17), which reacts in its acylic hydroxy aldehyde form. In one of the Merck Frosst syntheses of LTB$_4$, C_3 of arabinose is incorporated into a key synthon **113** (Scheme 20), which becomes C_{11} to C_{14}, to provide the chirality at C_{12}.

2. A cyclic chiral precursor, the C-glycoside **102** (Scheme 19) was used advantageously in one of the Merck Frosst syntheses of LTB$_4$. In this case, the C-glycoside was derived from 2-deoxy-D-ribose, so that C$_4$ of the sugar became C$_{12}$ in LTB$_4$. Since the chirality at C$_5$ of LTB$_4$ came from C$_3$ of 2-deoxy-D-ribose, as described above, the two chiral centers of the sugar provided the two chiral centers of the target molecule, a particularly satisfying result.

3. Optical induction by an external agent. In a synthesis of LTB$_4$ and of LTBx, Corey made good use of the method of chiral epoxidation developed by Sharpless. Because of the flexibility of the method he was able to prepare synthons with the 12(R) chirality required for LTB$_4$ (Scheme 18) or the 12(S) chirality present in LTBx (Scheme 23).

Introduction of the Olefinic Linkages

Synthesis of the multiple double bond structure is a problem common to the synthesis of all the leukotrienes. The solutions to this problem in the syntheses of LTB$_4$ represent a good variety of approaches and among them are some particularly novel reactions.

1. Sterically promoted intramolecular elimination. In one of the most elegant reactions in the leukotriene field, the setting up of a suitably substituted Z-double bond permitted the intramolecular generation of the conjugated triene system as the last step in one of Corey's LTB$_4$ syntheses (**101a**, Scheme 18). The required Z-double bond had been introduced by a Wittig reaction in this case.

101a   **96**
                             LTB$_4$

2. Opening of a cyclic structure. The novel concept of utilizing an "activated" furan ring was used to advantage in one of the Merck Frosst group's syntheses of LTB$_4$, wherein the E,E-diene unit and the chiral center for C$_{12}$ were efficiently delivered in one step from a cyclic precursor (**105 → 107**, Scheme 19).

105   **106**   **107**

3. Stereocontrolled Wittig reactions. This workhorse of the leukotriene synthetic effort was used to introduce E-double bonds from stabilized ylids and Z-double bonds from unstabilized ylids in a variety of steps in the various syntheses of LTB$_4$. The latter always retain the drawback that it is almost never possible to obtain 100% Z stereoselectivity in contrast to obtaining the thermodynamically more stable E geometrical isomers from stabilized ylids.

4. Reduction of acetylenes. This long-known procedure for obtaining Z-double bonds has also been widely used in leukotriene syntheses, including LTB$_4$ and related compounds.

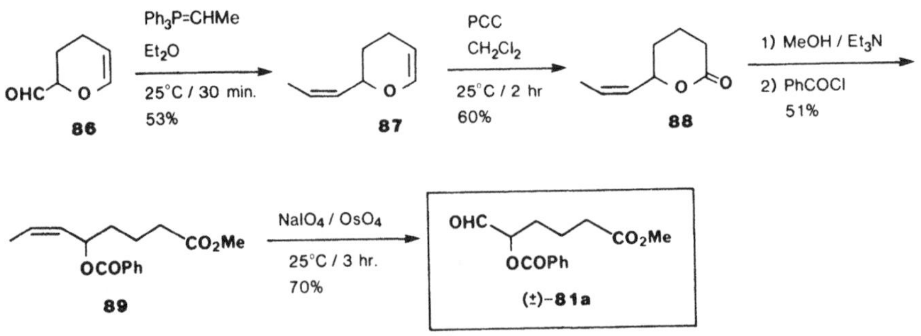

SCHEME 16. Synthesis of methyl (±)-5-benzoyloxy-6-oxohexanoate.[30]

Synthesis of Leukotriene B₄ (LTB₄)

Synthesis of (S)-, (R)-, and (±)-5-Benzoyloxy-6-oxohexanoic Acid Esters

The (S)-isomer of the title compound has been used as a synthon in all the reported syntheses of LTB₄ as well as the synthesis of LTBx and a synthesis of 5(R)- and 5(S)-HETE as discussed in an earlier section. Corey's group has described two syntheses of the methyl ester (S)-**81a**, both starting from 2-deoxy-D-ribose. In path A[27] (Scheme 14), 2-deoxy-D-ribose was first protected as the acetonide **77**, which was transformed to the epoxide **79**. The latter was also obtained in a second synthesis[28] (path B) which by-passed the initial protection of 2-deoxy-D-ribose as an acetonide.

Drawing on experience from their work on the synthesis of LTA₄ (to be described in a later section), the Merck Frosst group also developed a synthesis of the ethyl ester, (S)-**81b**, from 2-deoxy-D-ribose (Scheme 15).[29] The intermediate acetonide **83** served to protect the terminal diol unit and allow benzoylation to be effected on the desired hydroxyl group, thus serving the same protective purpose as the terminal epoxide in **79** (Scheme 14). In addition, the acetonide **83** served as a common intermediate to obtain the (R) isomer of **81b** used in the synthesis of 5(R)-HETE as shown previously in Scheme 5.[11]

Besides the above synthesis of the resolved isomers of **81**, an Italian group[30] has devised a novel synthesis of the racemic form of **81a** from the dimer of acrolein, **86** (Scheme 16). Worthy of note is the key conversion of the dihydropyran **87** to the δ-lactone **88** with PCC.

Synthesis of LTB₄

As mentioned earlier, all syntheses of LTB₄ reported to date have employed aldehyde esters **81a** (methyl ester) or **81b** (ethyl ester), derived from 2-deoxy-D-ribose, as the source of chirality at carbon 5. Two syntheses of LTB₄ have been reported from the Merck Frosst laboratories and two from the Harvard laboratory of Corey. In both cases, comparison of the synthetic material with natural LTB₄ established their identity.

The first synthesis of LTB₄, from Corey's laboratory, outlined in Scheme 17, starts from the derivative **90**, obtained from D-mannose.[27] In this synthesis, carbon 3 of D-mannose becomes carbon 12 in LTB₄. The triol tosylate resulting from deblocking of **91** underwent regioselective reaction with phenylchloroformate to give a 1,2-diol carbonate, allowing the remaining free alcohol to form epoxide **92** by internal displacement of the tosylate group. Reaction of the conjugated epoxydiene unit in **93** with one equivalent of HBr very cleanly gave a terminal bromo alcohol which was quaternized with triphenylphosphine to yield the Wittig salt **94**. Hydrolysis of **95** then yielded LTB₄, which was separated from an accompanying 15% of the 6E isomer by HPLC.

In a second synthesis of LTB₄,[28] Corey's group made use of the excellent method of chiral epoxidation developed by Sharpless and Katsuki[31] in order to obtain the optically active epoxy alcohol **98**[32] (Scheme 18). In this sequence, the Wittig reagent **100** seems to

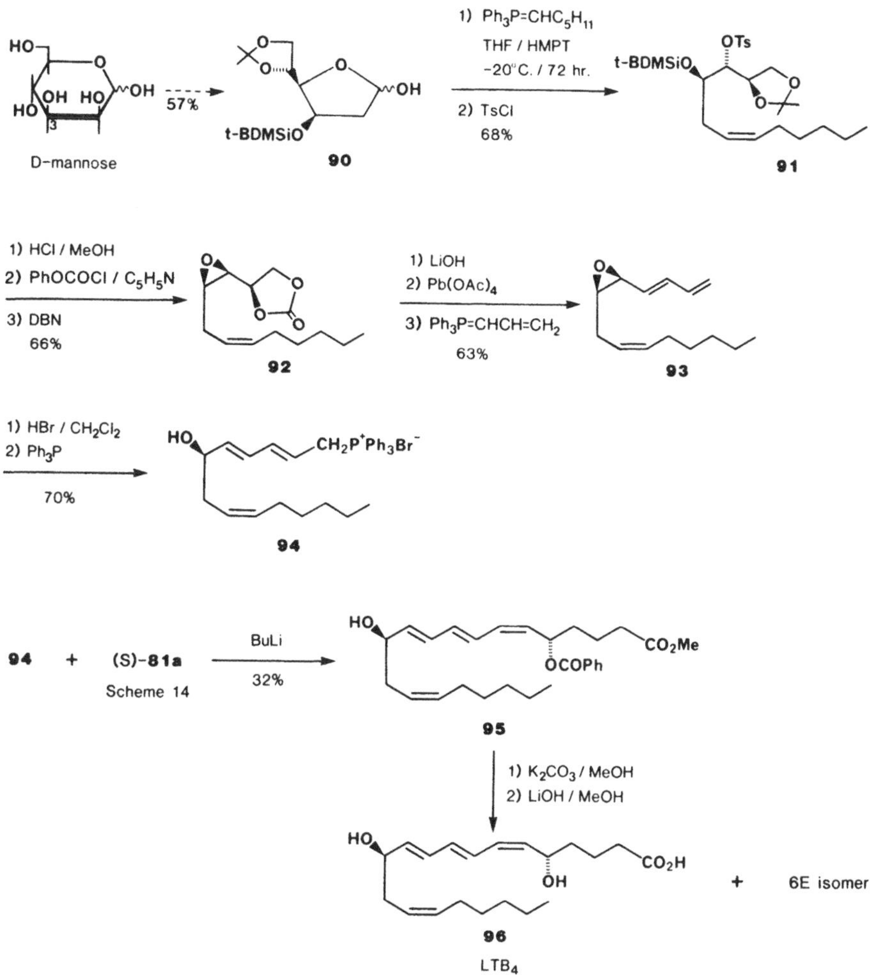

1) Ph$_3$P=CHC$_5$H$_{11}$
THF / HMPT
−20°C. / 72 hr.

2) TsCl
68%

57%

t-BDMSiO

D-mannose

90

t-BDMSiO

OTs

91

1) HCl / MeOH
2) PhOCOCl / C$_5$H$_5$N
3) DBN
66%

92

1) LiOH
2) Pb(OAc)$_4$
3) Ph$_3$P=CHCH=CH$_2$
63%

93

1) HBr / CH$_2$Cl$_2$
2) Ph$_3$P
70%

HO

CH$_2$P$^+$Ph$_3$Br$^-$

94

94 + (S)-**81a**
Scheme 14

BuLi
32%

HO

OCOPh

CO$_2$Me

95

1) K$_2$CO$_3$ / MeOH
2) LiOH / MeOH

HO

OH

CO$_2$H

+ 6E isomer

96
LTB$_4$

SCHEME 17. Synthesis of LTB$_4$.[27]

condense much more cleanly with (S)-**81a** than did **94** (Scheme 17). The key and very elegant step in the synthesis was the hydrolysis and rearrangement of the intermediate **101**, which undergoes a very facile intramolecular eliminative opening of the epoxide ring, as depicted in **101a**, to yield LTB$_4$ directly. In simple terms, phosphonium salts **94** and **100** are equivalent, except that in the latter, the opening of the epoxide ring has been deferred to a more propitious moment.

The first synthesis of LTB$_4$ from the Merck Frosst laboratories[29] is oulined in Scheme 19. One key aspect of this synthesis is the fact that 2-deoxy-D-ribose serves as the source of chirality for both asymmetric centers of LTB$_4$, with C$_3$ becoming C$_5$ in LTB$_4$ ((S)-**81b**, Scheme 15) and C$_4$ becoming C$_{12}$ (**107**, Scheme 19). The critical chemical step which made this synthesis possible was the novel finding that certain C-glycosides, containing a suitable leaving group in the tetrahydrofuran ring, unravel completely upon base treatment, giving rise to a conjugated diene unit in high yield, as depicted in the sequence **105** → **106** → **107**. The epoxide **106** is considered to be a true intermediate since in other cases it could be isolated.

The second synthesis developed at Merck Frosst for LTB$_4$[33] started from the known intermediate **111** (Scheme 20) which is easily prepared from L-arabinose, in which C$_3$ becomes C$_{12}$ in LTB$_4$. The synthesis of Scheme 20 converges with that of Scheme 19 with

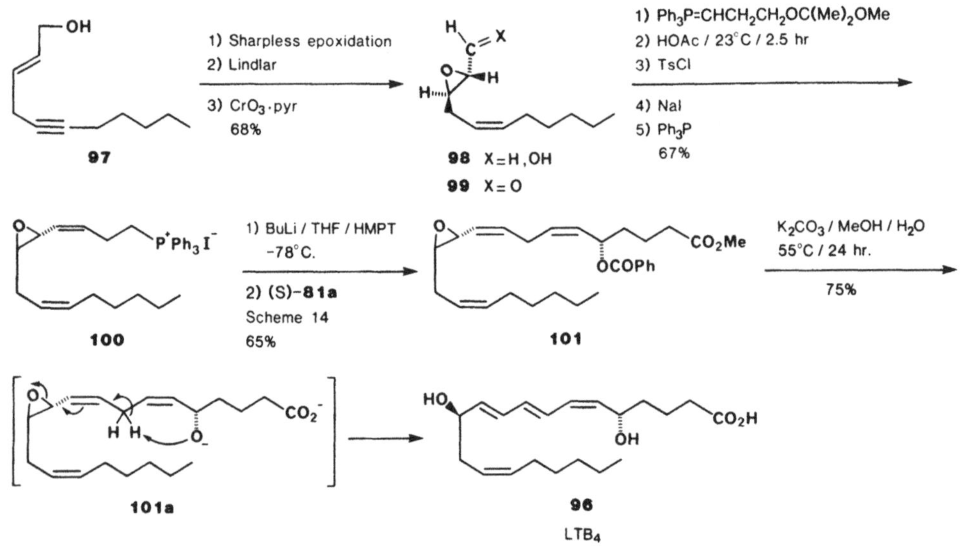

SCHEME 18. Synthesis of LTB$_4$.[28]

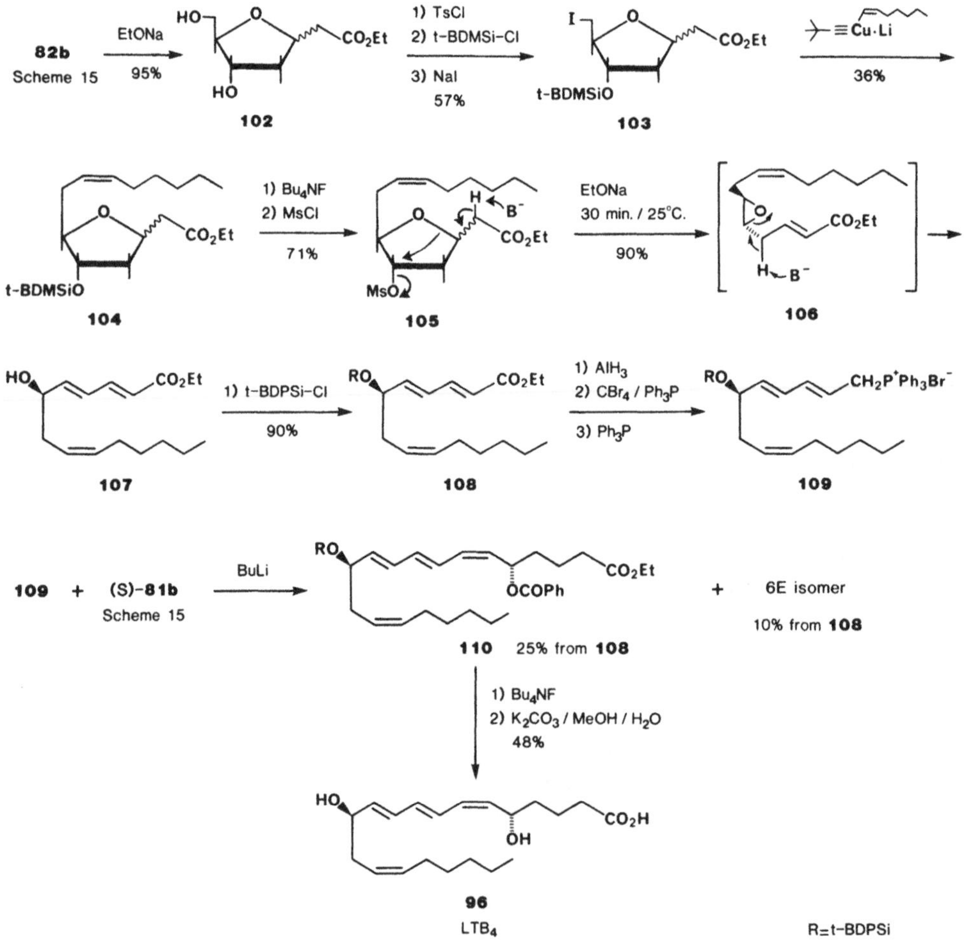

SCHEME 19. Synthesis of LTB$_4$.[29]

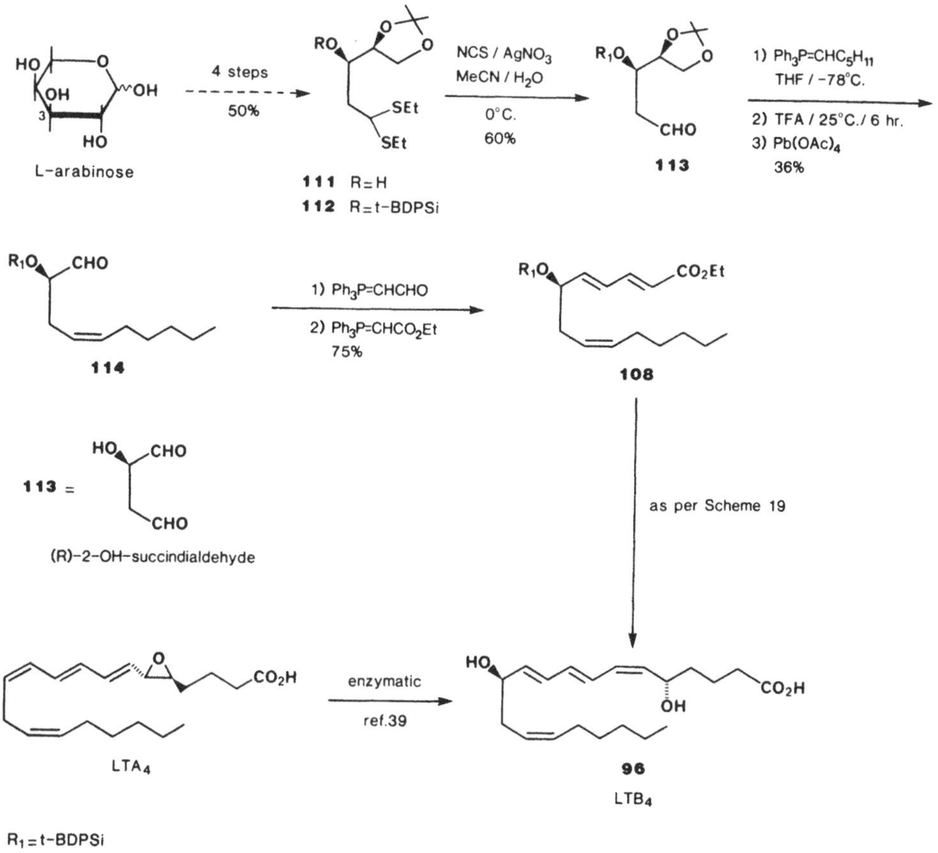

SCHEME 20. Synthesis of LTB$_4$.[33]

obtaining the protected ester **108.** It should be noted that the two-step chain elongation of **114** by four carbons to **108** was found to be superior to several one-step reactions investigated (two equivalents of Ph$_3$P=CHCHO, Ph$_3$P=CHCH=CHCHO, or (EtO)$_2$P(O)–CH$_2$CH=CH-CO$_2$C$_2$H$_5$). It can be seen that the Wittig reagent **109** (Scheme 19) obtained in both these syntheses is the protected equivalent of **94** (Scheme 17), and that for reasons which are not obvious, it appears to react more cleanly than the latter with aldehyde **81**. The particular advantages of this synthesis lie in the fact that starting from D-arabinose, the same sequence of reactions has allowed a synthesis of the 12(S) isomer of LTB$_4$;[33] and from synthon **113**, the terminally oxygenated compounds, 20-hydroxy-LTB$_4$ and 20-carboxy-LTB$_4$ have been prepared as described in later sections. Compound **113** thus serves as the synthetic equivalent of (R)-2-hydroxysuccindialdehyde in which the aldehyde groups are easily differentiated.

A clean biochemical preparation of LTB$_4$ from LTA$_4$ has also been developed by Maycock et al.[39] as indicated in Scheme 20.

Synthesis of 20-Hydroxy Leukotriene B$_4$ (20-OH-LTB$_4$)

As seen in Chart 1, LTB$_4$ is further metabolized by terminal oxidation, first to the primary alcohol, 20-OH-LTB$_4$, then to the dicarboxylic acid, 20-CO$_2$H-LTB$_4$. The terminal aldehyde is presumably an intermediate in the formation of the latter, but it has not yet been found in biological systems.

A synthesis of 20-OH-LTB$_4$ has now been carried out in the Merck Frosst laboratories,[34] using as a key intermediate the aldehyde **113** (Scheme 20), previously employed in one of the syntheses of LTB$_4$. The use of this aldehyde permits the ready introduction of various

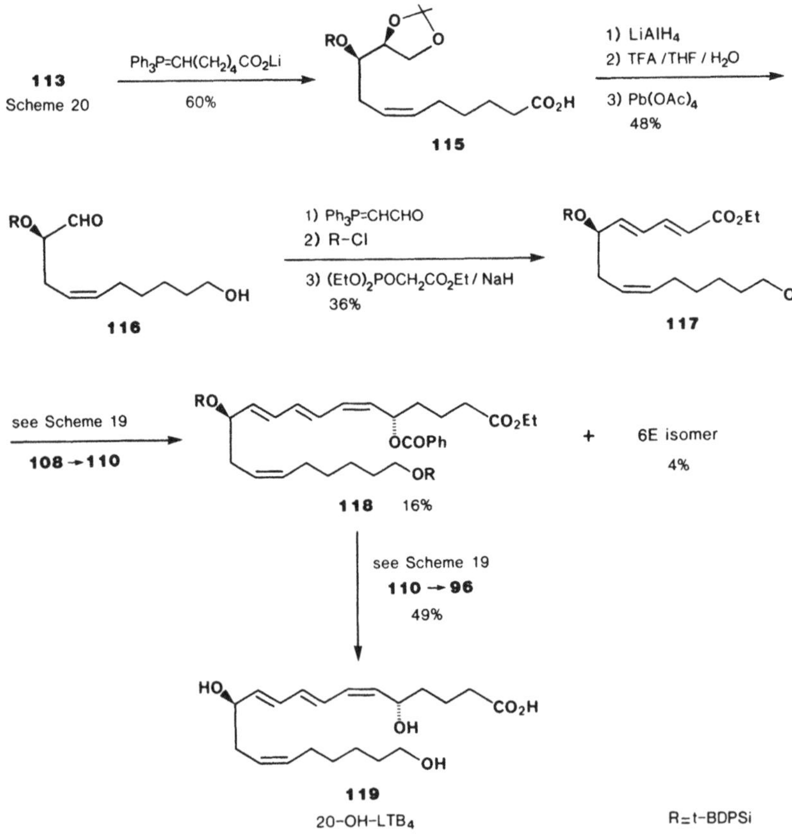

SCHEME 21. Synthesis of 20-hydroxy-LTB$_4$.[34]

functionalities into the terminal six carbons of eicosatetraenoic acid structures, and as outlined in Scheme 21, provided the basis for the synthesis of 20-OH-LTB$_4$. For the conversion of **116** to **117**, a two-step sequence for adding the four-carbon unit was again found advantageous, as in the case of homologation of the analogous aldehyde **114** (Scheme 20). The remainder of the synthesis through **118** to **119** (20-OH-LTB$_4$) was completed using methodology very similar to that shown in Scheme 19 for the synthesis of LTB$_4$.

Synthesis of 20-Carboxy Leukotriene B$_4$ (20-CO$_2$H-LTB$_4$)

Utilizing the acid **115** (Scheme 21) as the starting material, the Merck Frosst group has also realized a synthesis of the C$_{20}$ terminal acid, 20-CO$_2$H-LTB$_4$ (**123**, Scheme 22).[34] In this case the four-carbon homologation of **120** was carried out in one step, using formylmethylene triphenylphosphine, in spite of the modest yield obtained (25%) of **121**. This was done in order to differentiate the carbonyl groups in **121** and so preserve the desired terminal ester functionality. As described earlier (Scheme 2), advantage was again taken of the work of Gemal and Luche[12] to cleanly reduce the aldehyde group in **121** to the primary allylic alcohol. In a slight variation from related hydrolyses, it was found necessary to use LiOH instead of K$_2$CO$_3$ to convert **122** to **123** (20-CO$_2$H-LTB$_4$).

Synthesis of LTBx (5(S), 12(S)-DHETE)

A dihydroxyeicosatetraenoic acid, isomeric with LTB$_4$, was reported simultaneously by two groups[35,36] as being formed from AA in leukocytes. In contrast to LTB$_4$, it is formed by two successive lipoxygenase reactions, and the trivial names LTBx or 5(S), 12(S)-DHETE (from dihydroxyeicosatetraenoic acid) have been used as practical designations (Chart 1).

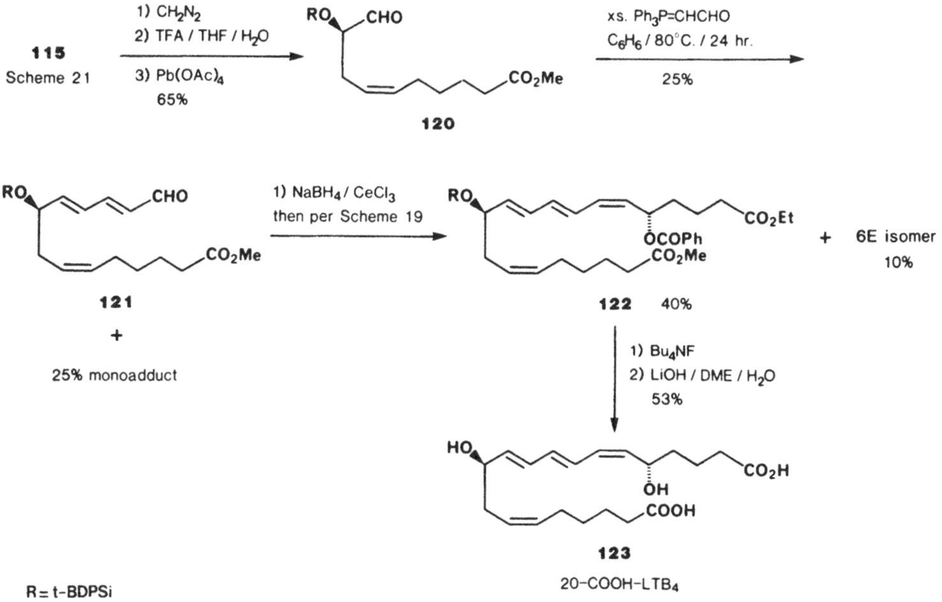

SCHEME 22. Synthesis of 20-carboxy-LTB$_4$.[34]

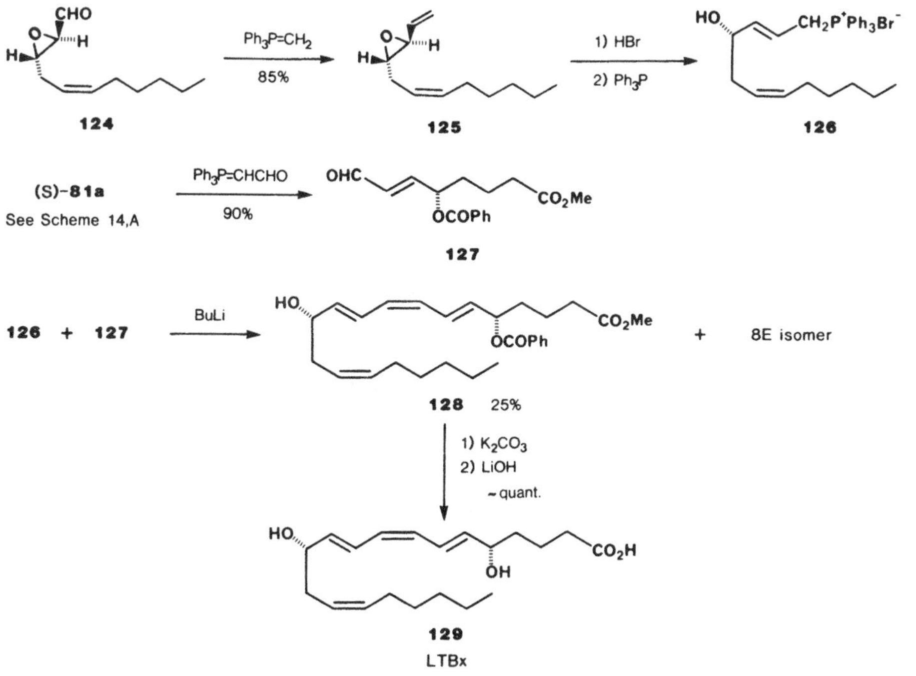

SCHEME 23. Synthesis of LTBx.[37]

A total synthesis of LTBx has been reported by Corey et al.[37] as outlined in Scheme 23. By using the Sharpless chiral oxidation,[31] the enantiomer of epoxy alcohol **98** (Scheme 18) was obtained, oxidation of which gave aldehyde **124**, the enantiomer of **99**. Epoxy olefin **125** was transformed to the desired Wittig salt **126** (see Scheme 17, **93** → **94**) which was condensed with the homologated aldehyde **127**, giving a modest yield of the desired 8Z

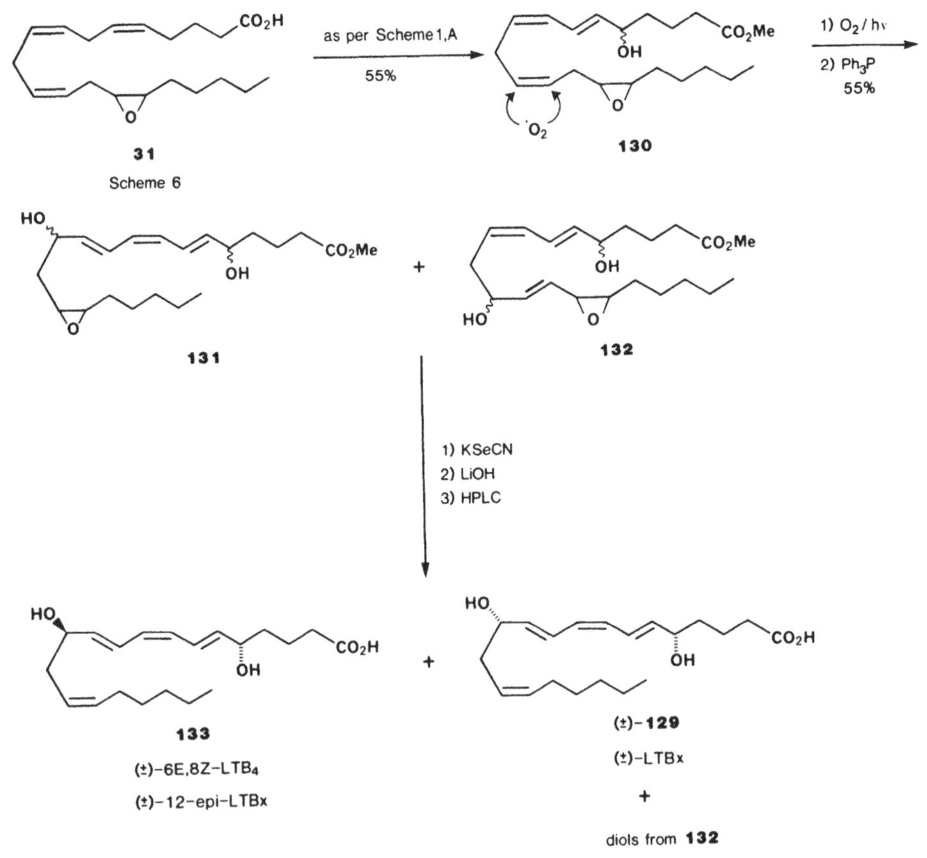

31

Scheme 6

130

131   +   **132**

1) KSeCN
2) LiOH
3) HPLC

133

(±)-6E,8Z-LTB₄

(±)-12-epi-LTBx

(±)-**129**

(±)-LTBx

+

diols from **132**

SCHEME 24. Synthesis of (±)-6E, 8Z-LTB₄ and (±)-LTBx.[32]

isomer, **128.** Again, the benzoyloxy aldehyde (S)-**81a** was the source of chirality for C_5 as in the syntheses of LTB₄ and its more highly oxygenated derivatives.

Synthesis of Isomers of LTB₄ and LTBx

Because of uncertainties remaining in the stereochemistry of the double bonds of LTB₄ after its structure determination, a number of double bond isomers were synthesized in order to provide definitive proof of structure. There was also considerable interest in synthesizing a number of closely related compounds in order to study the biological effects of controlled variation of the structure of these natural products.

Synthesis of (±)-6E, 8Z-LTB₄, (±)-5,12-LTBx, and 6E,10Z-LTB₄

The preparation of several isomers of LTB₄ as carried out by Corey et al.[32] is outlined in Schemes 24 and 25. Singlet oxygenation of the hydroxy epoxide **130** led to a mixture of dihydroxy epoxides containing the desired conjugated triene **131.** After reductive removal of the epoxide, the four resulting diols were separated and the conjugated trienes **129** and **133** were found not to be identical with LTB₄. Once the structure of LTBx (Scheme 23, **129**) was secured,[37] it was found that one of these was in fact racemic LTBx and the other was racemic 12-epi-LTBx (**133**).

The synthesis of another double bond isomer, 6E,10Z-LTB₄ (**139**) is outlined in Scheme 25.[32] The synthesis began with epoxy alcohol **98** (Scheme 18), the phenylcarbamate of which (**134**) underwent acid catalyzed rearrangement to a 1,2-diolcarbonate **135.** This was converted to the protected hydroxy aldehyde **136,** an alternative synthesis of which from arabinose has

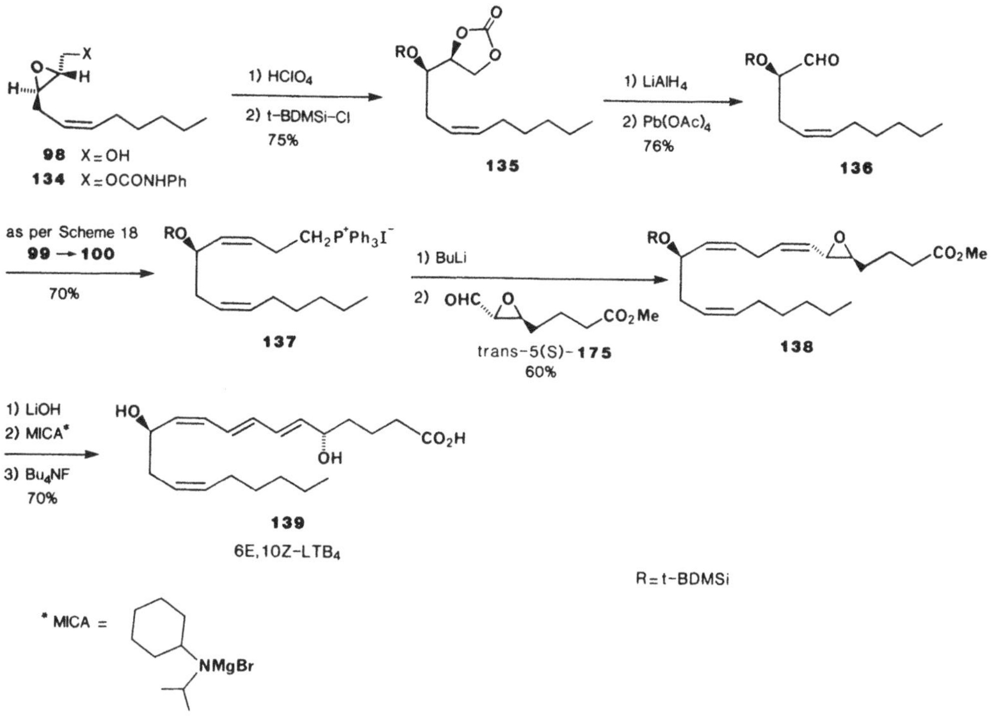

SCHEME 25. Synthesis of 6E,10Z-LTB₄.[32]

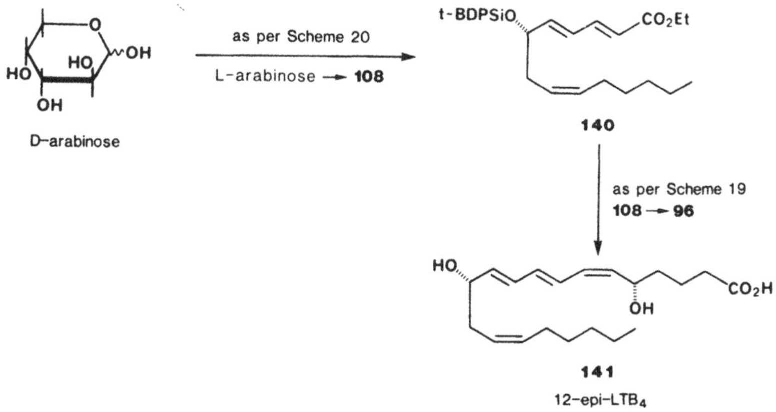

SCHEME 26. Synthesis of 12(S)-LTB₄ (12-epi-LTB₄).[33]

been outlined (Scheme 20, **114**). The epoxyaldehyde, trans-5(S)-**175,** used in the conversion of **137** to **138** has been widely used in syntheses of LTA_4 and will be discussed in detail in a later section. The MICA reagent (see Scheme 6) was again found useful in converting an epoxy olefin to a diene alcohol unit (**138 → 139**).

Synthesis of 12(S)-LTB₄ (12-epi-LTB₄)

As mentioned earlier in one of the Merck Frosst syntheses of LTB_4, by the simple expedient of using D-arabinose in place of L-arabinose as the starting material, it was a straightforward matter to prepare **140** (Scheme 26), which is the epimer of **108** (Scheme 20). This was then transformed to **141**, 12-epi-LTB₄ by a previously worked out sequence of reactions.[33]

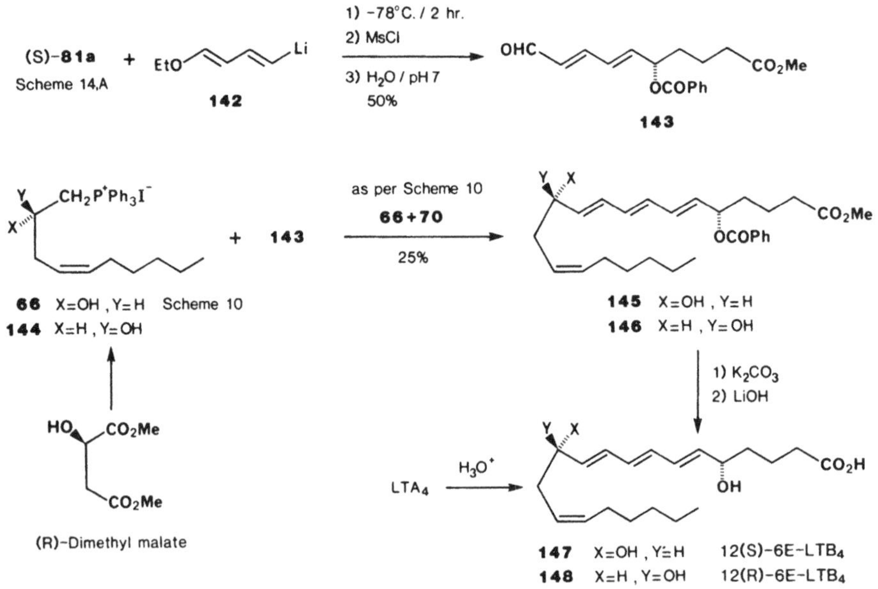

SCHEME 27. Synthesis of 12(S)- and 12(R)-6E-LTB$_4$.[38]

Synthesis of 12(R)- and 12(S)-6E-LTB$_4$)

The synthesis of these isomers of LTB$_4$ by Corey et al.[38] is outlined in Scheme 27. Intermediate **143**, derived from (S)-**81a**, was used as the source of C$_1$ to C$_{10}$ for both compounds. Wittig coupling of this with **66** (Scheme 10) gave **145** which was hydrolyzed to the 12(S) isomer **147**. The enantiomer (**144**) of **66**, which was required for the synthesis of the 12(R) isomer **148**, was prepared from the unnatural (R)-dimethyl malate. An efficient synthesis of the latter from (R) (R)-dimethyl tartrate is given by the authors. Isomer **148** had been previously obtained as a by-product in an earlier synthesis of LTB$_4$ (Scheme 17), and **147** and **148** were found to be identical with the pair of diastereomeric 5,12-diols obtained by acid-catalyzed epoxide ring opening of LTA$_4$.

SYNTHESIS OF LEUKOTRIENE A$_4$ (LTA$_4$) AND RELATED EPOXIDES

As can be seen from Chart 1, LTA$_4$ occupies a pivotal biochemical position, being the precursor of both the important chemotactic agent LTB$_4$ and of the potent contractile components of slow-reacting substance of anaphylaxis (SRS-A), LTC$_4$, LTD$_4$, and LTE$_4$. Because LTA$_4$ (as its methyl ester) has also been the widely used precursor of synthetic LTC$_4$, LTD$_4$, and LTE$_4$, a number of syntheses of it have been developed, including the racemic and optically active forms, *cis-* and *trans-*epoxy isomers, and a variety of double bond geometrical isomers.

Strategies of Synthetic Approaches to LTA$_4$

The various strategies for the synthesis of LTC$_4$, LTD$_4$, and LTE$_4$ all converge to the use of LTA$_4$ as their immediate synthetic precursor. A discussion of the strategies of approach to the synthesis of the components of SRS-A thus involves primarily a discussion of strategies for the synthesis of LTA$_4$ as its methyl ester **149**, which is then converted to the various components of SRS-A.

Three broad strategies have been employed in the various syntheses of LTA$_4$ as illustrated schematically in Chart 3. The *linear syntheses*, detailed in the section, ''Linear Synthesis of (±)-LTA$_4$'' all have started with the distal carbons of the molecule and introduce the

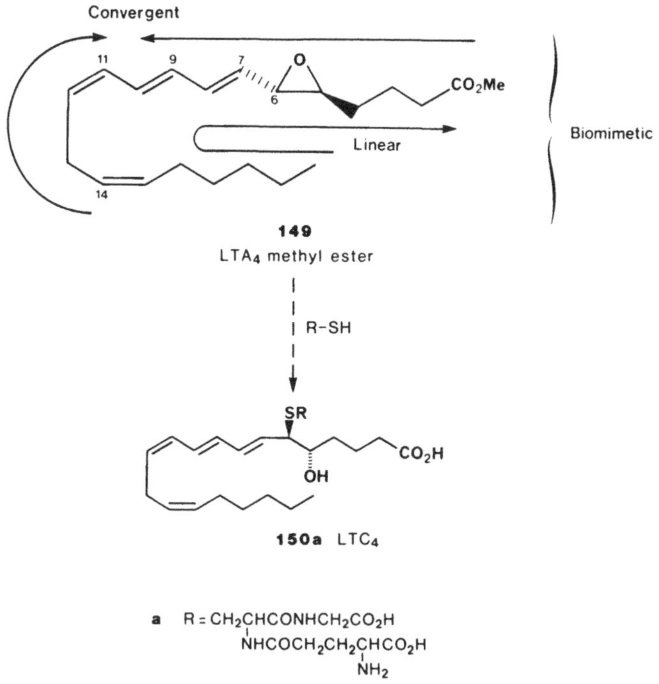

CHART 3. Strategies for the synthesis of LTA$_4$ methyl ester.

epoxide and methyl ester units at or near the end of the synthesis. The approaches used in the linear syntheses reported to date yield only racemic **149,** as a mixture of the separable *cis-* and *trans-*epoxides.

Convergent syntheses (section "Convergent Synthesis of LTA$_4$ and its Isomers") have been used for the majority of the reported approaches to LTA$_4$, principally because they allow the stereospecific synthesis of the four chiral forms of the compound. In all these cases a key chiral synthon was prepared which contained the epoxide unit in the desired optical form.

Finally, several *biomimetic syntheses,* outlined in the section "Biomimetic Synthesis of LTA$_4$ and 14,15-LTA$_4$", have been reported. The starting material is the purported biochemical precursor, 5-HPETE (section "Synthesis of the 5-HETEs and 5-HPETEs"), and thus contains all the carbons of the target molecule. This approach does not, however, lend itself readily to large-scale syntheses because of the instability of the peroxy intermediates.

With regard to the more specific strategies employed in the syntheses of LTA$_4$ methyl ester, they can be analyzed in terms similar to those used in the discussion concerning LTB$_4$.

Chiral Epoxide Synthesis

In this case there have been four basic approaches to preparing the appropriate chiral centers at C$_5$ and C$_6$ of LTA$_4$. The first two are based upon being able to discern, in an appropriate, readily available and optically active starting material, the necessary chirality which, by manipulation, would allow the synthesis of the desired chiral epoxide unit.

1. Acyclic chiral precursors. In the first synthesis of chiral LTA$_4$, Corey and co-workers employed D-ribose as their chiral starting material from which the key intermediate **208** (Scheme 37) was derived as the immediate precursor to the epoxide unit. Corey subsequently used D-mannose to obtain **213** (Scheme 39) from which the 6-epi isomer of LTA$_4$ was prepared.

The Merck Frosst group developed a comprehensive approach to the LTA$_4$ series starting from 2-deoxy-D-ribose. By appropriate manipulation of functionality, they prepared, from a single chiral starting material, **186**, the key intermediates **187**, **190**, **192**, and **193** (Scheme 35) as precursors to the four possible optical isomers of LTA$_4$ methyl ester.

2.   Cyclic chiral precursors. The C-glycoside **102** (Scheme 36), derived from 2-deoxy-D-ribose, was transformed by the Merck Frosst group into the intermediates **195**, **198**, and **201** (Scheme 36) from which three of the four possible optical isomers of LTA$_4$ were obtained. The cyclic structure in these syntheses provided the advantages of good control of stereochemistry and a novel and effective way of regioselectively masking one of the two secondary hydroxyl groups of the parent acyclic precursors.

3.   Optical resolution by an external agent. Sharpless has demonstrated the practical utility of his method of chiral epoxidation by two syntheses of the key epoxy alcohol trans-5(S)-**174** (Scheme 42) which has been widely used as a synthon in LTA$_4$ syntheses.

4.   Optical resolution by an internal agent. In this approach, utilized by Merck Frosst in syntheses of the four epoxy aldehydes **175** (Schemes 40 and 41), an optically active center in a precursor (**217**) gave rise to a separable pair of diastereoisomers during an epoxidation step. The original optical center was then removed to yield the desired optically pure epoxy aldehydes.

Introduction of the Olefinic Linkages

1.   Opening of a cyclic structure. This approach permitted the Merck Frosst group to generate either E,E or E,Z butadiene units by the retrocyclic ring opening of cyclo-propafurans such as **10** (Scheme 2) and **161** (Scheme 30) and has proved to be a useful and reliable approach to valuable synthons.

2. Stereocontrolled Wittig reactions. The Wittig reaction has again been widely used in the synthesis of LTA$_4$ and related structures, and has been applied to obtain E or Z double bonds as the case required.

3. Reduction of acetylenes. Controlled catalytic hydrogenation of acetylenes has also been applied in several syntheses of LTA$_4$. Thus, the widely used synthon **156** (Scheme 30) is readily available from an inexpensive acetylenic starting material, and catalytic hydrogenation was used to introduce two of the four double bonds in LTA$_4$ as illustrated in Schemes 32 and 33.

4. Vinyl organometallic compounds. Corey has made extensive use of the Wollenberg reagent, 1-lithio-4-ethoxybutadiene (**142**) in the LTB$_4$ field (Scheme 27) and in syntheses of LTA$_4$ (Schemes 33, 37, 39) to introduce the (E,E) conjugated diene aldehyde unit. He has also used a Gilman reagent to generate the Z double bond in the precursor to the synthon **156** (Scheme 29).

Evaluation of Approaches to the HETEs, LTB$_4$, and LTA$_4$

For the large-scale synthesis of (±)-5-HETE and (±)-5-HPETE, the iodolactonization route developed by Corey and associates (Scheme 1) is the most direct route to these compounds. Corey's group then developed a resolution of (±)-5-HETE (Scheme 3) based on the ready availability of the racemic compound. The total synthesis of 5(S) and 5(R)-HETE by the Merck Frosst group (Schemes 4 and 5) has made the corresponding optically active HPETEs available as well. Just's synthesis of 11-HETE (Scheme 9) makes available both the (R) and (S) isomers of this material. The enzymatic synthesis of 15(S)-HPETE and 15(S)-HETE by Baldwin and Davies (Scheme 11) makes these natural products readily available in one step in gram quantities from AA.

In a broad synthetic program it is very desirable to be able to take advantage of intermediates developed for a specific synthesis in as many other syntheses as possible. Thus, Corey's group, which established the utility of epoxyaldehyde trans-5(S)-**175** for synthesis of LTA$_4$ (Scheme 37), also utilized it for the synthesis of 11E-LTA$_4$ (Scheme 37) and 6E,10Z-LTA$_4$ (Scheme 25).

In the synthesis of LTB$_4$-like compounds, Corey's group used the aldehyde benzoate, (S)-**81a,** in syntheses of LTB$_4$ (Schemes 17 and 18), LTBx (Scheme 23), and 6E-LTB$_4$ (Scheme 27).

The most general approach to the synthesis of the lipoxygenase-derived products has been developed by the Merck Frosst group, using the triol **82b,** derived in one step from 2-deoxy-D-ribose, as a key intermediate to a wide variety of compounds. Thus, the epoxyaldehyde trans-5(S)-**175,** as well as its three unnatural epimers, were prepared from **82b,** en route to syntheses of LTA$_4$ and the three epimers of LTA$_4$ (Schemes 35 and 36). From **82b** were also prepared the (R) and (S) isomers of aldehyde benzoate **81b** (Scheme 15), from which were obtained 5(S)-HETE and 5(R)-HPETE (Scheme 4), 5(R)-HETE and 5(S)-HPETE (Scheme 5), LTB$_4$ (Schemes 19 and 29), 20-OH-LTB$_4$ (Scheme 21), and 20-CO$_2$H-LTB$_4$ (Scheme 22). The breadth of applicability of this approach has made it possible to supply the scientific community at large with the key leukotrienes, thus facilitating investigations of their biological properties in many laboratories.

The Wollenberg reagent **142** was used by Corey's group in syntheses of 12(S)- and 12(R)-6E-LTB$_4$ (Scheme 27), LTA$_4$ (Schemes 33 and 37), and 6-epi-LTA$_4$ (Scheme 39).

The synthon **90,** derived from D-mannose was used by Corey in the synthesis of LTB$_4$ (Scheme 17) and 6-epi-LTA$_4$ (Scheme 39), two quite different target structures.

D-mannose → **90** (t-BDMSiO) → LTB₄ / 6-epi-LTA₄

Synthon **113** was employed in three syntheses in the LTB$_4$ field by the Merck Frosst group since it permitted the introduction of varying functionality into the last six carbons of the structure, allowing syntheses of LTB$_4$ (Scheme 20), 20-OH-LTB$_4$ (Scheme 21), and 20-CO$_2$H-LTB$_4$ (Scheme 22).

L-arabinose → **113** → LTB₄ / 20-OH-LTB₄ / 20-CO₂H-LTB₄

Finally, the phosphonium salt **156** has been used extensively by a number of workers to introduce the terminal nine carbons in a variety of LTA$_4$ syntheses (Schemes 29, 30, and 34 to 40).

156 →
LTA₄
5-epi-LTA₄
6-epi-LTA₄
5-epi-6-epi-LTA₄
7Z-LTA₄
(±)-LTA₄
(±)-cis-LTA₄
(±)-11E-LTA₄

One last remark regarding the obtaining of the optically active centers in the various leukotrienes is in order. Classical resolution has been used in only one case (Scheme 3), and in the remaining syntheses, with two or three exceptions, sugars have served as readily available sources of naturally occurring optically active starting materials, once again pointing up their usefulness in syntheses of this type.[71]

The Problem of Double Bond Stereochemistry in the Leukotrienes: [1,7]-Hydrogen Migrations

The history of the structure assignments for the leukotrienes is somewhat unique in modern natural products chemistry as the complete structures were only assigned after comparison with totally synthetic material of known relative and absolute stereochemistry. In his original proposal of structures for LTA$_4$ and LTC$_4$, Samuelsson's group[5] pointed out that there remained several areas of uncertainty in the structures. The structures he suggested were as shown in Chart 4, and the areas of uncertainty were (1) the relative and absolute stereochemistry at C$_5$ and C$_6$, (2) whether or not the cysteine unit was further derivatized, and (3) the geometry of the double bonds at C$_7$ and C$_9$.

Despite these uncertainties, several groups set out to prepare the proposed leukotriene structures, and during these syntheses some interesting observations were made with respect to the double bond stereochemistry which eventually led to a reassignment of the proposed structures for LTA$_4$ and LTC$_4$. These observations are outlined here in a historical context to indicate their contribution to the final structure proof of the leukotrienes.

The Merck Frosst group first noted something unexpected in the behavior of compounds

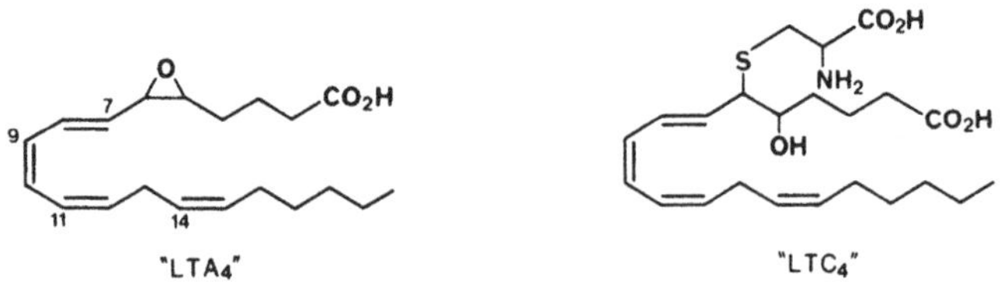

CHART 4. Originally proposed structures for LTA_4 and LTC_4.[5]

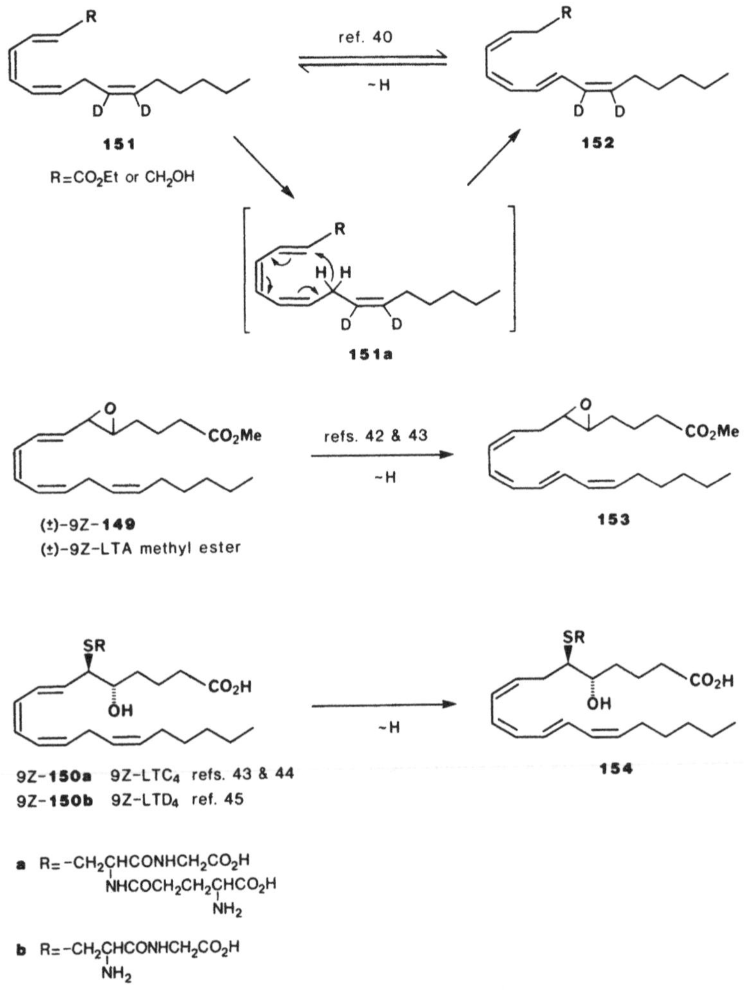

SCHEME 28. [1,7]-Hydrogen migrations in leukotriene-like structures.

in which the conjugated triene unit in the leukotrienes possessed the E-Z-Z stereochemistry, as depicted in Chart 4. In the early stages of their first synthesis of the leukotrienes,[40] compound **151** (R = CO_2Et, Scheme 28) was prepared as the intended precursor for the C_6 to C_{20} portion of the target structures. However, it was unexpectedly found that **151** (R = CO_2Et or CH_2OH) rearranged spontaneously at room temperature over a 24-hr period to a 1:2 mixture of **151** and **152**. This rearrangement appears to be a relatively facile example of an allowed sigmatropic antarafacial [1,7]-hydrogen migration as depicted in **151a**.[41] The Merck Frosst workers concluded that as there was no obvious reason why the structures of LTA_4 and LTC_4 as depicted in Chart 4 should not undergo a similar rearrangement, Samuelsson's tentatively proposed stereochemistry required revision. Samuelsson's evidence for the Z geometry of the C_{11} and C_{14} double bonds being based on solid grounds,[5] it was most reasonable to suppose that the C_9 double bond must have E stereochemistry rather than Z. With E geometry at $C_{9,10}$ it becomes sterically impossible for the molecule to assume a cyclic transition state, such as **151a**, which is necessary for the [1,7]-hydrogen migration to occur. This still left open the question of the correct geometry at the C_7 double bond, but it was assumed that the thermodynamically more favorable E configuration was most probable. Subsequent synthesis and comparison with the natural products then confirmed the proposed revision of the stereochemistry of the C_9 double bond.

Subsequently, a group at the Lilly Research Center[42] and Atrache and co-workers[43] prepared the ($\pm$)-LTA$_4$ isomer ($\pm$)-9Z-**149** (Scheme 28) with the originally proposed 9Z geometry (Chart 4) and both groups found that it rearranged readily at room temperature to the conjugated tetraene **153** as predicted by Rokach and co-workers.[40] Both groups also prepared the 9Z isomers of LTC$_4$[43,44] (9Z-**150a**) and LTD$_4$[45] (9Z-**150b**) and found that these fully functionalized leukotrienes also readily undergo the same [1,7]-hydrogen migration to produce the conjugated tetraene isomers **154**. The Lilly group also concluded that an earlier report[46] of the synthesis of 9Z-**150a** was in error and that the 9Z,11E isomer had actually been obtained.

It is important to note that all of these [1,7]-hydrogen migrations proceed in a matter of hours at or below room temperature, which, as was pointed out,[40,44] eliminated the 9Z structures (9Z-**149** and 9Z-**150**) for the natural products.

Linear Synthesis of ($\pm$)-LTA$_4$

There have been several linear (as opposed to convergent) syntheses of racemic LTA$_4$ methyl ester. All of these syntheses employ a sulfonium ylid, at or near the last step, in order to generate the epoxide by condensation with an aldehyde. It is conceivable that an optically active ylid would induce optical activity in the epoxide-forming step, but to date only optically inactive sulfonium salts have been used for this reaction.

The first synthesis of racemic LTA$_4$ methyl ester, ($\pm$)-**149**, as a mixture with ($\pm$)-cis-**149**, was reported by Corey et al.[47] as outlined in Scheme 29. Alcohol **155** was used as the synthon for the C_{12} to C_{20} portion of the target structure and (E,E)-2,4-hexadien-1,6-diol became C_6 to C_{11}. The functionality in the diol was differentiated by its transformation to **157**, which then underwent a Wittig reaction with **156** to yield the 15 carbon alcohol **158**. The formation of the mesylate of **158** and the sulfonium salt **159** pose delicate problems because of their reactivity. The condensation of **159** with methyl 4-formylbutyrate using LDA as a base is also a difficult step and only modest yields of the target compounds are obtained.[40,47]

A synthesis of racemic LTA$_4$ methyl ester (($\pm$)-**149**) was also carried out by Rokach and co-workers[40] during their first synthesis of LTC$_4$. During the course of this work, they have also synthesized ($\pm$)-cis-LTA$_4$ methyl ester (($\pm$)-cis-**149**) and ($\pm$)-11E-LTA$_4$ methyl ester (($\pm$)-11E-**149**). The preparation of the key intermediates for these syntheses is shown in Scheme 30, where alcohol **155** was prepared by the simple procedure of hydrogenation of the commercially available acetylene **160**, thus making the widely used phosphonium salt **156** cheaply and readily available. Synthons **162** and **163**, for carbons 6 to 11, were readily available by the photochemical addition of ethyl diazoacetate (EDA) to furan,[48] but a major practical improvement on this reaction was made with the finding that the addition was catalyzed efficiently and rapidly by rhodium(II) acetate.[11,49] These muconic acid semialdehyde derivatives (**162** and **163**) have readily differentiated functionality on either end of the diene unit and the broader potential utility of such 1,4-dicarbonyl-1,3-butadienes has been indicated earlier in Scheme 2. The synthesis of the alcohols **158** and **166** was straightforward, except to note that in the Wittig reaction the chloride salt of **156** gave only 10 to 15% of

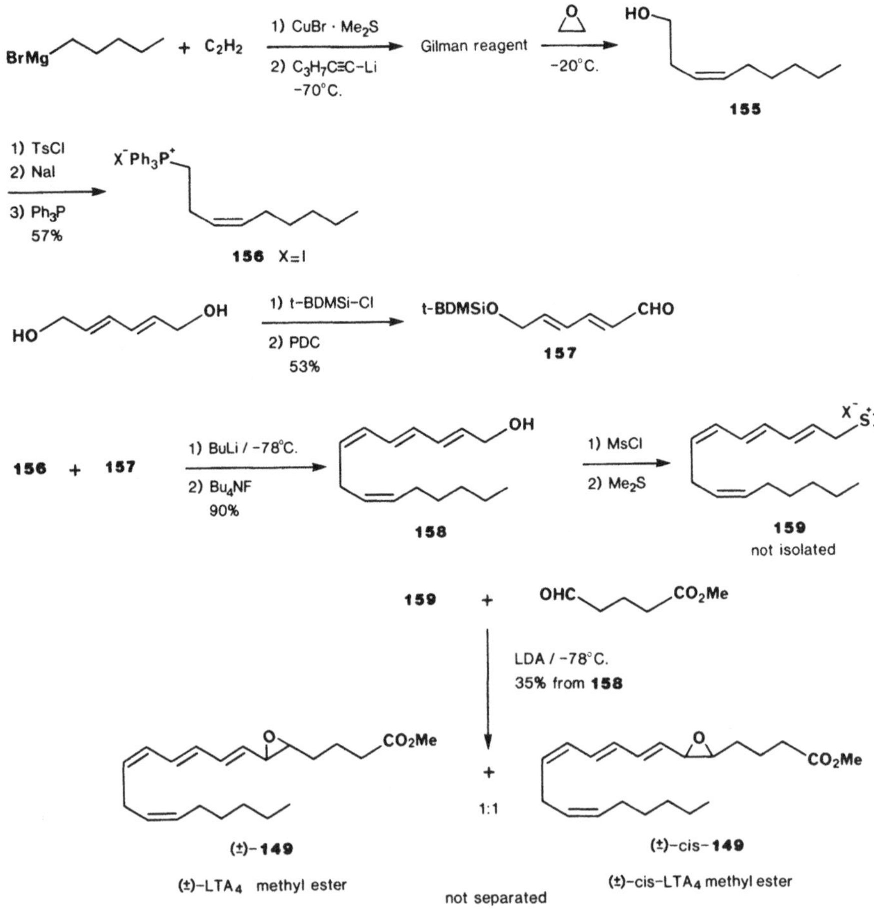

SCHEME 29. Synthesis of (±)-LTA$_4$ methyl ester.[47]

the undesired E isomer **165** as compared with 25 and 35% from the bromide and iodide, respectively.[49] Compound **151** (Scheme 28) was prepared by a Wittig reaction between **156** and aldehyde **162**.

The syntheses of the three racemic isomers of LTA$_4$ methyl ester, (±)-**149**, (±)-cis-**149**, and (±)-11E-**149** were then completed as indicated in Scheme 31. In the original report,[40] the use of LDA to form the sulfonium ylid from **159** gave a yield of only 10 to 20% of the epoxide mixture. A major improvement in this step was subsequently realized with the finding that it could be reliably effected with Triton B in 85% yield,[50,51] and an indication of the usefulness of the overall synthesis (Schemes 30 and 31) can be seen in the fact that as much as 35 g of the mixture of the epoxides (±)-**149** and (±)-cis-**149** has been prepared in a single reaction from 28 g of **158**.[49] Finally, two minor by-products resulting from alternative base-catalyzed reactions of **159** have been identified as the tetraene sulfide **168**, arising from a sigmatropic rearrangement of the ylid **167**, and the conjugated pentaene **169**.

Workers from Hofmann-LaRoche have also devised a linear synthesis of racemic LTA$_4$ methyl ester during the course of a synthesis of LTE$_4$.[52] They too employed a sulfonium salt, **172** (Scheme 32), in the epoxide forming step to obtain **173**. This step was made reasonably efficient by the use of a two-phase reaction system with a phase transfer catalyst. Their sequence is also amenable to reasonable scale-up as they describe the preparation of 8 g of **173** in a single reaction. A major strategic difference from the syntheses outlined in Schemes 29 to 31 is the deferral of reduction of acetylene bonds until the last step. It is not

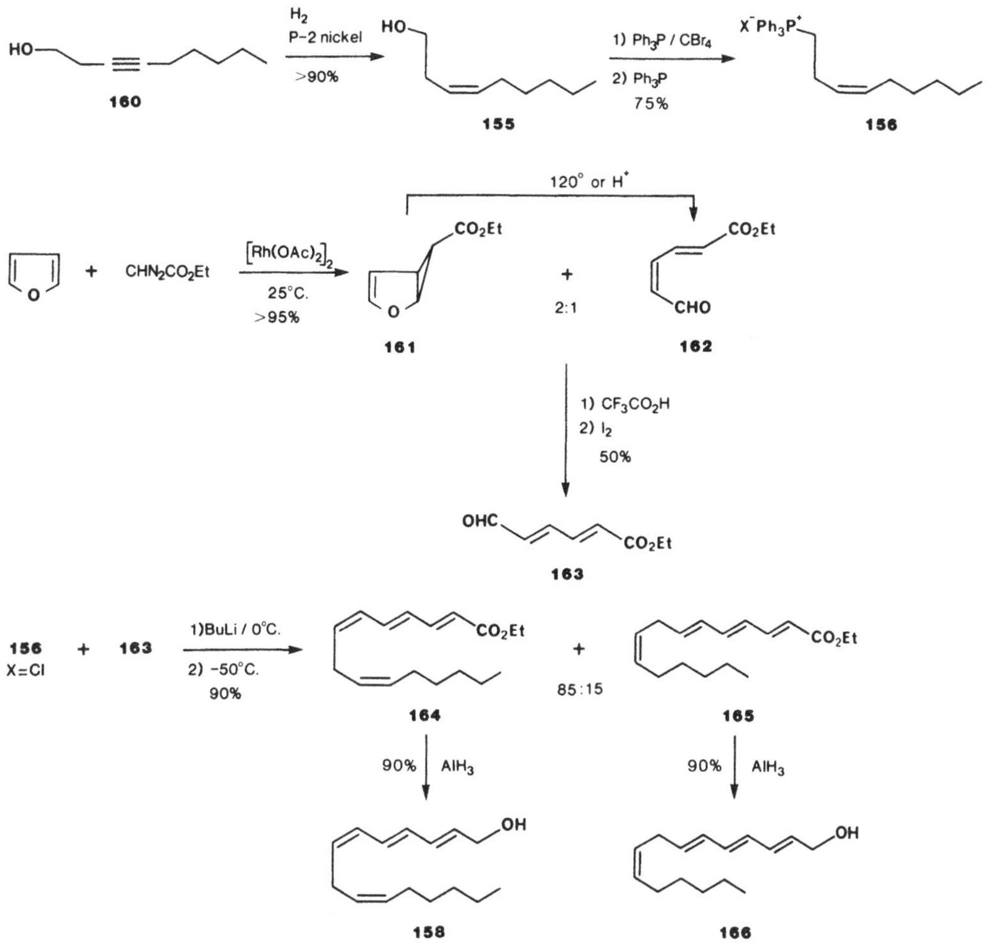

SCHEME 30. Preparation of intermediates for (±)-LTA$_4$ synthesis.[11,40,49]

clear whether this is advantageous as it is usually preferable to take a loss in yield at earlier, more accessible intermediates than at the last step of a synthesis. A unique aspect of this synthesis is that it is the only one reported which does not employ a Wittig reaction to obtain any of the four double bonds in the molecule.

Convergent Synthesis of LTA$_4$ and its Isomers

In all the reported convergent syntheses of LTA$_4$, including the racemic form, the natural and the three possible unnatural chiral forms, and various double bond geometrical isomers, the epoxy aldehydo ester **175** has been found to be a key intermediate. Consequently, a number of ingenious syntheses of the various stereo- and optical isomers of **175** have been devised. As outlined schematically in Chart 5, **175** has often been obtained from the corresponding alcohol **174**, although several other syntheses from various precursors have also been developed. Compound **175** has often been extended by various methodologies to the advanced synthons **176** and **177**.

Convergent Synthesis of Racemic LTA$_4$ and Double Bond Isomers

Several convergent syntheses of racemic LTA$_4$ methyl ester ((±)-**149**) have been reported. The first of these was by Gleason and co-workers,[53] who developed a very short, efficient synthesis of the key intermediate (±)-*trans*-**175** (Scheme 33). Corey and co-workers[54] also

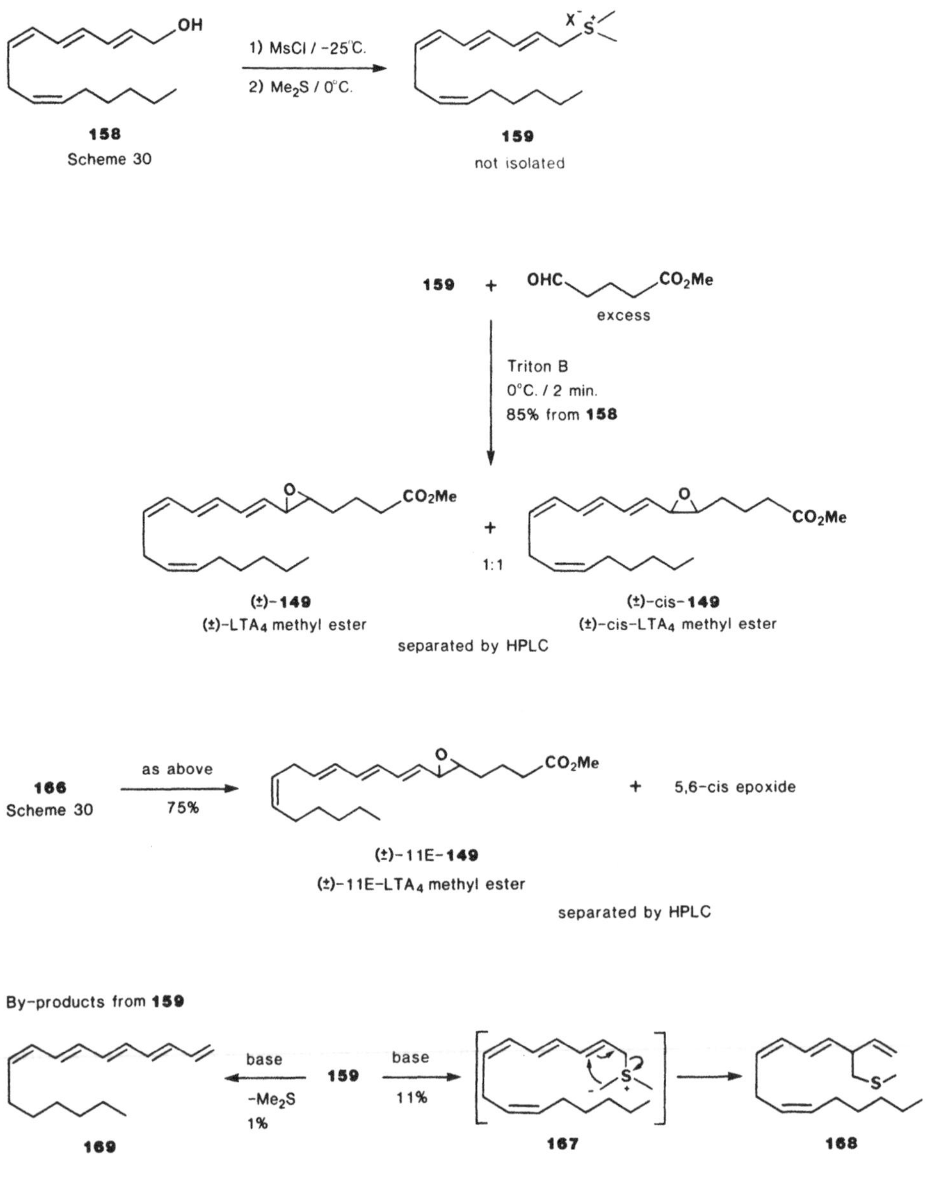

SCHEME 31. Synthesis of ($\pm$)-LTA$_4$ methyl ester and isomers.[11,40,49]

prepared the same racemic intermediate, albeit by a slightly longer route. Scheme 33 summarizes the synthesis of the various intermediates and Scheme 34 indicates the various isomers of ($\pm$)-**149** obtained. The group at the Lilly Research Center has examined these syntheses in detail[42,44,55] in order to characterize the four principal products indicated in Scheme 34. Gleason et al.[53] reported obtaining a mixture of two geometric isomers from the Wittig reaction between ($\pm$)-*trans*-**176** and **182**, but did not separate them. Based on the results of the Lilly group[42] it seems most probably that they obtained ($\pm$)-**149** and ($\pm$)-9Z,11E-**149**. The Lilly workers also isolated four minor isomers of LTA$_4$ methyl ester,[55] but did not obtain sufficient material to assign their double bond stereochemistry, although they report the UV and mass spectra, as well as HPLC behavior.

Convergent Synthesis of Chiral LTA$_4$ and its Opitcal Isomers

As would be expected, considerable effort has been put into developing syntheses for the

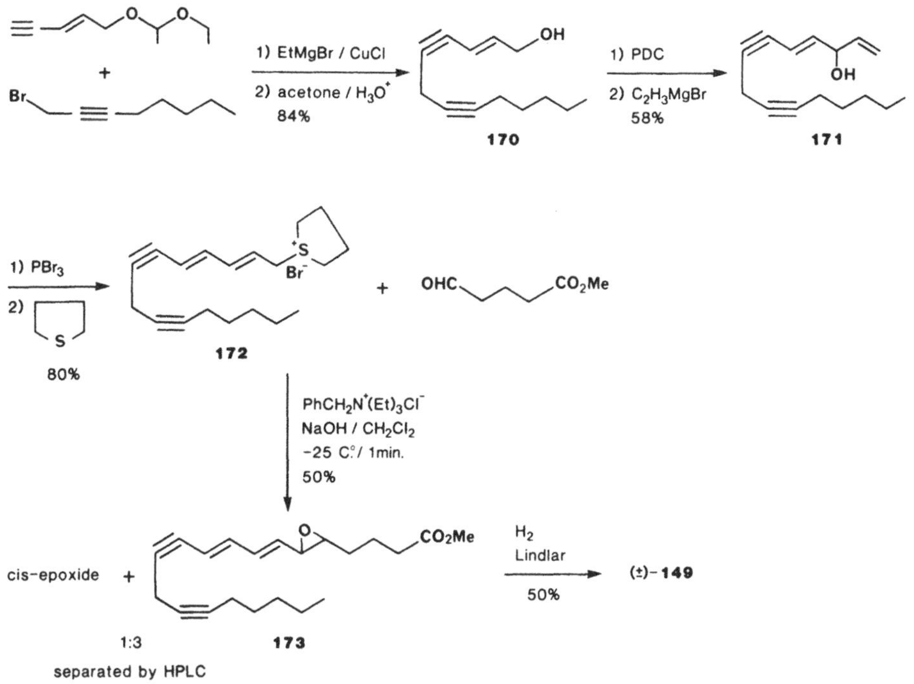

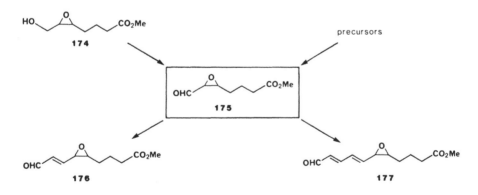

SCHEME 32. Synthesis of (±)-LTA$_4$ methyl ester.[52]

CHART 5. Key intermediates in convergent syntheses of LTA$_4$.

naturally occurring, chirally pure form of LTA$_4$, both for the purpose of preparing LTC$_4$, LTD$_4$, and LTE$_4$, and in order to study its enzymatic conversion to these same compounds and to LTB$_4$ (Chart 1). In order to study the effect of the stereochemistry at C$_5$ and C$_6$ on biological activity, there has also been a need to obtain the three unnatural stereoisomers (5-epi, 6-epi, and 5,6-diepi) of LTC$_4$, LTD$_4$, and LTE$_4$. As a result, considerable effort has gone into developing syntheses for the three unnatural stereoisomers of LTA$_4$ which are the immediate precursors of the desired leukotriene stereoisomers.

Synthesis of LTA$_4$, 5-epi-LTA$_4$, 6-epi-LTA$_4$, and 5-epi-6-epi-LTA$_4$ from 2-Deoxy-D-Ribose

The first complete study of the synthesis of LTA$_4$ and its three optical isomers (as their methyl esters) has been carried out by the group at Merck Frosst[56,57] which has resulted in two syntheses of this series of key compounds. The particularly successful and satisfying aspect of these syntheses is that from a single chiral starting material, 2-deoxy-D-ribose, by

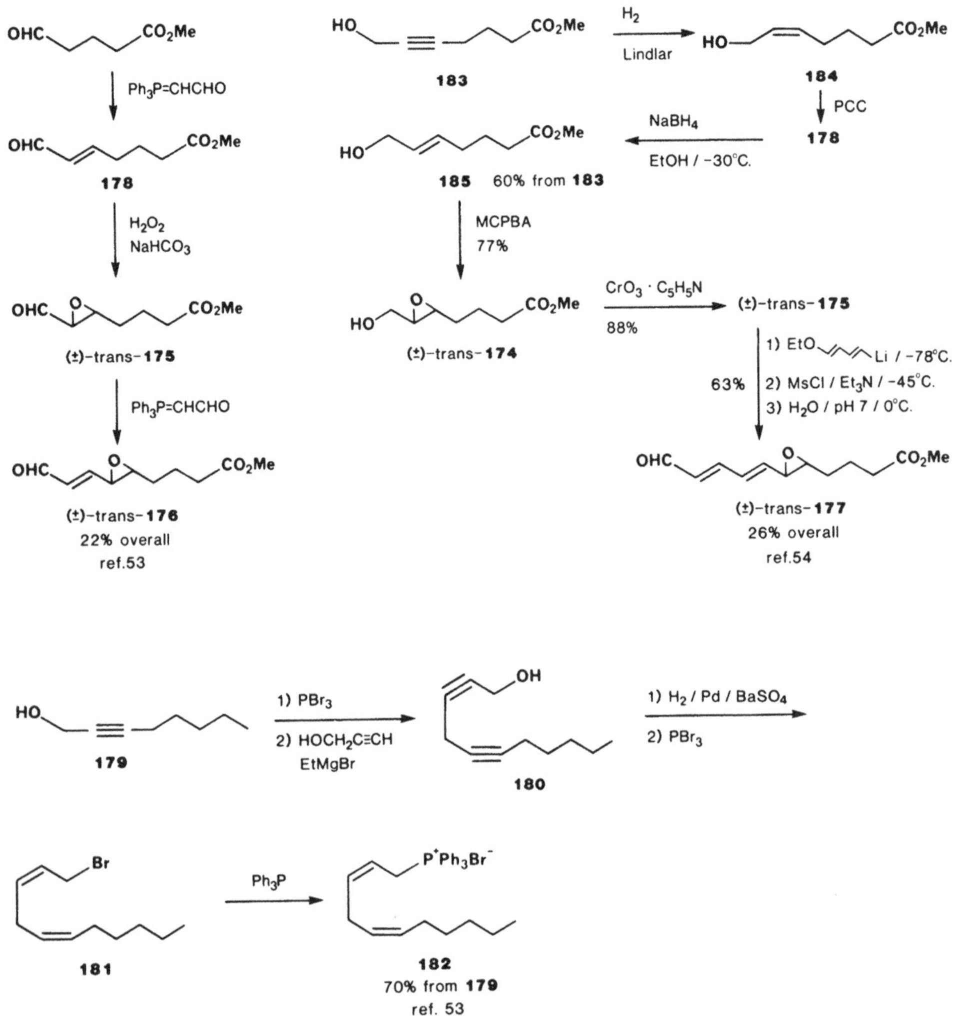

SCHEME 33. Intermediates for racemic LTA₄ methyl ester.

appropriate manipulation of its chiral centers, all four of the optically active isomers of LTA₄ were obtained. The overall stereochemical relationship between the starting material and the various isomers of LTA₄ is illustrated schematically in Chart 6. In each case, one of the series of epoxy alcohols **174** was first obtained and then elaborated to the corresponding LTA₄ isomer.

Scheme 35 outlines the stereospecific total synthesis of all four of the optical isomers of LTA₄ from 2-deoxy-D-ribose.[56] The readily available triol ester **186** was regioselectively functionalized on the primary hydroxyl group using the sterically hindered mesitylene sulfonyl chloride (Met-Cl). Treatment with sodium methoxide in methanol then yielded the desired epoxy alcohol *trans*-5(S)-**174**. The terminal epoxide **188** is an intermediate in the reaction and undergoes epoxide transposition under the basic reaction conditions. The asterisk in **188** and subsequent structures indicates the point of inversion, which is effected in order to achieve the desired stereochemistry as schematically indicated in Chart 6. The epoxy alcohol was then oxidized to the epoxy aldehyde *trans*-5(S)-**175**, which was extended by a double Wittig reaction with formylmethylenetriphenylphosphorane to *trans*-5(S)-**177**. The final step of the synthesis was a Wittig reaction between *trans*-5(S)-**177** and the widely used

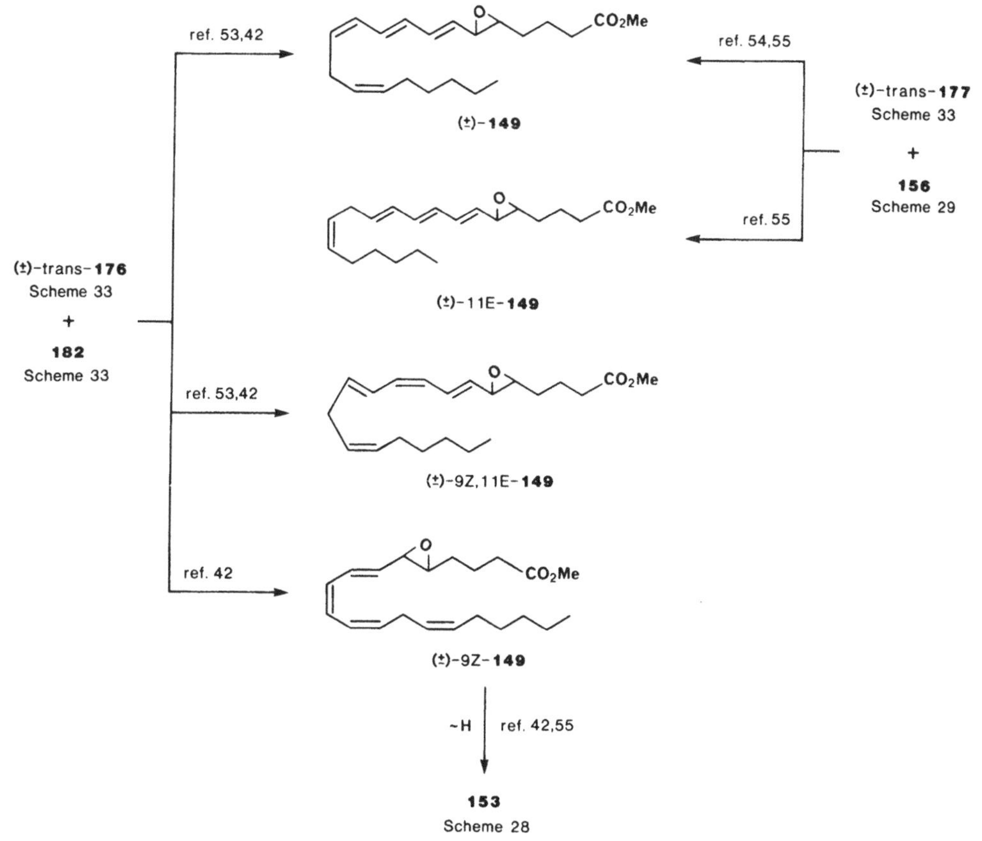

SCHEME 34. Convergent syntheses of racemic LTA$_4$ methyl ester and double bond isomers.

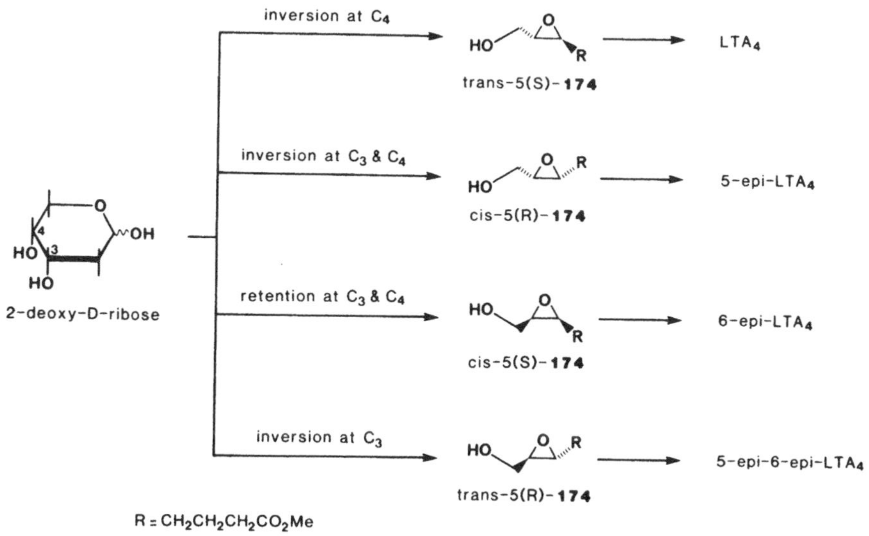

CHART 6. Stereochemical relationship between 2-deoxy-D-ribose and LTA$_4$ optical isomers.[56]

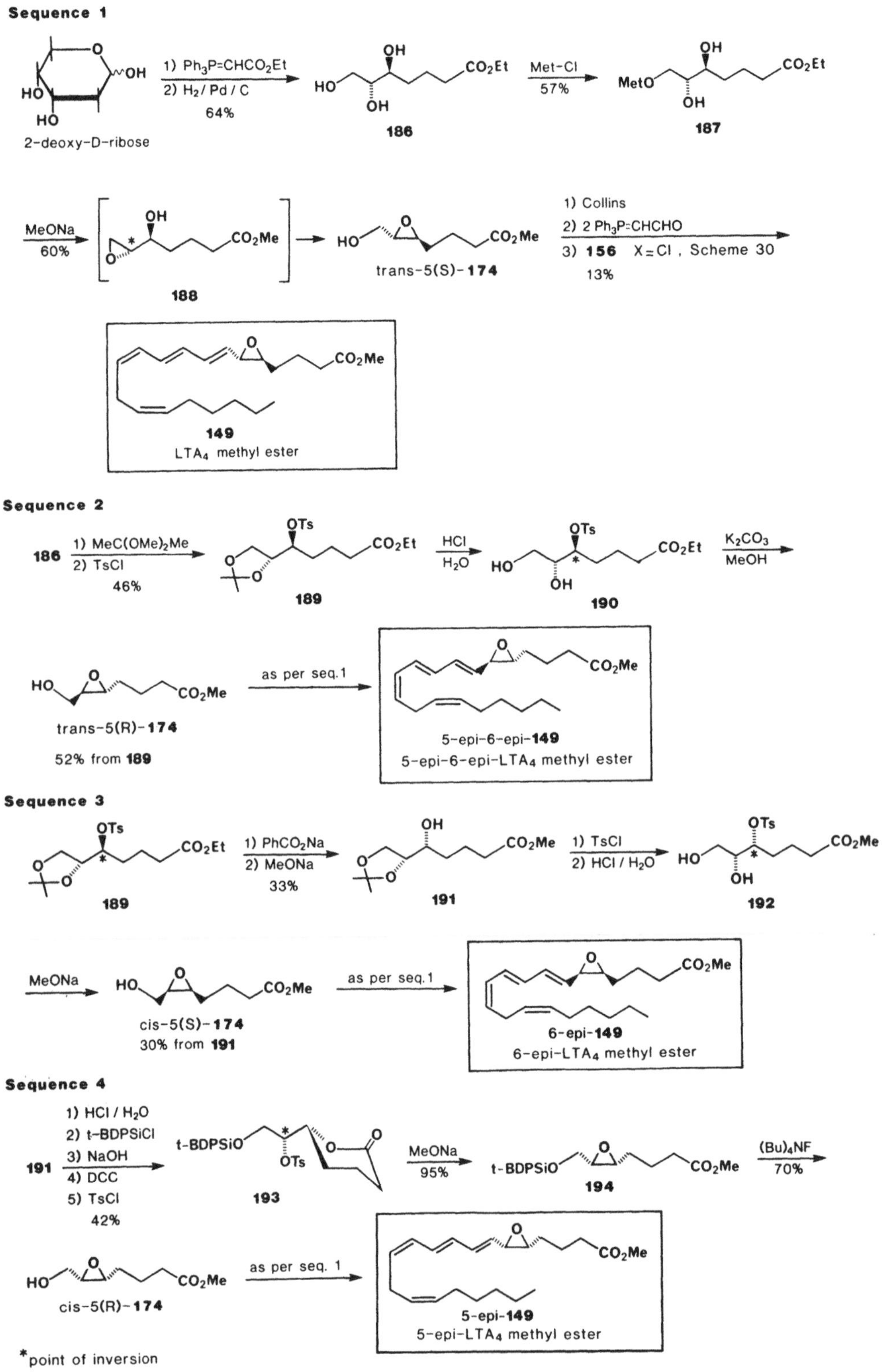

Sequence 1

Sequence 2

Sequence 3

Sequence 4

*point of inversion

SCHEME 35. Synthesis of LTA_4 and its optical isomers from 2-deoxy-D-ribose.[56]

triphenyl[(Z)-non-3-en-1-y]phosphonium chloride (**156**, X = Cl). In Sequence 2, in order to obtain the enantiomer of **149**, inversion was effected at the former C_3 hydroxyl group of 2-deoxy-D-ribose by selectively blocking the terminal diol unit of **186** followed by tosylation of the remaining secondary alcohol to obtain **189**. Removal of the acetonide blocking group and treatment with base then gave the epoxyl alcohol *trans*-5(R)-**174**, which was converted to 5-epi-6-epi-LTA$_4$ methyl ester by the same sequence of reactions as for LTA$_4$ methyl ester.

To obtain 6-epi-LTA$_4$ methyl ester (6-epi-**149**) it was necessary to retain the original stereochemistry at C_3 and C_4 of 2-deoxy-D-ribose. Operationally, this was carried out by effecting a double inversion at one center, as illustrated in Sequence 3, on compounds **189** and **192**. In contrast, in order to obtain 5-epi-LTA$_4$ methyl ester, inversion at both centers was required. This was achieved as illustrated in Sequence 4 in which the mono-inverted compound **191** was converted to the tosyl lactone **193**. Treatment of the lactone with methoxide opened the ring to the methyl ester, and the alkoxide intermediate displaced the tosylate group, with inversion at the second center, to give the desired *cis*-5(R)-**174**, which was converted to 5-epi-LTA$_4$ methyl ester.

In the second synthesis,[57] 2-deoxy-D-ribose was used to obtain the C-glycoside **102** as described previously in sections "Synthesis of (S)-, (R)-, and (±)-5-Benzoyloxy-6-Oxo-hexanoic Acid Esters" and "Synthesis of LTB$_4$". The key to the C-glycoside synthesis was the finding that with a suitably located leaving group, these structures open upon base treatment, as illustrated in **195** (Scheme 36), to yield unsaturated epoxy alcohols (**196**). In Sequence 1, the intermediate **188** (Scheme 35) was isolated prior to conversion to *trans*-5(S)-**174**. Sequence 2 of Scheme 36 illustrates the complete unraveling of the C-glycoside **198** to yield the conjugated diene alcohol **199a** as a reaction competing with the formation of the desired **199**. The formation of compounds such as **199a** formed the basis for one of the syntheses of LTB$_4$ described in section "Synthesis of LTB$_4$" and Scheme 19.[29] In Sequence 3, one of the necessary inversions was effected by a highly stereoselective reduction of ketone **200** with the hindered lithium perhydro-9B-boraphenalylhydride to obtain alcohol **201**. Conversion of the terminal epoxide **202a** to the desired *cis*-5(R)-**174** was unexpectedly difficult as compared with the corresponding reaction of **188** (Sequence 1). The difficulty was ascribed to steric interference between the epoxide ring and the ester side chain in the transition state necessary for the formation of the *cis*-epoxide. Among the attempts to improve the reaction **202b** was treatment with lithium iodide in hot methanol (Sequence 4), having the rather amusing result that a reasonable yield of the *trans*-epoxy alcohol was obtained. The proposed mechanism for this unexpected result implicates the formation of a small amount of **204**, which can either revert to starting material or be drained off to the more stable *trans*-5(R)-**174**.

Synthesis of LTA$_4$ and Double Bond Isomers from D-Ribose

The first stereospecific total synthesis of LTA$_4$ was carried out by Corey and co-workers,[46] and they established the general utility of the key intermediates of the general structures shown in Chart 5. Scheme 37 outlines Corey's synthesis of LTA$_4$ methyl ester (**149**) starting from the 2,3,5-tribenzoyl derivative of D-ribose. The clean reductive removal of the allylic benzoate (**206** → **207**) proved to be a particularly effective method to remove what had been the C_2 oxygen function of ribose. The selective removal of the acetate group from the saturated derivative of **207** using 0.005% HCl in methanol to uncover only the central oxygen as a hydroxyl group was a striking example of chemoselectivity. Tosylation then yielded **208**, which upon mild base treatment in methanol was debenzoylated and transformed directly to the desired *trans*-5(S)-**174**.

This conversion of **208** to *trans*-5(S)-**174** is a somewhat surprising result. The primary benzoate in **208** would be expected to hydrolyze first and form the terminal epoxide. Hy-

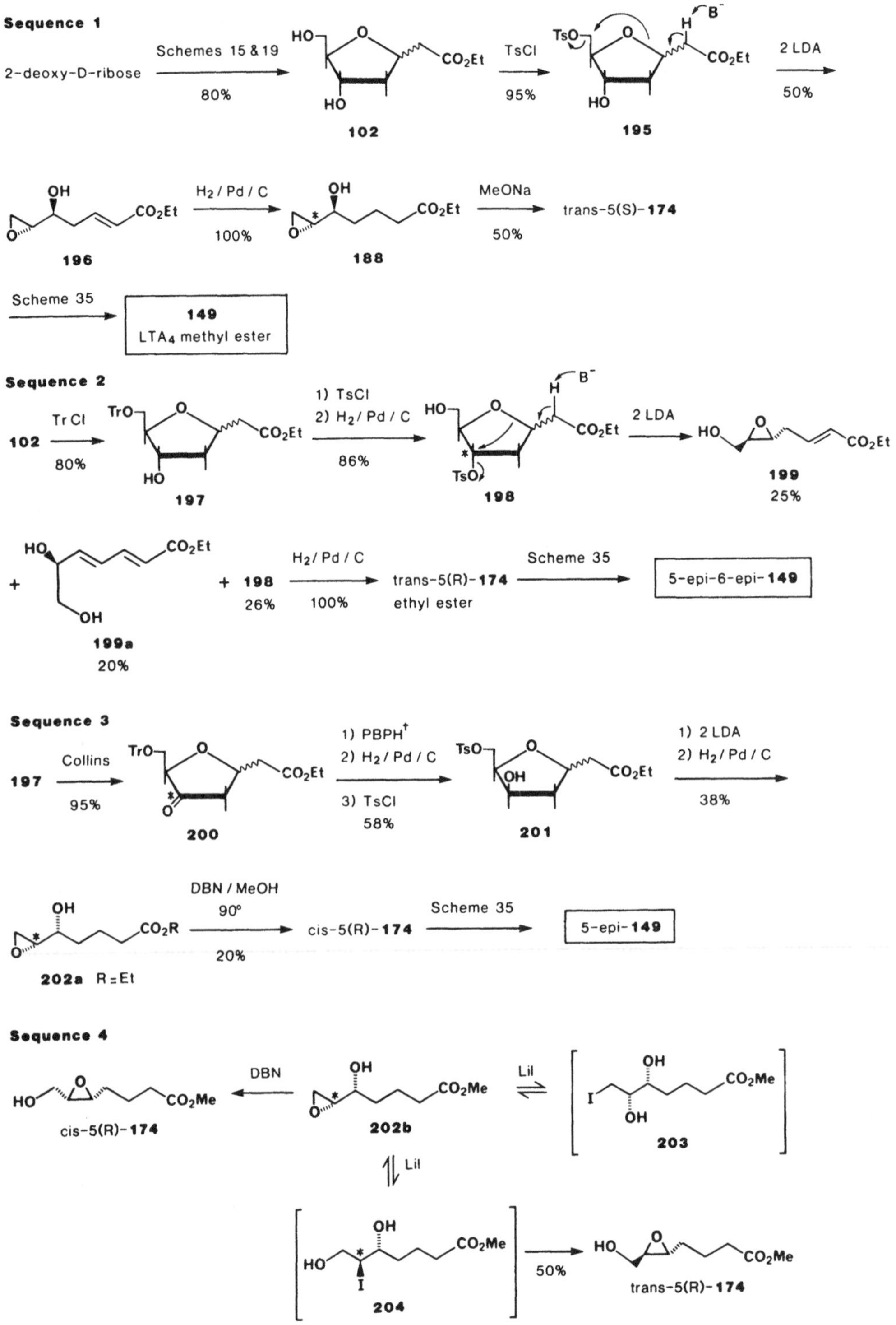

†PBPH = perhydro-9B-boraphenalyl hydride

*Point of inversion

SCHEME 36. C-Glycoside route to LTA_4, 5-epi-LTA_4, and 5-epi-6-epi-LTA_4.[57]

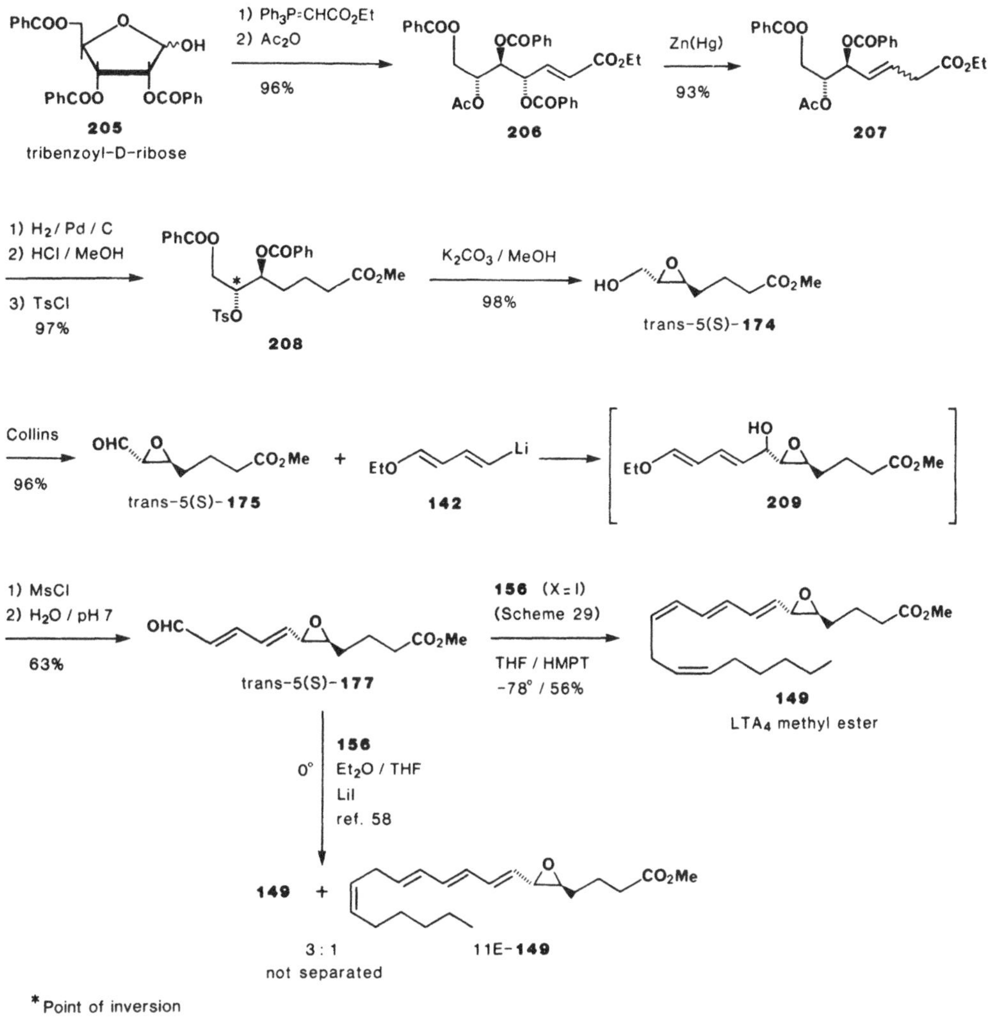

SCHEME 37. Synthesis of LTA₄ and 11E-LTA₄ from D-ribose.[46]

drolysis of the secondary benzoate, followed by epoxide rearrangement, should then give the internal epoxide, but it would be a *cis*-epoxide as a result of two inversions at the starred carbon. Since there seems to be no reason to doubt the *trans* nature of the epoxide, one must postulate a migration of one of the benzoate groups in **208** during hydrolysis, or during the initial Wittig reaction to form **206**, or during the acid methanolysis of the acetate group in the conversion of **207** to **208**. The last two possibilities could yield an isomer of **208** with a primary tosylate and two secondary benzoates, base hydrolysis of which would lead to the observed *trans*-epoxide.

After oxidation to the aldehyde, *trans*-5(S)-**175**, the latter was reacted with the Wollenberg reagent, **142**, to effect a four-carbon chain extension to *trans*-5(S)-**177**. A Wittig reaction with **156** (Scheme 29) then yielded LTA₄ methyl ester (**149**).

The last Wittig reaction also yielded a certain amount of the 11E-isomer of LTA₄ methyl ester, and conditions were varied in order to increase the yield of the latter[58] as shown in Scheme 37. Although 11E-**149** was not separated from **149**, the mixture was used to prepare the 11E isomer of LTC₄.[58]

Ernest and co-workers,[59] using Corey's methodology[46] to prepare the key aldehyde in-

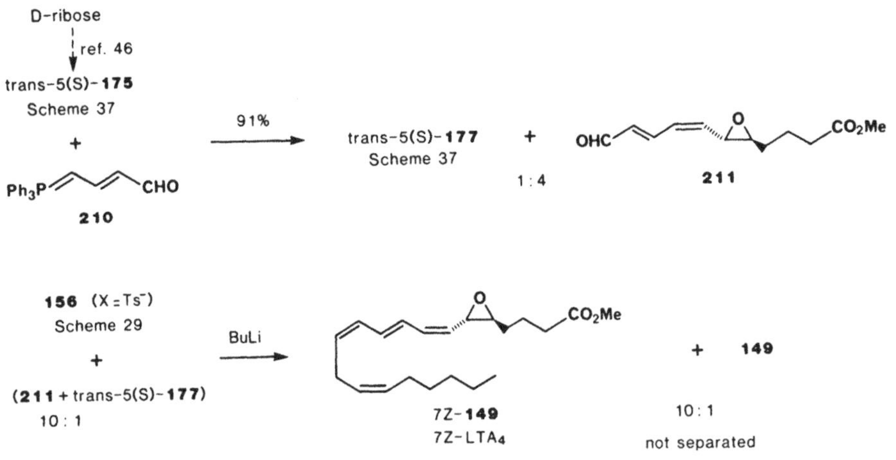

SCHEME 38. Synthesis of 7Z-LTA$_4$ from D-ribose.[59]

termediate *trans*-5(S)-**175**, developed a synthesis of the 7Z isomer of LTA$_4$ methyl ester (Scheme 38). These workers used the ylid **210** containing the crotonaldehyde unit as an alternative to the Wollenberg reagent to add four carbons to *trans*-5(S)-**175**. However, they found that the major product possessed the Z,E-diene structure **211**. Using a fraction enriched in **211**, they prepared 7Z-**149** which still contained about 10% of **149**. They also converted **211** to the E,E-diene *trans*-5(S)-**177** from which they also prepared LTA$_4$ methyl ester (**149**) and obtained it as a crystalline compound for the first time (m.p. 28 to 32°C).

Synthesis of 6-epi-LTA$_4$ from D-Mannose

As indicated earlier, the variation of stereochemistry at C$_5$ and C$_6$ of the leukotrienes has been of considerable interest in order to ascertain the effects of such changes on biological activity. Corey's group reported the first synthesis of 6-epi-LTA$_4$ as part of their efforts in this field,[60] and their synthesis is outlined in Scheme 39. They started with the hemiacetal **90**, readily derived from D-mannose (and which they subsequently also used in the first synthesis of LTB$_4$,[27] see section "Synthesis of LTB$_4$"), to obtain **212** by a Wittig reaction. Hydrogenation and extensive manipulation of functionality converted **212** to **213**. Although no explanation was given for changing the terminal diol protecting group from the acetonide in **212** to the orthoester in **213**, it was presumably done because of the acid sensitivity of the epoxide and the fact that much milder acid conditions are required to remove the orthoester group than the acetonide in **214**.

Synthesis of LTA$_4$ from L-Arabinose

Starting with the acetonide of L-glyceraldehyde (**215**), readily obtained from L-arabinose, Rokach and co-workers[61] have reported another type of synthesis of LTA$_4$ methyl ester as outlined in Scheme 40. The synthesis involves the epoxidation of **217** which gives rise to a 1:2 mixture of the diastereomeric epoxides **218** and **219** in favor of the desired **219**. The mixture was separable by chromatography, and **219** was hydrolyzed to the epoxy diol and cleaved to *trans*-5(S)-**175**. This transformation of **219** was carried out as a one-pot reaction because of the sensitivity of the epoxy diol to even quite mildly acidic conditions. Thus, the asymmetric center originally present in **215** served as an internal resolving agent (separation of **218** and **219**) and allowed the preparation of optically pure *trans*-5(S)-**175** from **219** (and *trans*-5(R)-**175** from **218**).

Chain extension by four carbons was carried out with formylmethylene triphenylphosphorane in which *trans*-5(S)-**176** was first obtained in 84% yield. The second Wittig reaction

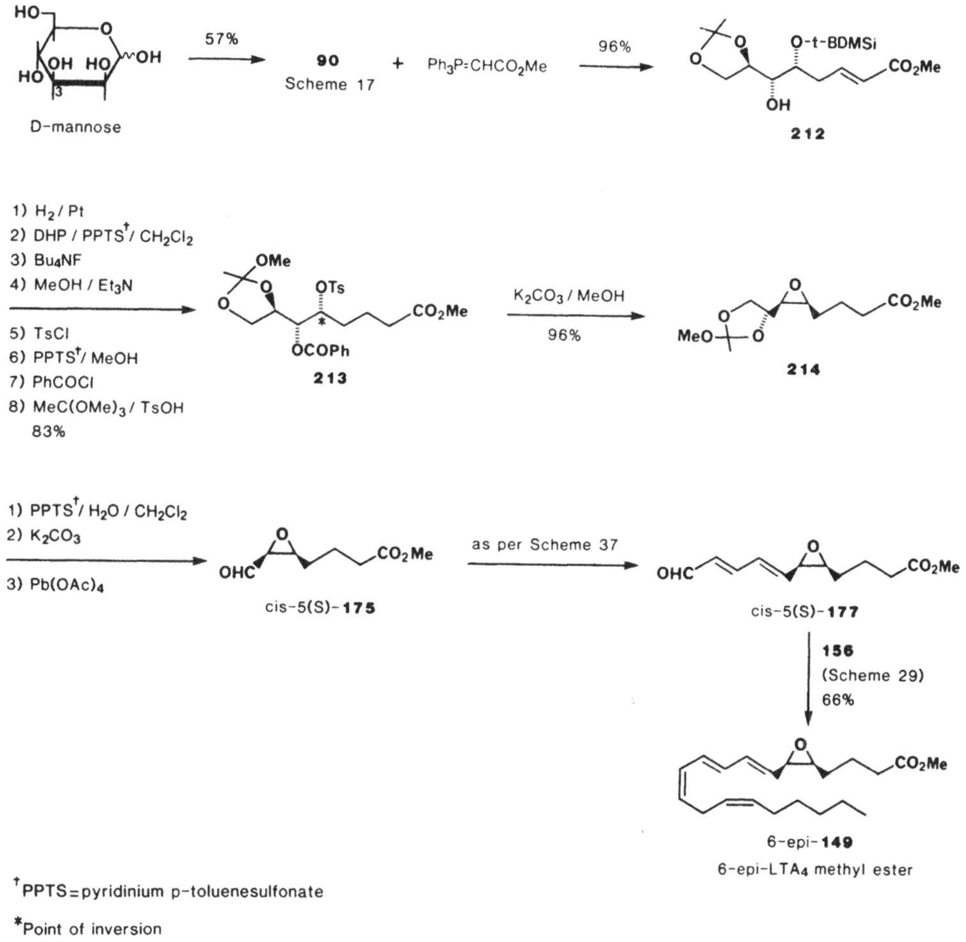

SCHEME 39. Synthesis of 6-epi-LTA₄ from D-mannose.[60]

to obtain *trans*-5(S)-**177** was much slower, and it was found more practical to carry out a one-pot double Wittig reaction with two equivalents of reagent as shown in Scheme 40.

Synthesis of LTA₄ Intermediates

Several other interesting syntheses of the key intermediates **174** and **175**, the elaboration of which to LTA₄ has just been detailed, have been reported, and are summarized in this section.

Concurrently with the previously described (section "Synthesis of LTA₄ from L-Arabinose") synthesis of LTA₄ from L-glyceraldehyde acetonide, Rokach and co-workers[61] prepared the three unnatural isomers of aldehyde **175**. The syntheses were based on the same reactions as outlined in Scheme 40, and are illustrated briefly in Scheme 41. From D-glyceraldehyde acetonide, **220**, epoxidation of the derived **222** again gave a diastereomeric mixture of separable epoxides analogous to **218** and **219**, but in which the unnatural enantiomer *trans*-5(R)-**175** was now the major component. Epoxidation of Z-ester **221** (R = Me), gave a diastereomeric mixture of *cis*-epoxides, which after separation and oxidative cleavage gave the enantiomeric *cis*-aldehydoepoxides **175**.

In a very different approach to the epoxy alcohol *trans*-5(S)-**174**, Sharpless and co-workers[62] demonstrated the practical effectiveness of their previously mentioned[31] chiral epoxidation technique based on the use of *t*-butyl hydroperoxide, titanium alkoxide, and

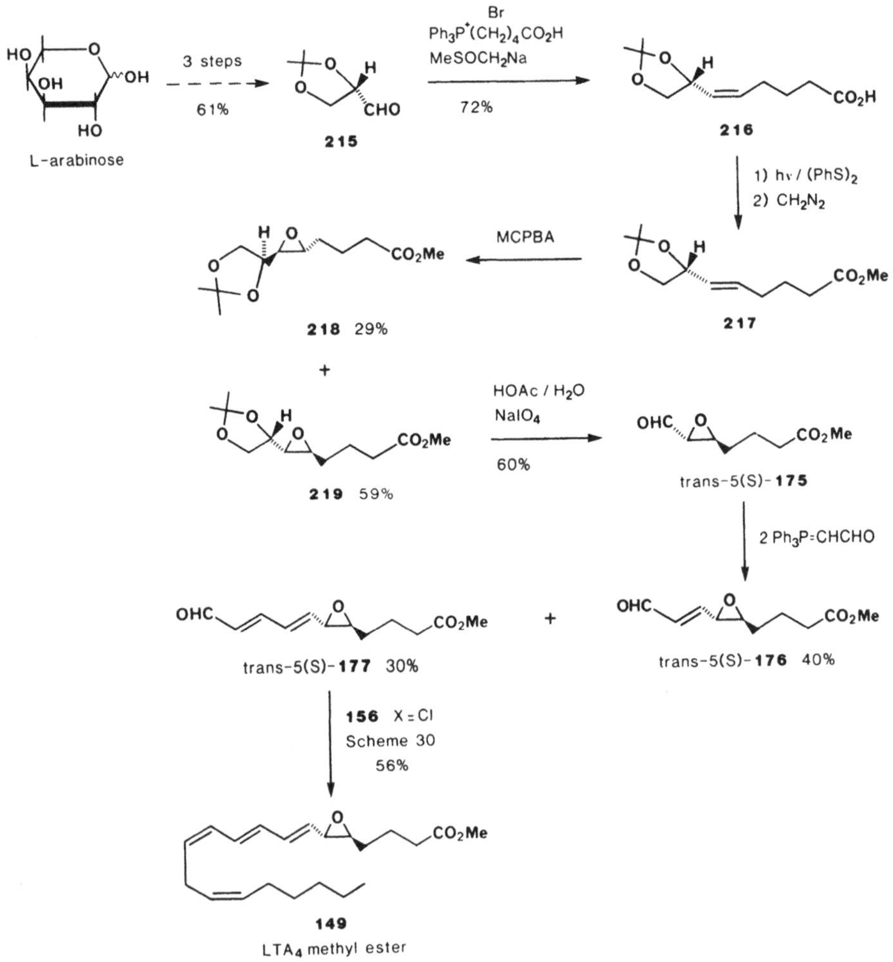

SCHEME 40. Synthesis of LTA₄ from L-arabinose via L-glyceraldehyde.[61]

tartrate esters as the asymmetry-inducing agents. As outlined in Scheme 42, starting with the inexpensive diene alcohol **223**, available in high yield in two steps from butadiene, they obtained *trans*-5(S)-**174** in 60% overall yield and excellent optical purity. As pointed out by Sharpless, the original procedure[31] gave poor yields of relatively water-soluble epoxides such as **174**, but they describe[62] a modified work-up procedure which largely circumvents the previous difficulty, allowing the direct oxidation of **225** to *trans*-5(S)-**174**, again with excellent optical purity. For large-scale preparations, **223** is undoubtedly a preferable starting material as it is much less expensive to prepare than **225** and the overall yield is equivalent.

Using the Sharpless procedure, Corey et al.[63] have also described an analogous, though somewhat longer synthesis of *trans*-5(S)-**174** from (E)-8-methyl-2,7-nonadien-1-ol.

Cohen et al.[64] have described a synthesis of *trans*-5(S)-**174** starting from erythrose acetonide **226** as outlined in Scheme 43. Acid treatment of **228** brought about deprotection of the diol unit followed by intramolecular lactonization to yield the crystalline lactone **229**. This lactone had the desired hydroxyl group free to be converted into a suitable leaving group (**230**) which was smoothly transformed to the target structure. Aldehyde intermediate **231**, epimeric at C_3 with respect to **226**, was obtained in a straightforward sequence from L-(+)-diethyl tartrate, and was carried through the same reactions as **226** to obtain the *cis*-epoxy alcohol *cis*-5(S)-**174**.

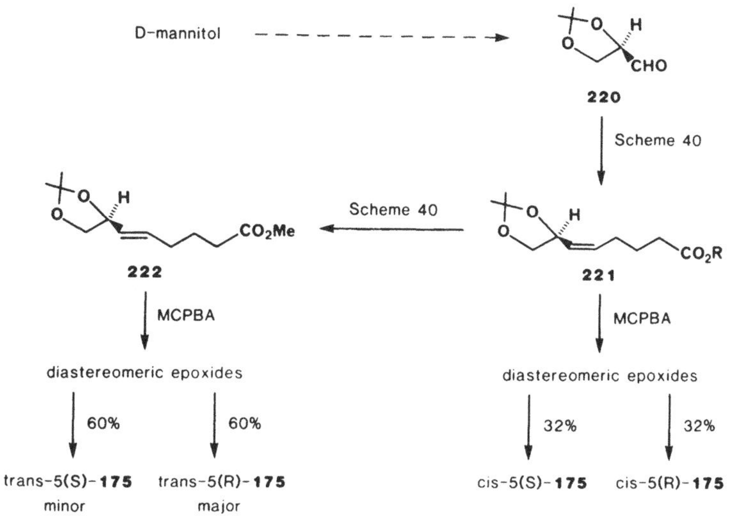

SCHEME 41. Synthesis of LTA$_4$ intermediates from D-mannitol via D-glyceraldehyde.[61]

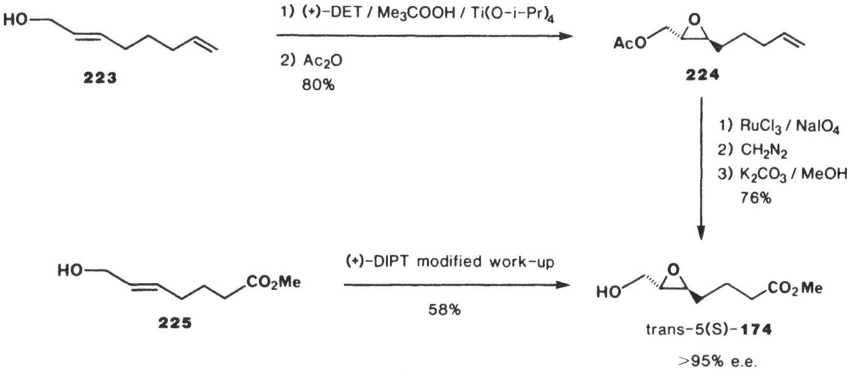

SCHEME 42. Synthesis of LTA$_4$ intermediates by asymmetric epoxidation.[62]

Starting with 3,5-dibenzoyl-2-deoxy-D-ribose, Marriott and Bantick[65] developed an abbreviated synthesis of Corey's tosylate intermediate, **208** (Scheme 37). Their sequence is similar to that shown in Scheme 37, but by starting with the deoxy sugar there is, of course, no need to remove the oxygen function (**206** → **207**), nor was it necessary to acetylate the hydroxyl group as in **206**, thus eliminating a protection-deprotection sequence.

Biomimetic Synthesis of LTA$_4$ and 14,15-LTA$_4$

As indicated in Charts 1 and 2, 5(S)-HPETE and 15(S)-HPETE are generally considered to be the biochemical precursors of LTA$_4$ and 14,15-LTA$_4$, although solid biochemical evidence on this point has not yet been obtained. A reasonable biochemical mechanism for this conversion would be the formation of a phosphate or sulfate ester of the hydroperoxide which would undergo elimination of the anion by nucleophilic attack of the adjacent π-electrons of the double bond on the peroxide oxygen, with concomitant loss of a proton from the methylene group at C_{10}.

The groups of Corey and of Sih have reported studies on chemical analogues of the proposed biochemical mechanism, in which they activated the hydroperoxide function by forming the corresponding mesylate or trifluoromethanesulfonate (triflate). In no case were

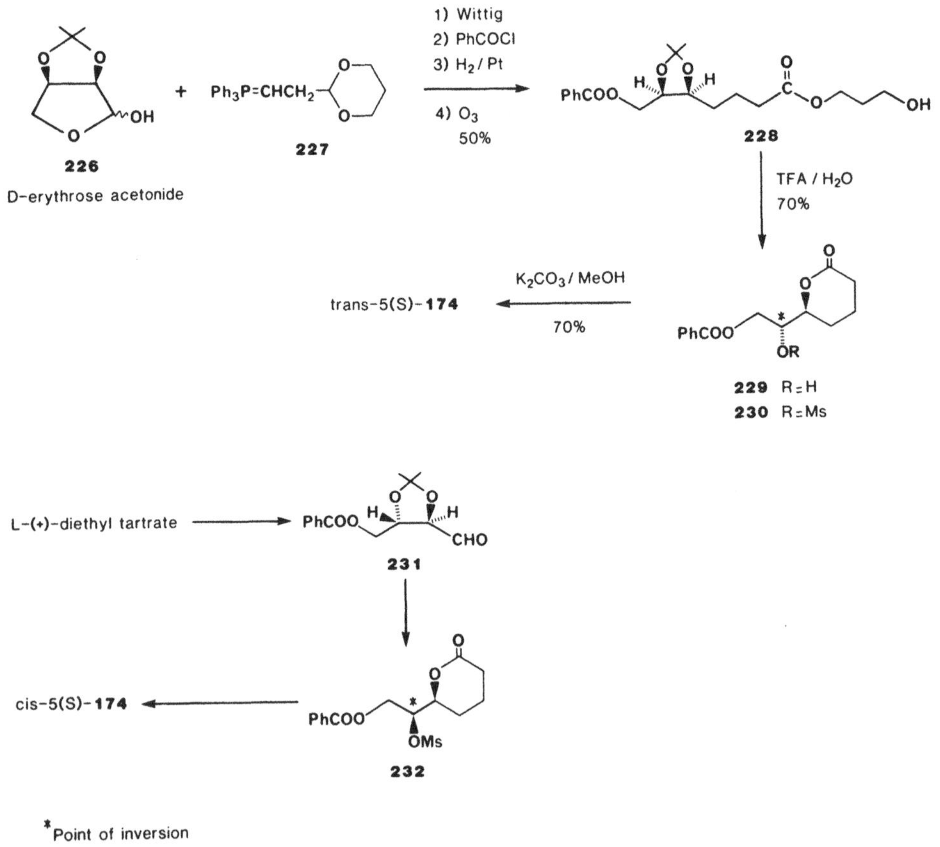

SCHEME 43. Synthesis of LTA$_4$ intermediates from D-erythrose and L-(+)-tartrate.[64]

the peroxy esters isolated as they underwent elimination under the conditions of formation, even at $-110°C$.

Corey's group[66] was the first to report the successful conversion of 5(S)-HPETE methyl ester (**233**, Scheme 44) into LTA$_4$ methyl ester in 25% yield via the biomimetic mechanism, using the triflate ester of the hydroperoxide. Shortly afterwards, Sih and co-workers[67] reported the same transformation in 15% yield using the mesylate ester. Corey emphasized the sensitivity of the reaction conditions in order to obtain a reasonably successful conversion of the HPETE into LTA$_4$, and he found that a significant amount of ketone **234** was also formed by a 1,2-elimination of the peroxy triflate. Sih subsequently confirmed this sensitivity to conditions, and in a detailed study of the reaction, found that a third product, 9Z-LTA$_4$ methyl ester (9Z-**149**), was also formed in significant amounts.[43] This compound was found to undergo a facile [1,7]-hydrogen migration as discussed earlier in the section "The Problem of Double Bond Stereochemistry in the Leukotrienes". In a subsequent report, Corey and Barton[68] provide an improved procedure for the reaction and confirmed the formation of 9Z-**149**.

Two points with regard to the stereochemistry of the epoxide-forming reaction are to be emphasized. Using optically active 5(S)-HPETE methyl ester, Corey et al. obtained optically pure LTA$_4$ methyl ester, indicating that the stereochemistry of the C$_5$-oxygen bond in the transformation of **233** to **149** is unaffected. Both groups of workers found that the *trans*-epoxide is formed exclusively, with no *cis*-epoxide being observed.

It should be pointed out that Scheme 44 depicts the reaction of optically active compounds,[66] but the data of References 43 and 68 were obtained using racemic 5-HPETE as starting material.

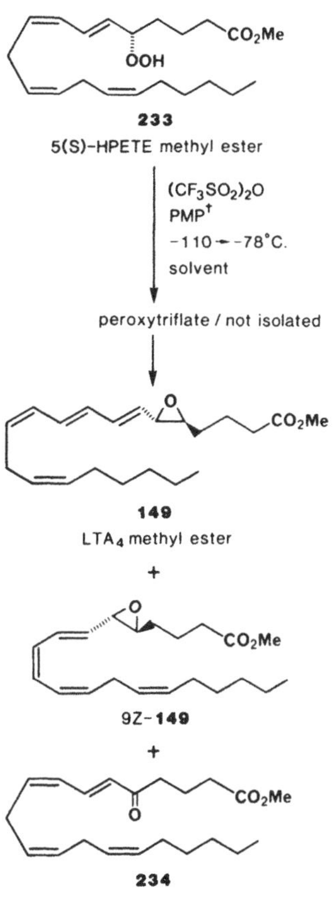

Reference and yields	66	43	68
149	25%	6 - 13%	33%
9Z-149	not observed	14 - 24%	~10%
234	~25%	7 - 14%	~4%

† PMP = 1,2,2,6,6-pentamethylpiperidine

SCHEME 44. Biomimetic synthesis of LTA_4 methyl ester.

This biomimetic route represents a convenient synthesis of small amounts of LTA_4 and hence of LTC_4, LTD_4, and LTE_4,[66,67] but the reactivity of the peroxide intermediates would probably make scale-up difficult.

As shown in Chart 2, 14,15-LTA_4 is also a proposed product of the lipoxygenase reaction of AA. Its preparation by a biomimetic pathway has also been effected in the laboratories of Corey[68,69] and Sih[43] (Scheme 45). The characteristics of the reaction and the results were very similar to those discussed for the formation of LTA_4, and again Sih observed the formation of an E,Z,Z-triene **237**, which underwent a [1,7]-hydrogen migration analogous to that observed with 9Z-**149**. Sih's group has provided evidence for the enzymatic conversion of 15(S)-HPETE to 14,15-LTA_4, and it is interesting to note that none of the E,Z,Z-triene **237** was formed under biochemical conditions.[70]

SYNTHESIS OF THE COMPONENTS OF SRS-A: LEUKOTRIENE C_4, D_4, E_4, AND RELATED COMPOUNDS

Of all the products derived from the lipoxygenase metabolism of AA, the group which has attracted the greatest attention biologically and chemically are those compounds which make up the material known as "slow-reacting substance of anaphylaxis" (SRS-A).[4(d),7] SRS-A is now known to be made up of varying amounts of the cysteine-containing eicosanoids LTC_4, LTD_4, and LTE_4 (Chart 1), and the synthesis of these and related compounds

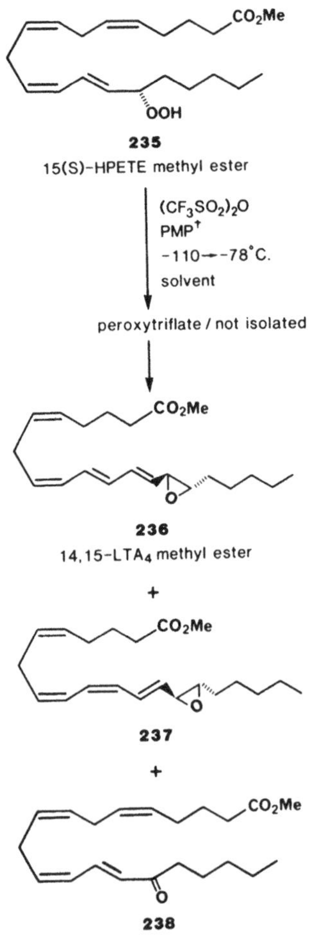

Compound	Reference and yields: 69	43	68
236	40%	8%	37%
237	not observed	23%	15%
238	20%	35%	not reported

SCHEME 45. Biomimetic synthesis of 14,15-LTA_4 methyl ester.

is detailed in this section. The section is broken into two major subsections, the first of which deals with naturally occurring leukotrienes, while the second presents the synthesis of closely related analogs.

Synthesis of Naturally Occurring Leukotrienes

Synthesis of Leukotrienes C_4, D_4, and E_4

As mentioned above, the title compounds are now known to constitute the active components of SRS-A, and as such have been the subject of intense biological and synthetic research. All the syntheses of these thioether structures have converged on LTA_4 methyl ester (**149**), the various syntheses of which were outlined in the third major section of this chapter. Corey at Harvard University, the Merck Frosst Canada group, and workers at Hoffmann-LaRoche independently developed synthetic routes to LTA_4 methyl ester and hence to various components of the SRS-A complex.

Harvard Synthesis of LTC_4 and LTD_4

The first synthesis of LTC_4 was achieved by Corey's group,[46] and as chiral intermediates were employed, it also established the full stereochemistry of the compound. As outlined in Scheme 46, the epoxide group in **149** was found to undergo a clean S_N2 reaction at C_6 under basic conditions with the thiol of **239a** (glutathione) or **239e** (*N*-trifluoroacetylgluta-

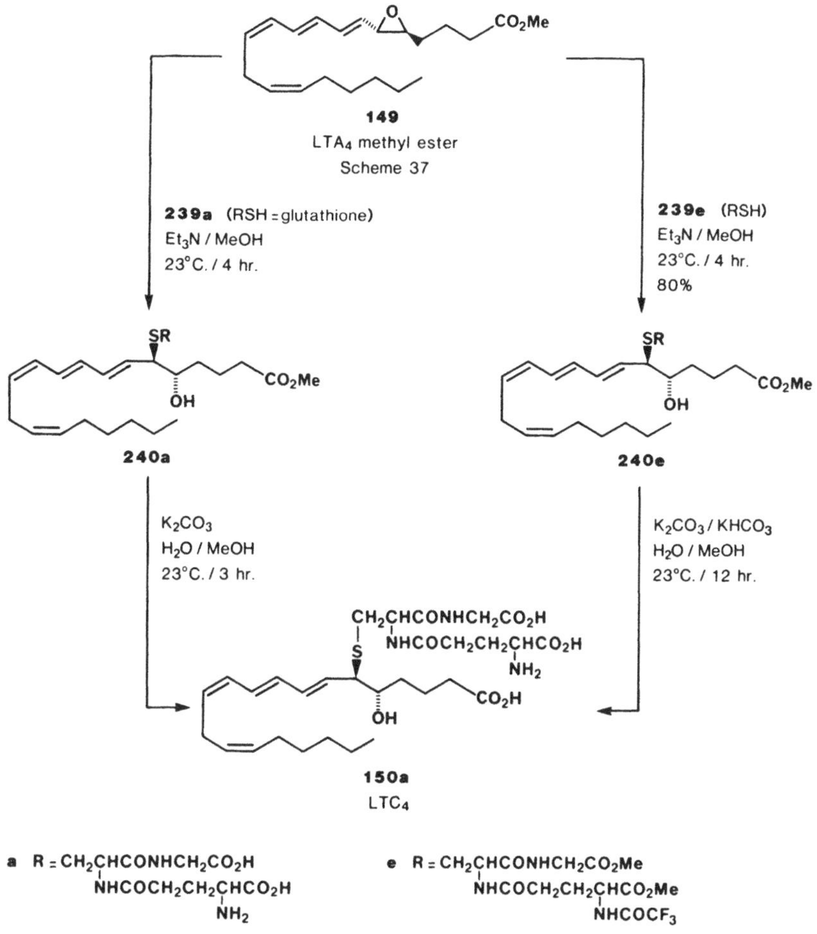

SCHEME 46. Synthesis of LTC$_4$ from optically active LTA$_4$.[46]

thione dimethyl ester, *N*-TFA-GSH-Me$_2$) to yield the adducts **240a** and **240e**, respectively. Mild basic hydrolysis of these adducts yielded LTC$_4$ (**150a**). The synthetic material was found to be identical with natural LTC$_4$ by comparison of their UV spectra, HPLC properties, biological activity and their reaction with soybean lipoxygenase to produce a conjugated tetraene.[46,72]

It should be noted that when the coupling reaction between a model epoxide and thiols was carried out under solvolytic conditions (LiClO$_4$ in MeOH), attack of the thiol occurred at the terminal double bond of the conjugated system rather than at the allylic position.

In a similar sequence of reactions, Corey and co-workers subsequently synthesized LTD$_4$ (**150b**) by first coupling **239f** (*N*-TFA-cys-gly-Me) with LTA$_4$ methyl ester (Scheme 47).[58] Hydrolysis of **240f** under the same conditions as used for the hydrolysis of **240e** (Scheme 46) was incomplete and gave a mixture of LTD$_4$ and **150j**, the *N*-trifluoroacetyl derivative of LTD$_4$, which was prepared separately for comparison by mild hydrolysis of **240j**. Under somewhat more basic conditions, for a longer period of time, the hydrolysis of **240f** was completed to yield LTD$_4$. Identity of the synthetic material with naturally occurring LTD$_4$ was secured by the same criteria as in the case of LTC$_4$.[73]

Merck Frosst Synthesis of LTC$_4$, LTD$_4$, and LTE$_4$

In a second independent synthesis, the Merck Frosst group disclosed the preparation of LTC$_4$ from racemic LTA$_4$ methyl ester as outlined in Scheme 48.[40] The initial synthesis

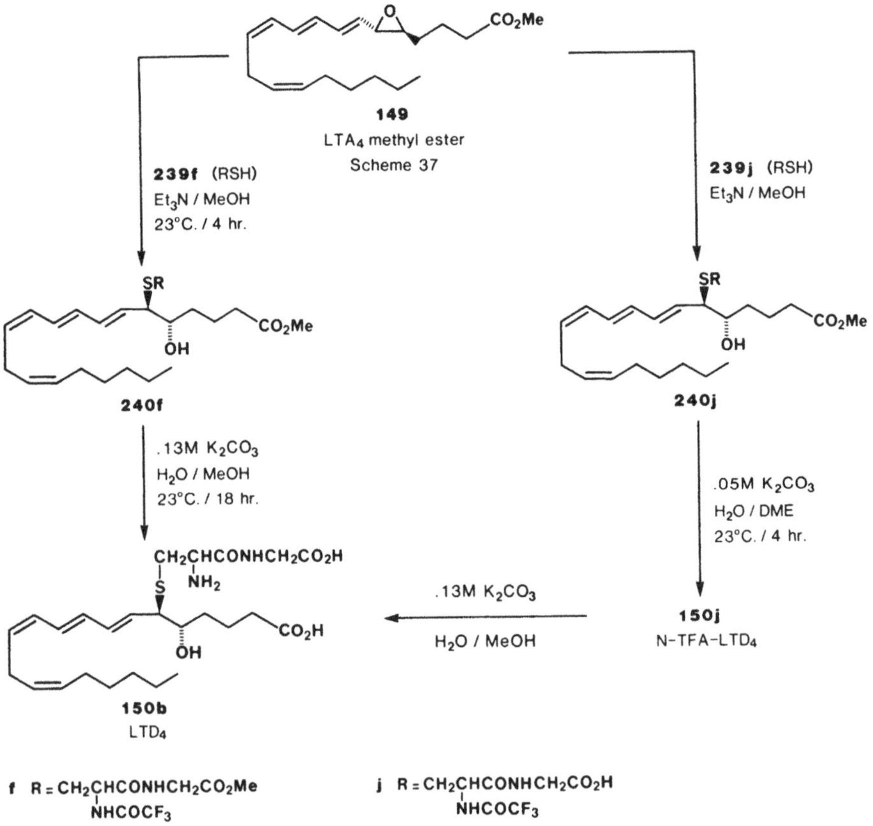

149

LTA$_4$ methyl ester

Scheme 37

239f (RSH)

Et$_3$N / MeOH

23°C. / 4 hr.

239j (RSH)

Et$_3$N / MeOH

23°C. / 4 hr.

240f

240j

.13M K$_2$CO$_3$

H$_2$O / MeOH

23°C. / 18 hr.

.05M K$_2$CO$_3$

H$_2$O / DME

23°C. / 4 hr.

150j

N-TFA-LTD$_4$

.13M K$_2$CO$_3$

H$_2$O / MeOH

150b

LTD$_4$

f   R = CH$_2$CHCONHCH$_2$CO$_2$Me
       NHCOCF$_3$

j   R = CH$_2$CHCONHCH$_2$CO$_2$H
       NHCOCF$_3$

SCHEME 47. Synthesis of LTD$_4$ from optically active LTA$_4$.[58]

(path A) utilized the novel reagent **241e**, the S-trimethylsilyl derivative of N-TFA-GSH-Me$_2$, to react with the epoxide of (±)-**149** to generate the diastereomers **240e** and 5-epi-6-epi-**240e** which were separated by HPLC. Hydrolysis of these separated derivatives led to two isomers of LTC$_4$, one of which showed the potent biological activity characteristic of natural LTC$_4$. The material was further characterized by its UV spectrum and conversion by lipoxygenase to a conjugated tetraene, a reaction also characteristic of natural LTC$_4$. The same pair of isomers of LTC$_4$ was also obtained by reacting free glutathione (**239a**) with (±)-**149** followed by hydrolysis of the separated diastereomers **240a** and 5-epi-6-epi-**240a**. With the more recent availability of fast atom bombardment MS, a comparison of the natural and synthetic LTC$_4$ served to further establish their identity.[74]

It was subsequently found that the trimethylsilyl reagents of structure **241** provided a general and mild synthesis of β-hydroxythioethers from a variety of epoxides,[75] but in the reaction of amino acid-type thiols with LTA$_4$, the best conditions turned out to be the base-catalyzed reaction of the free thiol with the epoxide (path B).[49,50]

A synthesis of LTD$_4$ from the reaction of racemic LTA$_4$ methyl ester with N-TFA-cys-gly-Me, **239f**, was carried out by the Merck Frosst group as shown in Scheme 49. A detailed comparison of the active diastereomer **150b**, thus obtained with naturally derived LTD$_4$ was carried out in collaboration with Morris and colleagues.[76] They were found to be identical in their UV spectra, HPLC behavior, biological activity, and reaction with soybean lipoxygenase. Furthermore, the mass spectra of the derivative **241l** prepared from both synthetic and natural LTD$_4$ were found to be identical. It should be noted that this work constituted the first positive structure identification of LTD$_4$.

The synthesis of LTE$_4$ was carried out as indicated in Scheme 50 by the condensation of

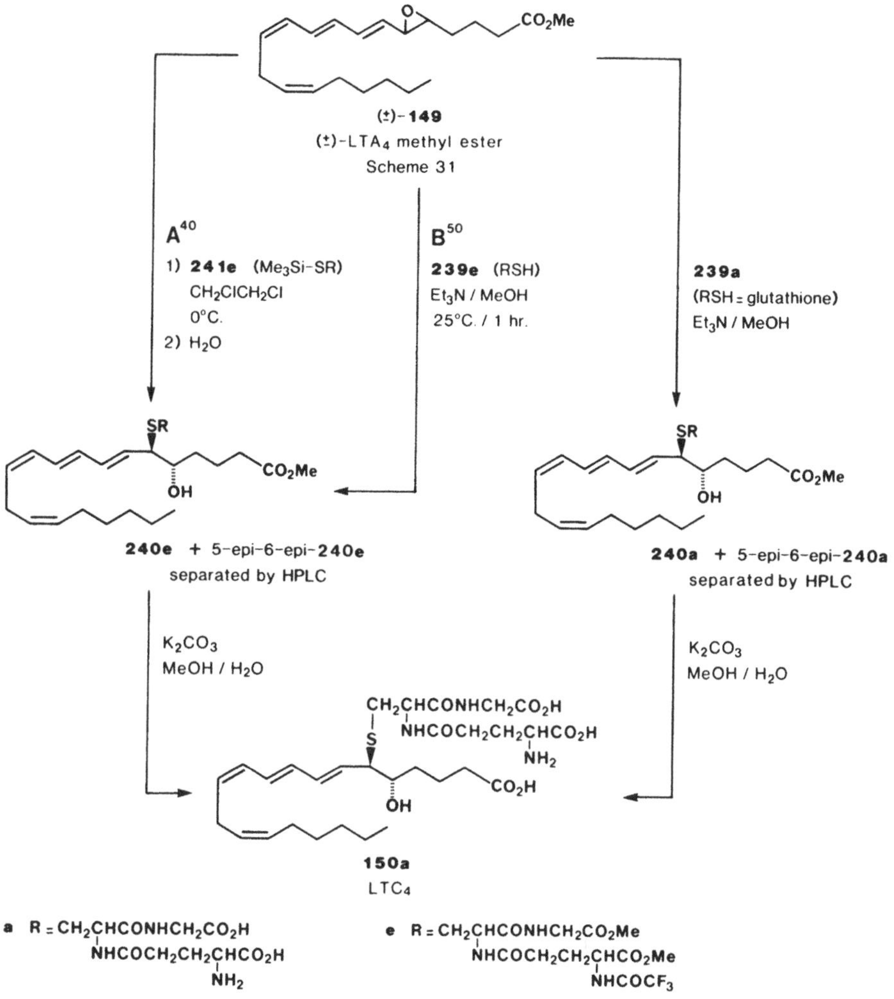

SCHEME 48. Synthesis of LTC$_4$ from racemic LTA$_4$.[40]

racemic LTA$_4$ methyl ester with *N*-TFA-cys-Me (**239g**) followed by separation of the diastereomers and hydrolysis.[51]

Finally, it may be noted that the ease of scale-up of the synthesis of racemic LTA$_4$ methyl ester (Schemes 30 and 31) allowed, in the early stages of the work at Merck Frosst, the preparation of gram quantities of LTC$_4$, LTD$_4$, and LTE$_4$ which have been made widely available to the scientific community.

The Merck Frosst group also carried out the synthesis of the pure natural isomers of LTC$_4$, LTD$_4$, and LTE$_4$ from optically active LTA$_4$ methyl ester (sections "Synthesis of LTA$_4$, 5-epi-LTA$_4$, 6-epi-LTA$_4$, and 5-epi-6-epi-LTA$_4$ from 2-Deoxy-D-Ribose" and "Synthesis of LTA$_4$ and Double Bond Isomers from D-Ribose").

As illustrated in Scheme 51, these preparations involve the same chemistry at that previously described.[61] Comparison of the biologically active leukotriene diastereomers obtained from racemic LTA$_4$ methyl ester (Schemes 48, 49, and 50) with the pure compounds prepared as in Scheme 51 established the identity of the compounds from the different syntheses.

Hoffmann-LaRoche Synthesis of LTE$_4$

The third independent synthesis of a leukotriene, LTE$_4$, was achieved by Rosenberger

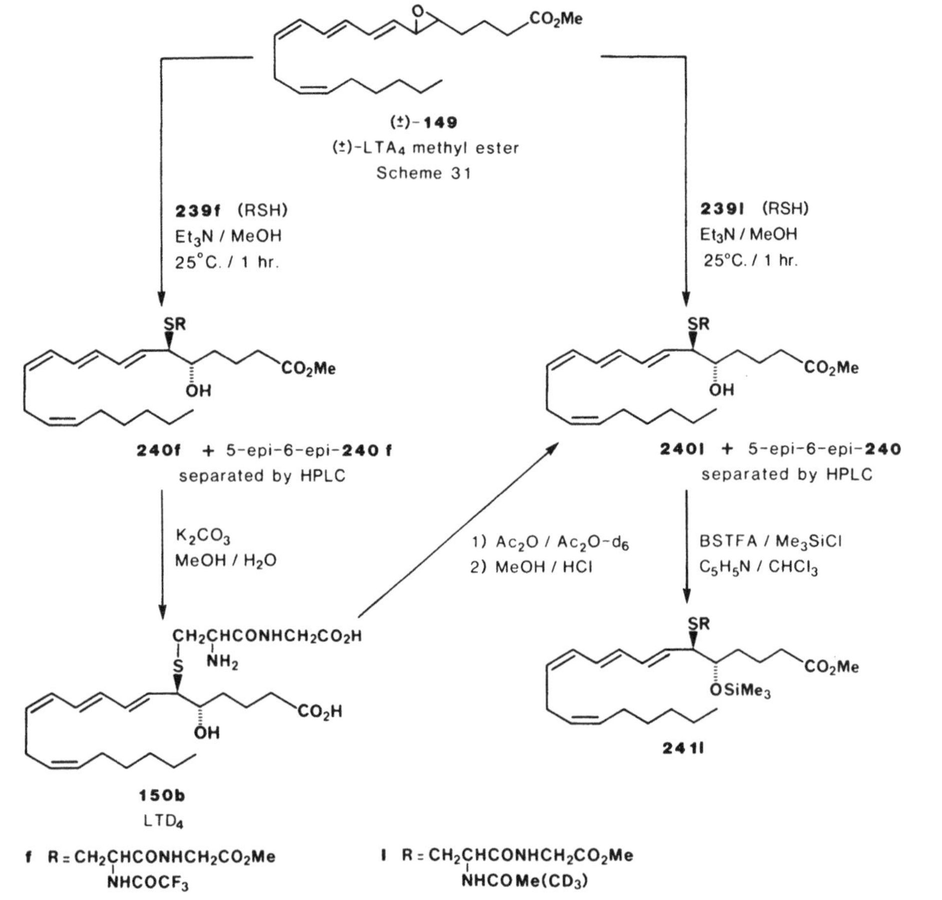

SCHEME 49. Synthesis of LTD$_4$ from racemic LTA$_4$.[76]

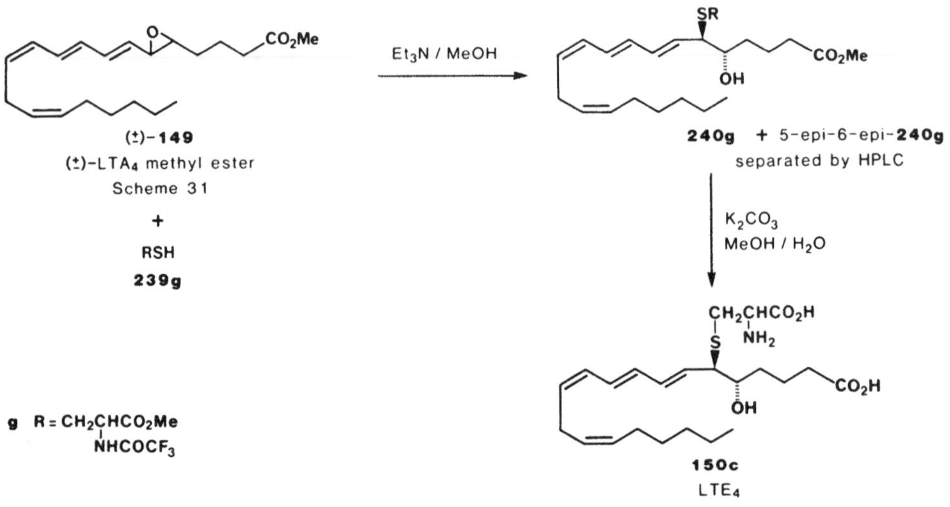

SCHEME 50. Synthesis of LTE$_4$ from racemic LTA$_4$.[51]

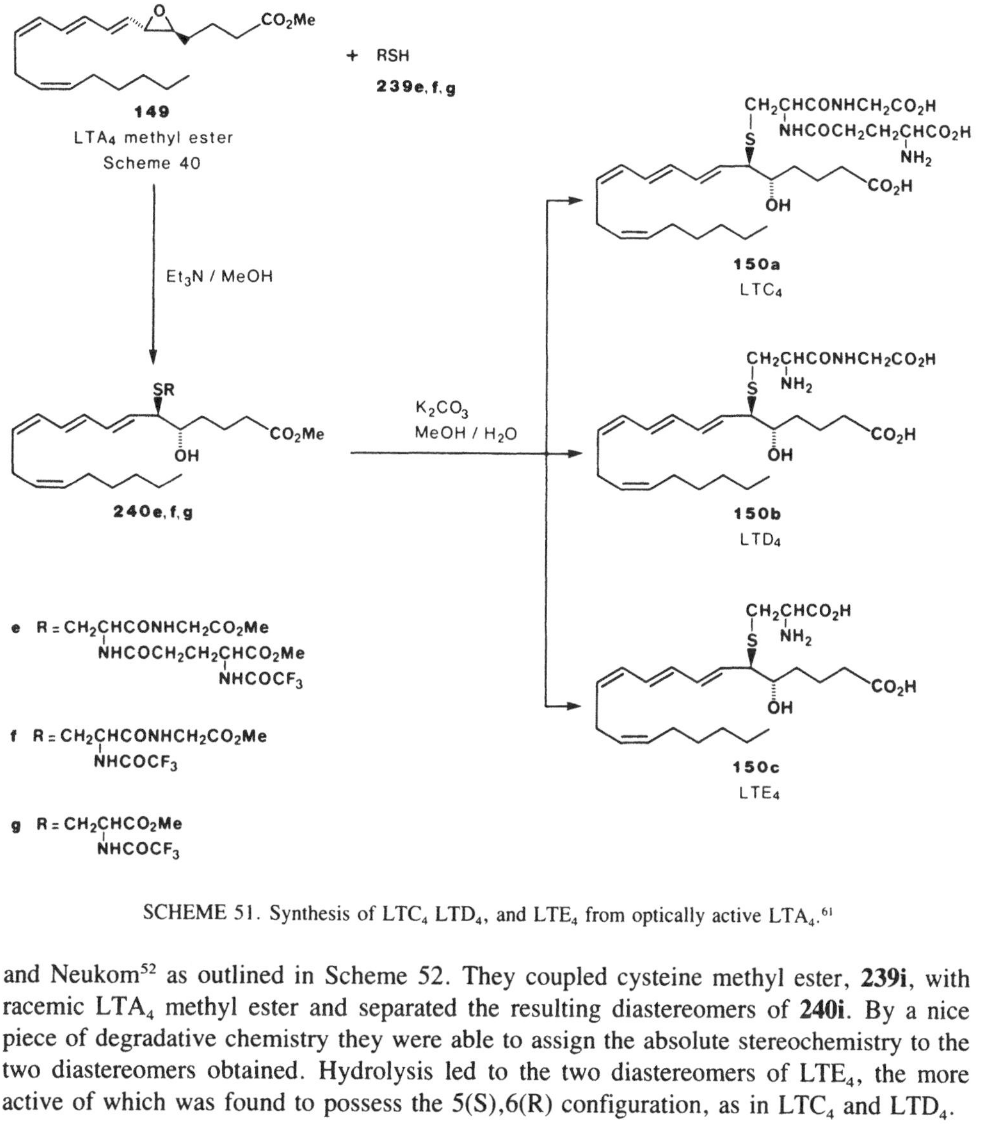

SCHEME 51. Synthesis of LTC_4 LTD_4, and LTE_4 from optically active LTA_4.[61]

and Neukom[52] as outlined in Scheme 52. They coupled cysteine methyl ester, **239i**, with racemic LTA_4 methyl ester and separated the resulting diastereomers of **240i**. By a nice piece of degradative chemistry they were able to assign the absolute stereochemistry to the two diastereomers obtained. Hydrolysis led to the two diastereomers of LTE_4, the more active of which was found to possess the 5(S),6(R) configuration, as in LTC_4 and LTD_4.

Synthesis of Biochemically Prepared Leukotrienes
Synthesis of LTF_4

This more recent member of cysteine-containing eicosanoids contains the γ-glutamylcys-teinyl dipeptide unit (see Chart 1); i.e., glutathione with glycine removed. Although it has not at this writing been isolated as a natural product, two groups have generated it bio-chemically from LTE_4,[77,78] indicating that it probably will be found in whole cell systems. The Merck Frosst group has synthesized this member of the leukotriene family by coupling the methyl benzyl ester of cys-γ-glu with LTA_4 methyl ester to generate **240k**, hydrolysis of which yielded **150d**, LTF_4 (Scheme 53).[79]

Synthesis of 14,15-LTC_4 and 14,15-LTD_4

As in the case of LTF_4, these positional isomers of LTC_4 and LTD_4 (Chart 2) have not yet been demonstrated to be natural products, but Sih and co-workers[70] have demonstrated the biosynthesis of both **243a** and **243b** (Scheme 54) when 14,15-LTA_4 was incubated with RBL-1 cells with added cysteine. The synthesis of both 14,15-LTC_4 and 14,15-LTD_4 (Scheme

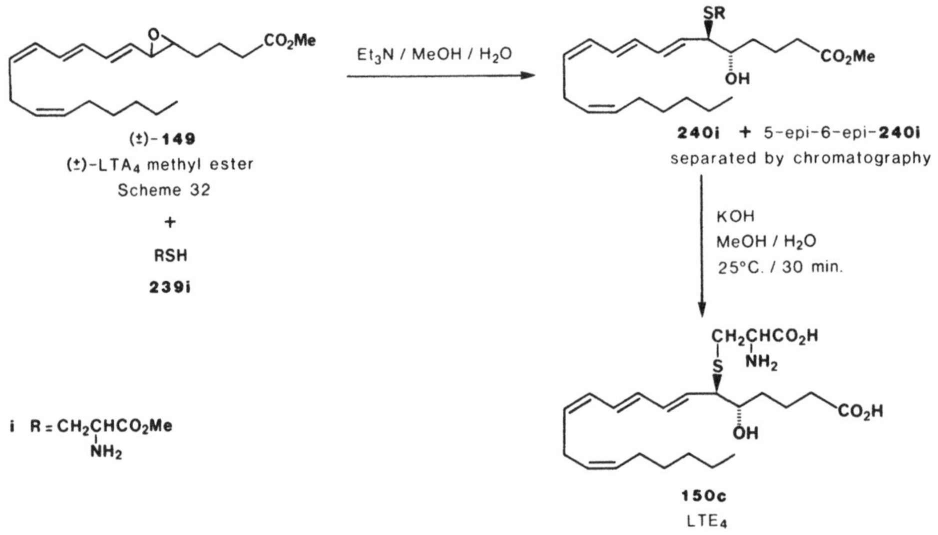

SCHEME 52. Synthesis of LTE$_4$ from racemic LTA$_4$.[52]

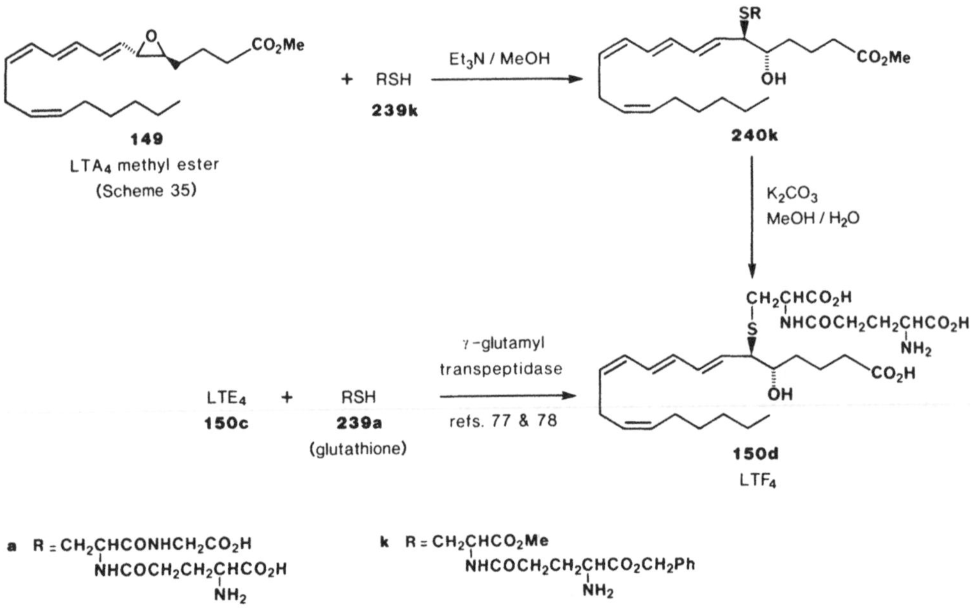

SCHEME 53. Synthesis of LTF$_4$ from optically active LTA$_4$.[79]

54) was actually carried out by Corey's group[69] prior to their biosynthetic preparation. This was effected by the now-familiar coupling of the epoxide **236** with the appropriate blocked thiol precursors to yield **242e** and **242f**, which were hydrolyzed to the target structures.

Synthesis of Leukotrienes with Higher Oxidation Levels of Sulfur

Although limited, evidence has been obtained that more highly oxidized levels of the sulfide unit in the leukotrienes are, or can be, formed under biological conditions. The two obvious analogs, the sulfones and the sulfoxides, have now been studied and their preparation is outlined in the following sections.

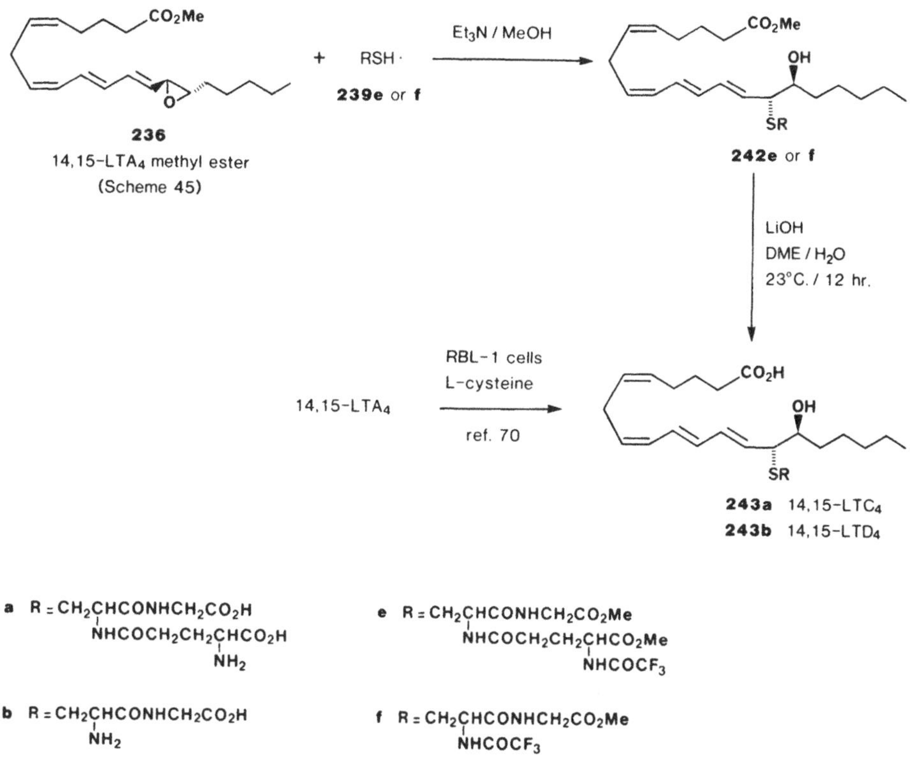
CO2Me
236
14,15-LTA4 methyl ester
(Scheme 45)
+ RSH
239e or f
Et3N / MeOH
CO2Me
OH
SR
242e or f
LiOH
DME / H2O
23°C. / 12 hr.
RBL-1 cells
L-cysteine
14,15-LTA4
ref. 70
CO2H
OH
SR
243a 14,15-LTC4
243b 14,15-LTD4
a R = CH2CHCONHCH2CO2H NHCOCH2CH2CHCO2H NH2
e R = CH2CHCONHCH2CO2Me NHCOCH2CH2CHCO2Me NHCOCF3
b R = CH2CHCONHCH2CO2H NH2
f R = CH2CHCONHCH2CO2Me NHCOCF3

SCHEME 54. Synthesis of 14,15-LTC_4 and LTD_4 from 14,15-LTA_4.[69]

Synthesis of the Sulfones of LTC_4, LTD_4, LTE_4, and LTF_4

Working with SRS-A from rat peritoneal cells, Ohnishi et al.[80] isolated a potent contractile substance which they characterized as the *sulfone* of LTC_4 (see Scheme 1). Although the sulfone structure was somewhat unexpected in the light of the fact that previous workers had identified only the sulfide structure in LTC_4, LTD_4, and LTE_4, its proposal stimulated efforts to synthesize the compound. This was achieved by the Merck Frosst group, as indicated in Scheme 55, by the direct oxidation of the sulfide precursor with potassium hydrogen persulfate.[81] This proved to be a unique reagent for sulfone formation in this series of compounds and permitted the preparation of the sulfones of LTC_4, LTD_4, and LTE_4,[81] as well as the sulfone of LTF_4.[79] Attempts to obtain the sulfones with a number of well-known oxidizing agents (*m*-chloroperbenzoic acid, H_2O_2, $NaIO_4$) were unsuccessful, yielding only the sulfoxides, and attempts to react the epoxide (LTA_4) with model sulfinates resulted in extensive side reactions. A second synthesis of the sulfones was carried out from the acetylenic analogs **245** in order to confirm that the isolated double bond in **150** had not been subject to reaction with the oxidizing agent. Biological evaluation of the sulfones **244a**, **b**, and **c** confirmed that they are indeed powerful contractile agents, and are qualitatively very similar to the parent sulfides, LTC_4, LTD_4, and LTE_4.[82]

Synthesis of the Sulfoxides of LTC_4 and LT_{D4}

It has recently been shown that human polymorphonuclear leukocyte cells (PMNs), when stimulated with phorbol myristate acetate, bring about the metabolism of LTC_4 to three different types of products, two of which have been identified as having the sulfoxide structure **247a** and the diol pair **147/148** (Scheme 56).[83] The sulfoxides (as diastereomeric pairs) **247a** and **247b** of LTC_4 and LTD_4 have now been prepared by Corey's group[84] (Scheme 56) using standard oxidizing agents. It was observed that LTC_4 sulfoxide undergoes an overall 1,7-

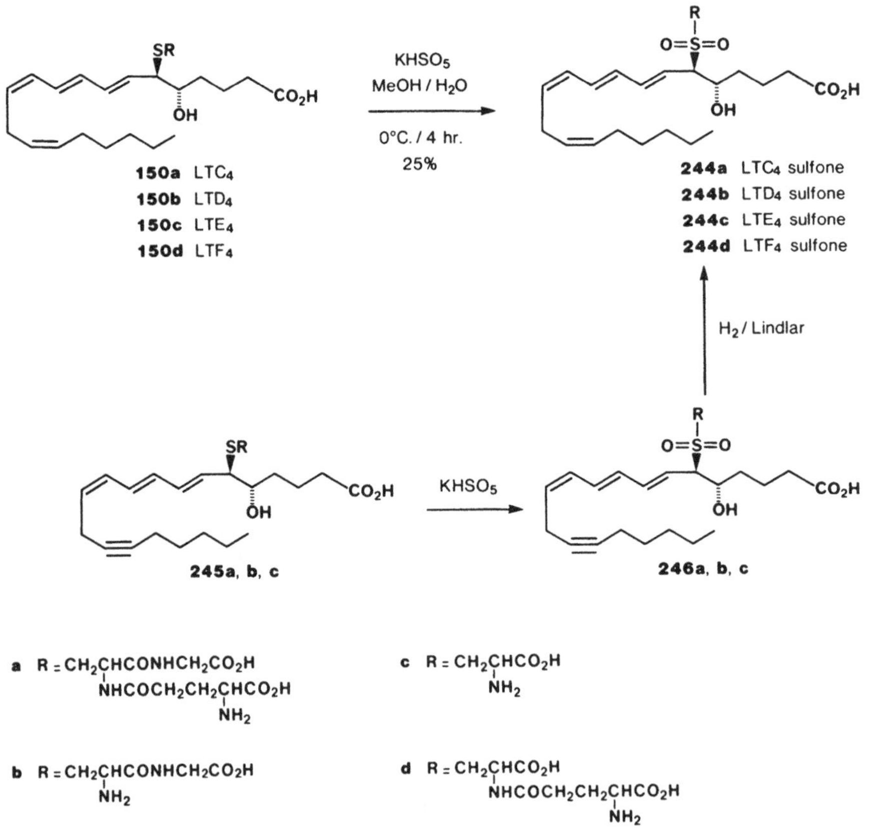

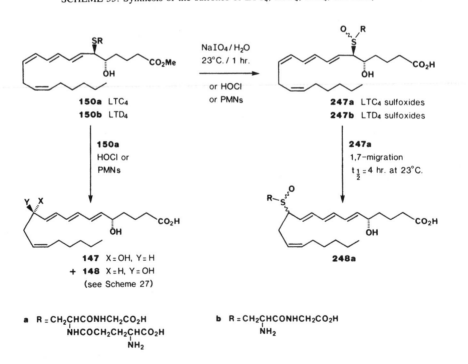

SCHEME 55. Synthesis of the sulfones of LTC$_4$, LTD$_4$, LTE$_4$, and LTF$_4$.[79,81]

SCHEME 56. Synthesis of the sulfoxides of LTC$_4$ and LTD$_4$.[84]

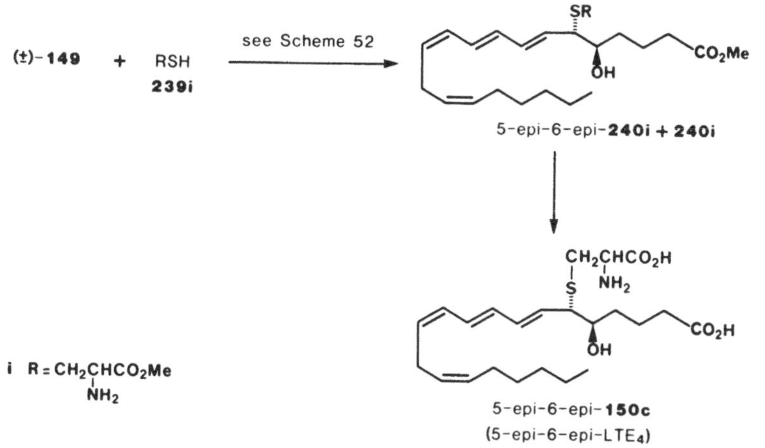

SCHEME 57. Synthesis of 5-epi-6-epi-LTE$_4$.[52]

rearrangement of the sulfur unit to produce **248a**, but curiously, LTD$_4$ sulfoxide is much more stable, and does not rearrange appreciably over 72 hr.

Close Structural Analogs of the Leukotrienes

During the course of synthesis of the various components of SRS-A, a number of closely related analogs have been prepared by the various research groups involved. During the early stages of the synthetic efforts several isomeric compounds were prepared in order to prove the structures of the different leukotrienes. In addition to structure proof, it was important to synthesize a number of closely related compounds for biological study, and thus gain insight into the structural features responsible for the biological activity. What constitutes "close structural analogs" can only be arbitrarily defined, and most of the compounds reported in this section are in fact isomers of LTC$_4$, LTD$_4$, or LTE$_4$. No compounds which have been made for the purpose of potential development as drugs are included, as this would more properly be the domain of a medicinal chemical review.

Synthesis of the 5,6-epimers of the Leukotrienes
Hoffmann-LaRoche Synthesis of 5-epi-6-epi-LTE$_4$

Since they started with racemic LTA$_4$ methyl ester, Rosenberger and Neukom[52] obtained a mixture of **240i** and 5-epi-6-epi-**240i**, which were separated by chromatography. Hydrolysis of the latter then yielded 5-epi-6-epi-LTE$_4$ (Scheme 57), which showed one half the biological activity of the natural isomer.

Harvard Synthesis of 6-epi-LTC$_4$ and 6-epi-LTD$_4$

As part of their work in the structure proof of the natural leukotrienes, Corey's group carried out a stereospecific synthesis of 6-epi-LTA$_4$ methyl ester as discussed in the section "Synthesis of 6-epi-LTA$_4$ from D-Mannose" and Scheme 39. Coupling of the epoxide with the appropriately blocked derivative of glutathione, **239e**, or cysteinylglycine, **239f**, followed by hydrolysis of the 6-epi-**240** derivatives yielded the desired 6-epi-LTC$_4$ and 6-epi-LTD$_4$ (Scheme 58).[60] On guinea pig parenchymal strips 6-epi-**150a** had 1/10 the activity of natural LTC$_4$ and 6-epi-**150b** had 1/500 the activity of LTD$_4$.

Merck Frosst Synthesis of 5-epi, 6-epi, and 5-epi-6-epi-LTC$_4$ and LTD$_4$

Following up on their stereospecific total syntheses of the four optical isomers of LTA$_4$ methyl ester, described in an earlier section and Schemes 35 and 36, the Merck Frosst group

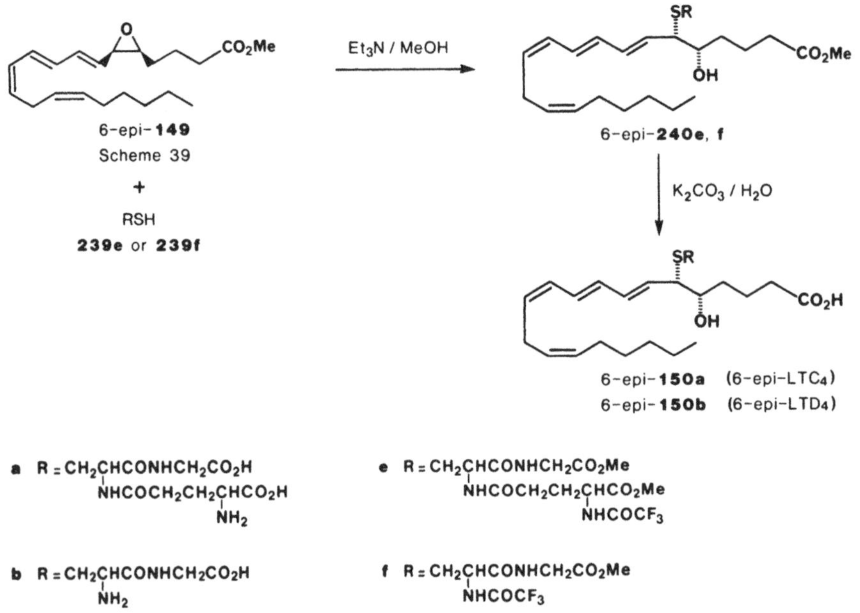

SCHEME 58. Synthesis of 6-epi-LTC$_4$ and 6-epi-LTD$_4$.[60]

has prepared the three possible unnatural 5,6-isomers of LTC$_4$ and LTD$_4$.[85] Scheme 59 summarizes the preparation of the six compounds thus obtained. The same compounds were also obtained starting from racemic **149** and racemic *cis*-**149**, and the structure of each pair of diastereomers so obtained was assigned by comparison with the individual isomers obtained from the optically pure epoxides.[49] In the LTC$_4$ series, on the guinea pig ileum (in Tyrode's solution containing 1 μM atropine and 0.5 $\mu g/m\ell$ of timolol), 5-epi-LTC$_4$, 6-epi-LTC$_4$, and 5-epi-6-epi-LTC$_4$ showed 1/20, 1/70, and 1/60 the potency of LTC$_4$, respectively; in the LTD$_4$ series, 5-epi-LTD$_4$ had 1/20 the activity of natural LTD$_4$ and 5-epi-6-epi-LTD$_4$ had 1/30 the activity.[86] Rather surprisingly, 6-epi-LTD$_4$ showed weak antagonist rather than agonist activity.

Synthesis of Double Bond Isomers of the Leukotrienes

With four double bonds, there are 16 possible E-Z double bond isomers of the leukotriene structure. There has been no systematic attempt to prepare all these isomers, and only four, derived from the corresponding LTA$_4$ (epoxide) precursors, have been studied.

Synthesis of 7Z-LTD$_4$

As outlined in an earlier section and Scheme 38, Ernest and co-workers described a synthesis of 7Z-LTA$_4$ methyl ester. This was converted to 7Z-LTD$_4$ (Scheme 60) by standard conditions and was found to exhibit about 1/10 the activity of LTD$_4$ on the guinea pig ileum.

Synthesis of 9Z-LTC$_4$ and 9Z-LTD$_4$

The synthesis and behavior of the 9Z-isomers of LTC$_4$ and LTD$_4$ has been studied by the group at the Lilly Research Center[44,45] and by Sih and co-workers.[43] As outlined in Scheme 61, condensation of glutathione (**239a**) with ($\pm$)-9Z-**149** (($\pm$)-9Z-LTA$_4$ methyl ester) yielded the monomethyl ester 9Z-**240a**, which was hydrolyzed to 9Z-LTD$_4$.[45] Since racemic epoxide was used in all cases, the 5-epi-6-epi diastereomers were also obtained and separated by HPLC. In all cases, these compounds underwent a facile [1,7]-hydrogen migration as discussed in an earlier section. The 9Z-LTD$_4$ isomer was found to be only 2.5 times less potent

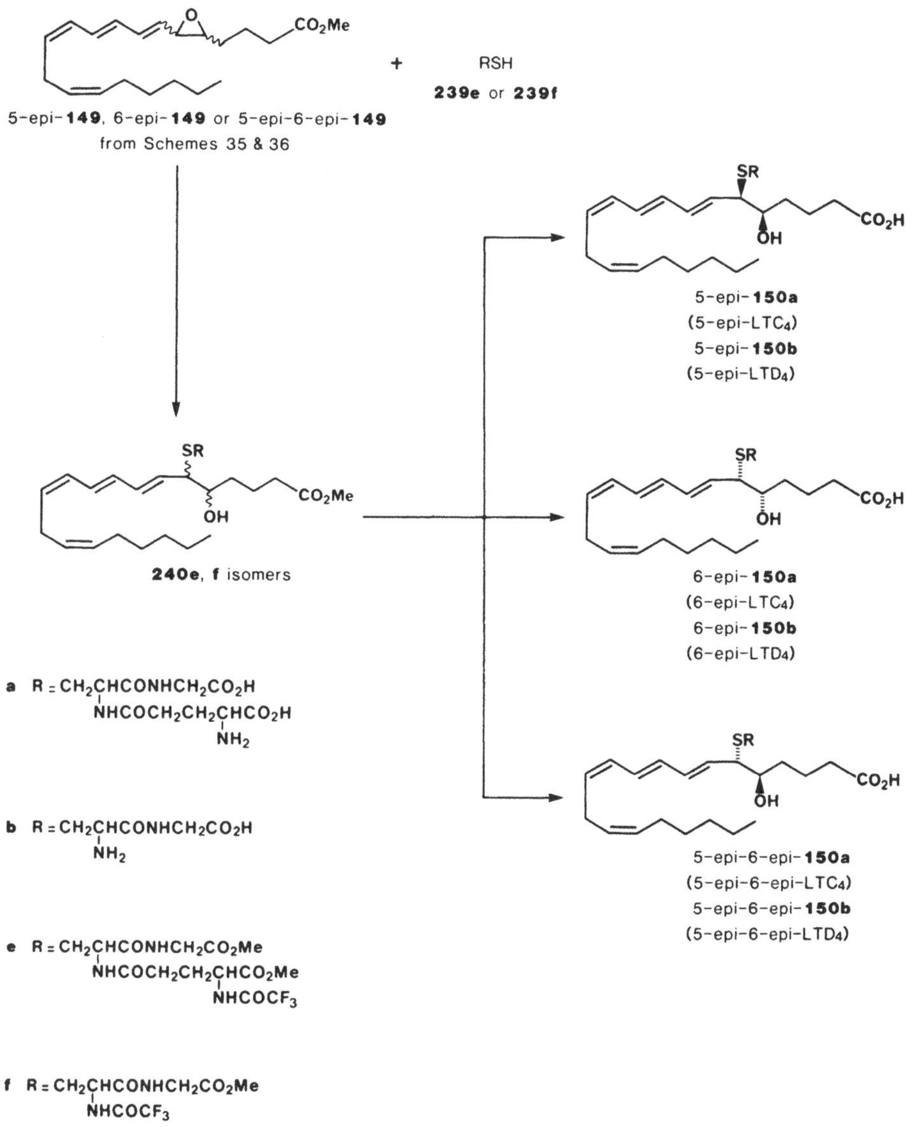

SCHEME 59. Synthesis of the 5-epi, 6-epi, and 5-epi-6-epi isomers of LTC$_4$ and LTD$_4$.[85]

than natural LTD$_4$ on the guinea ileum, whereas the conjugated tetraene **154b** derived from 9Z-LTD$_4$ was approximately 1000 times less potent than LTD$_4$.

Synthesis of 9Z,11E-LTD$_4$

With ($\pm$)-9Z,11E-LTA$_4$ methyl ester in hand, the Lilly group prepared the corresponding isomers of LTC$_4$[44] and LTD$_4$,[45] as illustrated in Scheme 62. Once again, the diastereomeric pairs of the structures **240** were separated, and the more active were assigned the natural 5(S), 6(R) configuration. The 9Z,11E isomer of LTD$_4$ was found to be quite potent, showing approximately 50% the contractile activity of LTD$_4$ on the guinea pig ileum.[45]

Synthesis of the 11E Isomer of LTC$_4$, LTD$_4$, and LTE$_4$

The 11E double bond isomer of the leukotrienes has been the object of several studies because it has been isolated as a minor component from several biological systems. Corey's

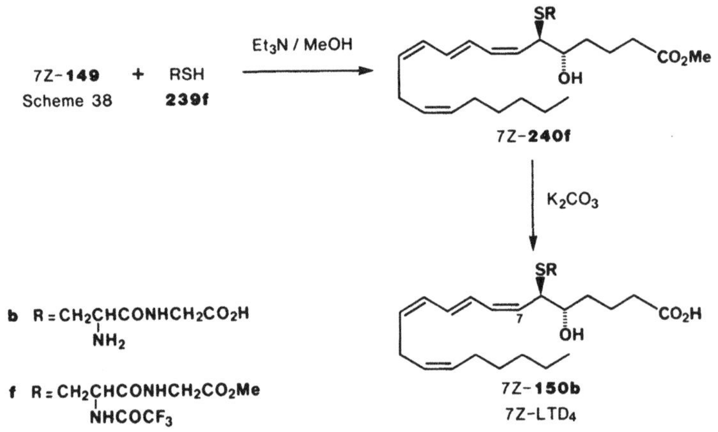

SCHEME 60. Synthesis of 7Z-LTD$_4$.[59]

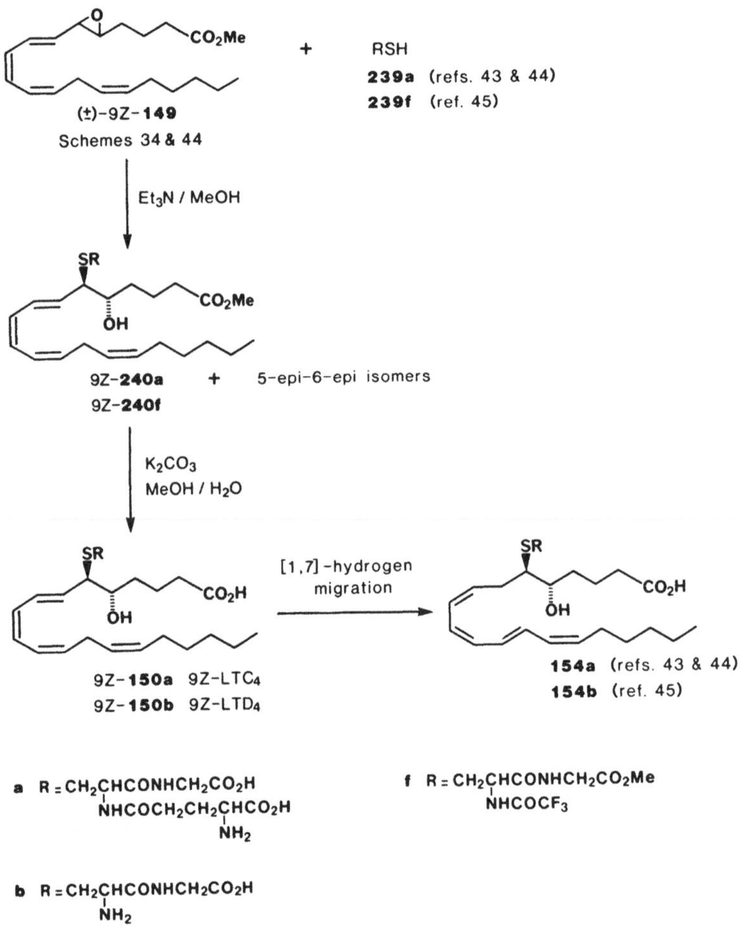

SCHEME 61. Synthesis of 9Z-LTC$_4$ and 9Z-LTD$_4$.

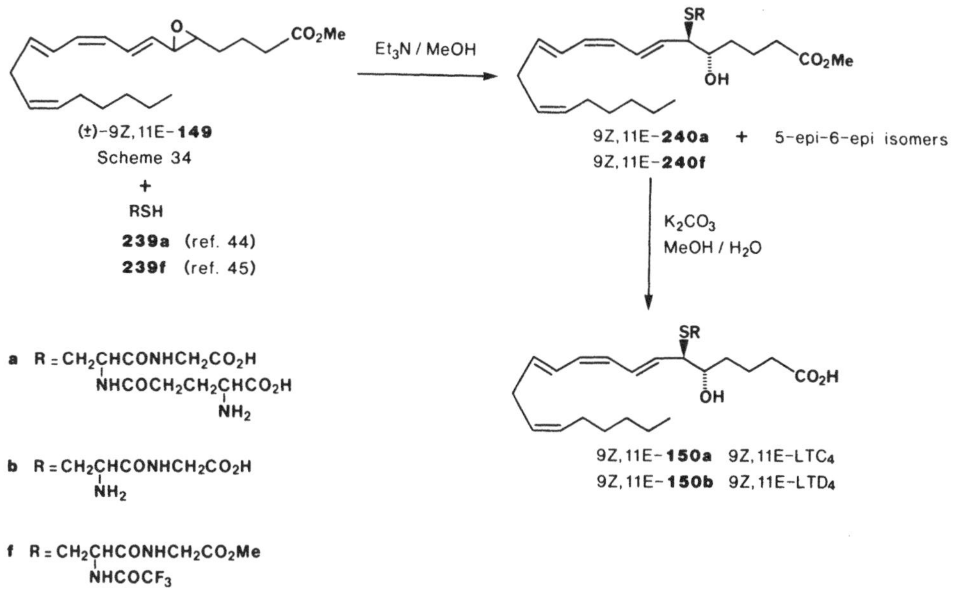

SCHEME 62. Synthesis of 9Z,11E-LTC$_4$ and 9Z,11E-LTD$_4$.

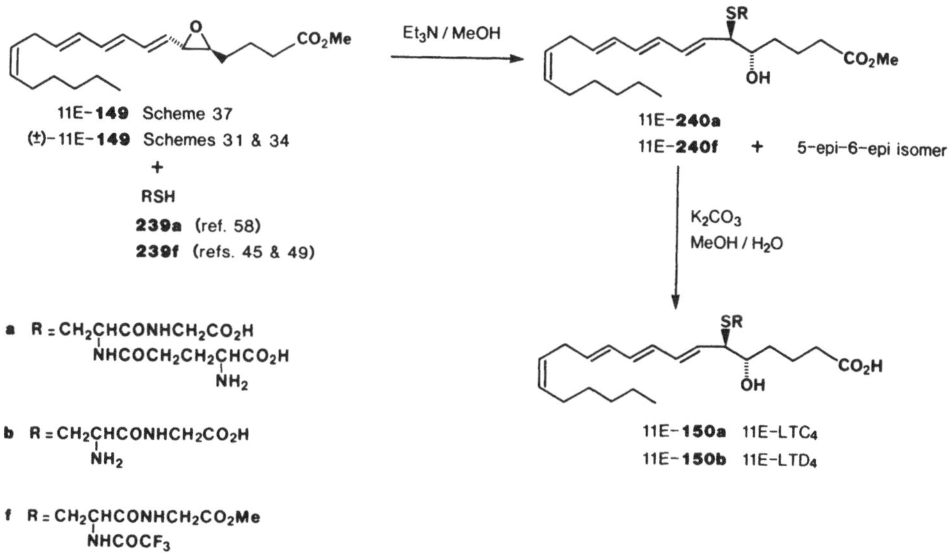

SCHEME 63. Synthesis of 11E-LTC$_4$ and 11E-LTD$_4$.

group[58] was the first to prepare an 11E isomer in this series with their synthesis of 11E-LTC$_4$, as shown in Scheme 63. The Lilly[45] and Merck Frosst[49] groups subsequently prepared 11E-LTD$_4$ from racemic 11E-LTA$_4$ methyl ester and separated the 5,6-epimers of 11E-**240f** prior to hydrolysis. The biological activities for contraction of the guinea pig ileum by the 11E isomers of LTC$_4$, LTD$_4$, and LTE$_4$ were found by Sih and co-workers[87] to be 1/3, 1/5 and 1/8 that of the natural compounds, respectively.[87]

Mechanism of Formation of the 11E Isomers of the Leukotrienes

As mentioned earlier, the 11E isomers of LTC$_4$, LTD$_4$, and LTE$_4$ have been obtained

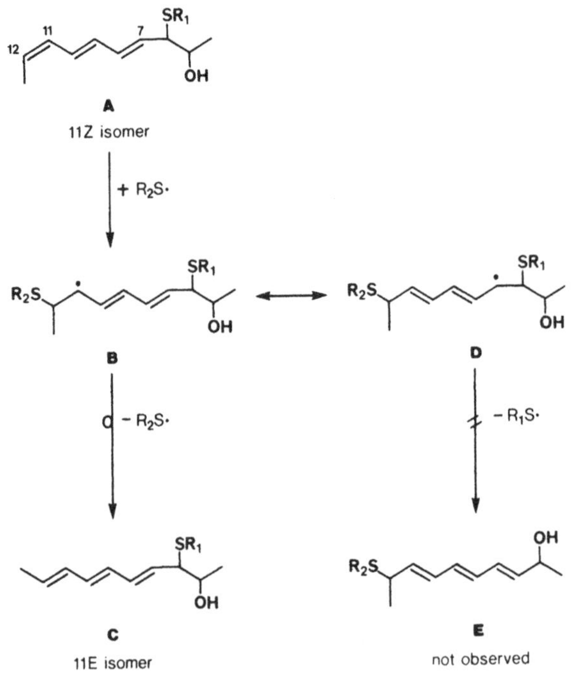

SCHEME 64. Proposed mechanism for formation of 11E leukotrienes.[58,87]

during the course of their isolation from various biological sources. Sih and co-workers[87] have carried out a very careful study of the formation of the 11E isomers of LTC_4 and LTE_4 under both biological and chemical conditions of preparation. Working in the absence and presence of free thiols and using the free radical inhibitor, 4-hydroxy-2,2,6,6-tetramethyl-piperidinoxy (HTMP), they demonstrated quite conclusively that the formation of the 11E isomers is a free radical catalyzed reaction requiring the presence of free thiol groups in the medium. They further conclude that their formation in biological systems is probably not an enzymatic process, but rather chemical, catalyzed by endogenously formed thiol radicals. Thus, at present, it would seem to be a matter of semantics whether the 11E isomers should be called natural products, since they are probably formed in biological systems, but by a nonenzymatic process.

With regard to the mechanism of the 11Z → 11E isomerization, both Corey et al.[58] and Atrache et al.[87] suggest a reversible addition of a thiol radical to C_{12} of the triene unit, as a result of which the more stable E isomer, C, is formed (Scheme 64, part structures **A**, **B**, **C**). Sih has quite conclusively shown that there is no incorporation of label in recovered 11E-LTE_4 when radioactive free cysteine is present as the thiol source, thus ruling out the sequence **A** → **B** → **D** → **E**, which would have yielded an isomeric LTE_4 containing radioactive cysteine, now attached to C_{12}. It is rather striking that an intermediate such as **B**, which should be in facile resonance equilibrium with **D**, should eliminate $R_2S\cdot$ (**B** → **C**) to the complete exclusion of the known [88] alternate possibility (**D** → **E**) with loss of $R_1S\cdot$. This leads us to suggest an alternative mechanistic possibility (Scheme 65) in which the external thiol radical ($R_2S\cdot$) adds to C_7 rather than to C_{12}, generating the resonance stabilized pair **G** ↔ **H**. Such a radical intermediate can then eliminate *only* $R_2S\cdot$, generating **C** exclusively, as is observed. Regioselective addition to C_7 to give rise to **G** may be explained by the formation of a σ-sulfuranyl radical similar to structures investigated by Perkins et al.,[89] which could then rearrange to **G**, or by a more conventional type of hydrogen bonding association between the polar groups of R_1 and R_2.

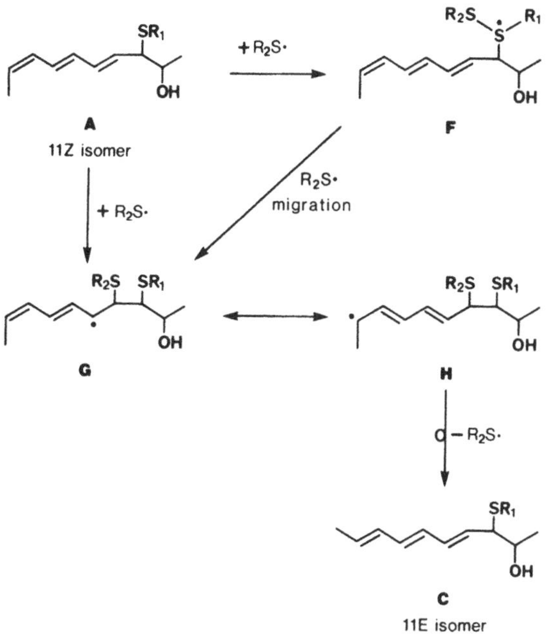

SCHEME 65. Alternative mechanism for formation of 11E leukotrienes.[49]

Hydro Derivatives of the Leukotrienes

In order to study the effect of the double bonds on the biological activity of the leukotrienes, several partially or fully saturated analogs of the leukotrienes have been studied. The biologial activity of the partially saturated analogs of LTC_4 and LTD_4 shown in Chart 7 has been studied by Drazen et al.,[90] but the details of their synthesis have not been published.

The fully saturated analogs of LTC_4, LTD_4, and LTE_4 have been prepared by Young et al.,[91] as outlined in Scheme 66. It is obvious that the epoxide opening reaction could not be utilized in the saturated series, as the regiochemistry is not subject to the control existing in allylic epoxides. Thus, the use of sulfenyllactonization was investigated as a means of introducing the required oxygen and sulfur functionality at C_5 and C_6 respectively, using 5E-eicosenoic acid (**249**) as substrate. Although there was no precedent for the use of polyfunctional sulfenyl chlorides such as **250**, it was found that the regio- and stereochemistry were satisfactorily controlled, and that the desired erythro δ-lactones **251** could be obtained in 30 to 40% yield. The diastereomeric pairs (5(S),6(R) and 5(R),6(S)) of lactones obtained were separated by chromatography and the absolute configuration of each determined by desulfurization to the corresponding 5-hydroxyeicosanoic acid δ-lactone. Starting with 5Z-eicosanoic acid, the threo analog of LTE_4 was also prepared.

These perhydro derivatives of LTC_4, LTD_4, and LTE_4 were found to retain contractile properties on the guinea pig trachea, but were about 1000 times less potent than the parent compounds. Taken together with Drazen's results, this indicates the important role of the double bonds, and particularly the $C_{7,8}$ double bond, in determining the biological activity of the natural leukotrienes.

Synthesis of 5-Desoxy LTD_4

For purposes of further structure-activity studies, Corey and Hoover[92] have carried out a total synthesis of the 5-desoxy analog of LTD_4. As outlined in Scheme 67, when **254**, formed *in situ* from **252** and the lithium derivative of **253**, was allowed to warm to room temperature, it underwent a double [3,2] sigmatropic rearrangement to the terminal sulfoxide **255**. After

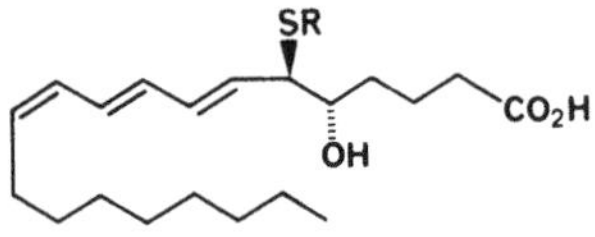

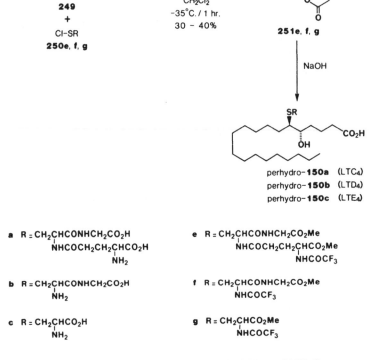

Potency on ileum
relative to parent cpd.

14,15-H$_2$-**150a** 14,15-H$_2$-LTC$_4$ 1
14,15-H$_2$-**150b** 14,15-H$_2$-LTD$_4$ 2/3

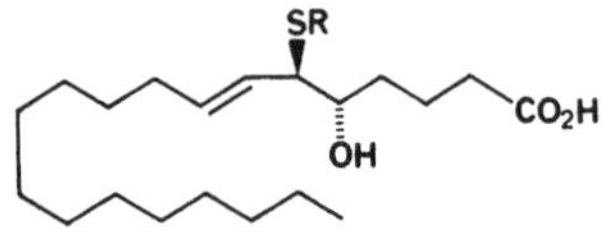

7E-H$_6$-**150a** 7E-H$_6$-LTC$_4$ 1/3
7E-H$_6$-**150b** 7E-H$_6$-LTD$_4$ 1/200

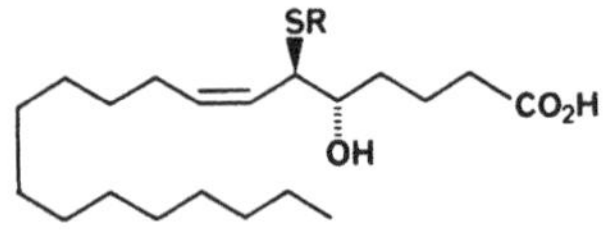

7Z-H$_6$-**150a** 7Z-H$_6$-LTC$_4$ 1/3
7Z-H$_6$-**150b** 7Z-H$_6$-LTD$_4$ 1/140

a R = CH$_2$CHCONHCH$_2$CO$_2$H
$\quad\quad$ NHCOCH$_2$CH$_2$CHCO$_2$H
$\quad\quad\quad\quad\quad\quad$ NH$_2$

b R = CH$_2$CHCONHCH$_2$CO$_2$H
$\quad\quad\quad$ NH$_2$

CHART 7. Partially saturated derivatives of LTC$_4$ and LTD$_4$.[90]

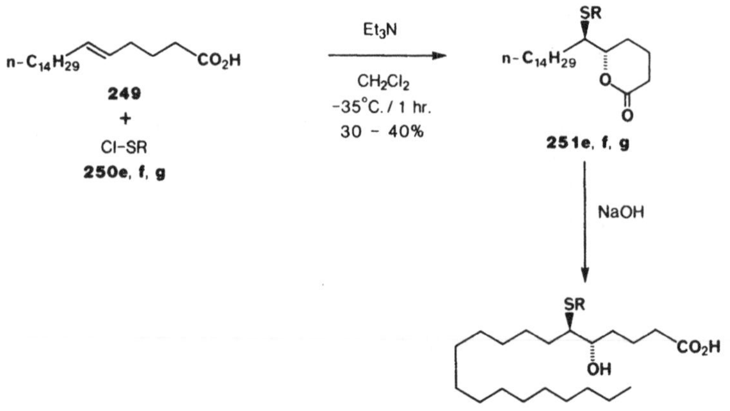

Et$_3$N

CH$_2$Cl$_2$
-35°C. / 1 hr.
30 – 40%

249
+
Cl-SR
250e, f, g

251e, f, g

NaOH

perhydro-**150a** (LTC$_4$)
perhydro-**150b** (LTD$_4$)
perhydro-**150c** (LTE$_4$)

a R = CH$_2$CHCONHCH$_2$CO$_2$H
$\quad\quad$ NHCOCH$_2$CH$_2$CHCO$_2$H
$\quad\quad\quad\quad\quad\quad$ NH$_2$

b R = CH$_2$CHCONHCH$_2$CO$_2$H
$\quad\quad\quad$ NH$_2$

c R = CH$_2$CHCO$_2$H
$\quad\quad\quad$ NH$_2$

e R = CH$_2$CHCONHCH$_2$CO$_2$Me
$\quad\quad$ NHCOCH$_2$CH$_2$CHCO$_2$Me
$\quad\quad\quad\quad\quad\quad$ NHCOCF$_3$

f R = CH$_2$CHCONHCH$_2$CO$_2$Me
$\quad\quad\quad$ NHCOCF$_3$

g R = CH$_2$CHCO$_2$Me
$\quad\quad\quad$ NHCOCF$_3$

SCHEME 66. Synthesis of perhydro LTC$_4$, LTD$_4$, and LTE$_4$.[91]

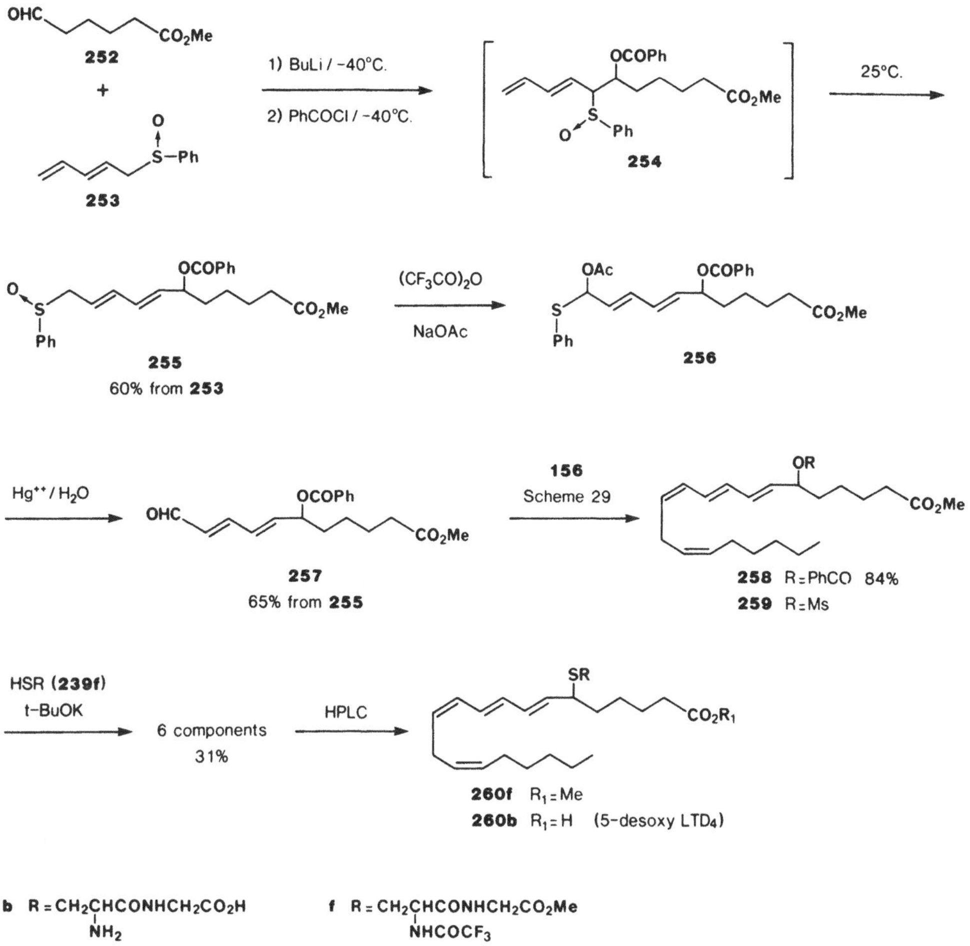

SCHEME 67. Synthesis of 5-desoxy LTD_4.[92]

conversion of **255** to **256** by a Pummerer reaction, the full carbon skeleton was assembled by a Wittig reaction with the ylid of **156**. Conversion of **258** to the mesylate **259** was effected by standard procedures; **259** was then reacted with the potassium salt of *N*-TFA-cys-gly-Me (**239f**). This reaction gave rise to six compounds in a combined yield of 31%, in sharp contrast to the numerous condensations already described with the epoxide as the leaving group, in which very clean $S_{N}2$ displacement occurs. Two of the compounds were found to be the diastereomeric pair of the desired structure **260f**, although the absolute configuration at C_6 was not established. Hydrolysis then yielded the two isomers of **260b**, 5-desoxy LTD_4. Both isomers showed less than 1% of the contractile activity of LTD_4 on the guinea ileum and parenchymal strips, demonstrating the importance of the C_5 hydroxyl group to biological potency.

CONCLUSION

The synthesis of the major lipoxygenase-derived products arising from AA has now been achieved. A number of the minor components and secondary metabolites of the leukotrienes will undoubtedly be synthesized in the near future, particularly those for which important biological properties may be found.

The synthetic availability of the compounds outlined in this review, especially of LTB_4,

LTC_4, LTD_4, and LTE_4, is already proving to be of great value in two major broad areas of investigation. Because of the difficulty of isolating the pure compounds in reasonable amounts, it has been slow work to study and evaluate the actions of these powerful agents in a variety of biological systems. However, with the pure synthetic compounds in hand, progress is being rapidly made toward a greater understanding of their functions.[4]

A second major area in which the ready availability of the leukotrienes will undoubtedly prove of great value is in the discovery and development of new drug entities for the treatment of a variety of disease states. The leukotrienes have long been implicated in allergy and inflammation,[4(c)] and it seems safe to say that they will be found to play important roles in a variety of other disorders.[4(e)] Thus, the development of drugs which stimulate, inhibit, or modulate their activities will hopefully prove useful in the treatment or cure of some of these disorders, and the ready accessibility of the pure leukotrienes by synthesis will facilitate the task of the pharmacologist and the medicinal chemist to discover agents which modify their actions.

ACKNOWLEDGMENTS

We wish to express our appreciation to a number of people who made major contributions to this manuscript: Miss Suzanne Longpré who skillfully and rapidly typed the whole text; Mr. Scott Williamson and Mr. Kevin Clark of Williamson Labs for an equally skillful execution of the drawings; and Dr. Yvan Guindon, Mr. Yves Girard, Dr. R. N. Young, and Dr. Robert Zamboni for many constructive discussions during the course of this writing.

ADDENDUM

A synthesis of LTF_4 as a mixture with its 5-epi-6-epi isomer has been published by workers from Glaxo Laboratories,[93] and Just et al.[94] have published further details of their synthesis (Scheme 9) of 11(R)- and 11(S)-HETE along with a detailed nuclear magnetic resonance study of the compound. A brief review of the synthetic work in the leukotriene field has now appeared.[95] The synthesis of 5(R)- and 5(S)-HETE and the corresponding HPETEs (Schemes 4 and 5) has now been published by Zamboni and Rokach.[96]

Workers in the Ono Laboratories have reported the preparation of LTF_4, several partially hydrogenated analogs of LTC_4 and LTD_4, and deamino LTC_4 and LTD_4, along with preliminary results of their biological activity.[97] Glaxo workers have also reported a new synthesis of racemic LTA_4 methyl ester.[98]

Just after this article was finalized, we learned that Dr. B. Samuelsson, along with Dr. S. Bergström and Dr. J. Vane had been awarded the 1982 Nobel Prize in Medicine for their research on the roles and actions of the prostaglandins and the leukotrienes. We extend our warmest congratulations to these outstanding researchers for having won this most distinguished of scientific awards for their achievements in this area.

SYNTHESIS OF THE MONOHYDROXY AND MONOHYDROPEROXY EICOSATETRAENOIC ACIDS (HETES AND HPETES)

Rokach and co-workers have now published their synthesis of racemic 5-HETE (Scheme 2),[99] as well as the synthesis of 5(R)- and 5(S)-HETE and the corresponding HPETEs (Schemes 4 and 5).[96] Gunn has also disclosed a new approach to the synthesis of racemic 5-HETE.[100] Using their diazo ketone-cyclopropafuran methodology, Adams and Rokach have achieved the first synthesis of 8- and 9-HETE in their racemic forms.[101]

Just has published further details of his synthesis (Scheme 9) of 11(R)- and 11(S)-HETE,[94] as well as a new synthesis of 11(S)- and 12(S)-HETE.[103] Corey has also published a new synthesis of 12-HETE.[102] Rokach and co-workers have synthesized 12(S)-HETE, as well as the more recently discovered 12(R) isomer.[104a] The same group has also prepared the 20-hydroxy metabolite of 12(S)-HETE.[104a] Falck and co-workers have also synthesized the 20-hydroxy metabolite of 12(S)-HETE, along with the putative 19(R)- and 19(S)-hydroxy metabolites.[104b]

The first total synthesis of 15(S)-HETE has recently been reported by Nicolaou et al.[105]

SYNTHESIS OF THE DIHYDROXY AND THE MORE HIGHLY OXYGENATED EICOSATETRAENOIC ACIDS (LTB$_4$, ETC.)

New syntheses of LTB$_4$ have been published by Italian workers,[106] by Nicolaou, et al.,[107] by the Glaxo group,[108] and a chemoenzymatic synthesis by Sih's group.[109] Young et al., have published work on the development of a radioimmunassay for LTB$_4$.[110] Rokach and co-workers have achieved a synthesis of LTBx, further exemplifying their cyclopropafuran technology,[111] and Fitzsimmons and Rokach have synthesized several isomers of 8,15-LTB.[112] Nicolaou and Webber describe syntheses of 5,15-DHETE and 8,15-DHETE,[113] and workers at the Karolinska Institute describe the isolation and synthesis of a 14,15-dihydroxyeicosatetraenoic acid.[114]

Syntheses of LTB$_3$[115] and LTB$_5$[116] have now been reported.

Pirillo, et al., have published improvements on their preparation of the LTB$_4$ synthon (±)-**81a** (Scheme 16).[117] Glaxo workers have developed a significantly improved synthesis of the synthon **94** (Scheme 17).[118] LeMerrer, et al., have reported a general route to optically active α-hydroxyaldehydes, including the synthons **114** (Scheme 20), **136** (Scheme 25), and the corresponding (S) enantiomers.[119]

SYNTHESIS OF LEUKOTRIENE A$_4$ AND RELATED EPOXIDES

New syntheses of racemic LTA$_4$ (as its methyl ester) have been reported by workers at Glaxo,[98] by a Russian group,[120a] and by researchers at Okayama University.[102b] A new synthesis of LTA$_4$ has recently been reported by a group from Shanghai.[121] Corey has reported further on the conversion of 5-HPETE to LTA$_4$ and 14,15-LTA$_4$. respectively.[122] Rokach and co-workers have reported total syntheses of 11(S), 12(S)-LTA$_4$[124] and 14(S), 15(S)-LTA$_4$.[125] The above-mentioned Russian group have also employed the Sharpless epoxidation methodology to prepare the synthons trans-5(S)-**174** and trans-5(S)-**175** (Charts 5 and 6, and Scheme 42).[126]

SYNTHESIS OF THE COMPONENTS OF SRS-A: LEUKOTRIENE C$_4$, D$_4$, E$_4$ AND RELATED COMPOUNDS

A synthesis of LTF$_4$ as a mixture with its 5-epi-6-epi isomer has been published by workers from Glaxo Laboratories,[93] and Ono Laboratories have reported the preparation of LTF$_4$, of several partially hydrogenated analogs of LTC$_4$ and LTD$_4$, and deamino LTC$_4$ and LTD$_4$, along with preliminary results of their biological activity.[97]

Rosenberger et al., have published details of their initial synthesis of LTE$_4$ (Scheme 52),[123] and Cohen et al., following up on their earlier synthesis of LTA$_4$ (Scheme 43), have prepared LTC$_4$, D$_4$, and E$_4$.[127] Interest in possible metabolic products of the leukotrienes has led the Merck Frosst group to prepare the 20-hydroxy and 20-carboxy derivatives of LTD$_4$.[128]

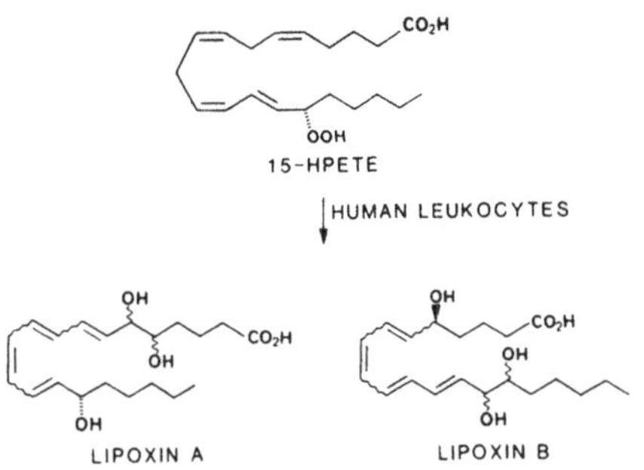

SCHEME 68.

THE LIPOXINS: SYNTHESIS AND IDENTIFICATION OF THE NATURAL PRODUCTS

Introduction

In the spring of 1984, Serhan, Hamberg, and Samuelsson reported the isolation of a new class of metabolites of arachidonic acid and coined the names lipoxin A and lipoxin B;[129] the lipoxins represent the first natural products containing a fully conjugated tetraene derived from arachidonic acid (Scheme 68). These novel trihydroxy tetraene eicosanoids were also reported to possess interesting biological properties.[130]

Incubation of 15-hydroperoxy eicosatetraenoic acid (15-HPETE) with human leukocytes produced lipoxins A and B, having characteristic UV absorptions for the conjugated tetraenes (λmax = 301nm). The Samuelsson group performed a very elegant structural elucidation to determine the skeletal connectivity of the trihydroxy tetraenoic acids. However, the geometric configuration of the conjugated double bonds and the relative stereochemistry of the vicinal alcohols at C_5 and C_6 for lipoxin A and the alcohol at C_{14} for lipoxin B had not been determined.

A program was therefore initiated by the Merck Frosst group to chemically and enzymatically synthesize various lipoxin isomers with three goals in mind. First, by comparing synthetic samples of unambiguous stereochemical origin with the natural product, the absolute configurations of the hydroxyl groups and the geometry of the double bonds could be assigned. Second, with this stereochemical information in hand, a biosynthetic route which accounts for the formation of lipoxins could be postulated. Third, since only minute quantities of lipoxins are available from natural sources, sufficient quantities of the lipoxins were prepared to allow a more extensive evaluation of their biological properties.

Upon analysis of the reported data, consideration of the known polyoxygenation products of arachidonic acid, and in an effort to define the biochemical origins of lipoxins A and B, we considered two possible biosynthetic routes (Schemes 69 and 70). Scheme 69 depicts successive enzymatic oxidations at C15 and C5 of arachidonic acid to yield the known 5(S),15(S)-diHPETE which, by analogy to the formation of leukotriene A_4 (LTA$_4$), could undergo a stereospecific enzymatic dehydration to produce a 5(S),6(S) epoxide with concomittant formation of a tetraene of 7E, 9E, 11Z, 13E geometry. This undergoes transformation to produce the lipoxins. An enzymatic hydrolysis of the epoxide at C6 would lead to lipoxin A, while nonenzymatic homo-conjugate addition of water at C14 would produce lipoxin B. Formation of lipoxin A (solid arrows), as shown, should leave the tetraene geometry intact and the stereochemistry at C6 would be R, assuming an S_N2 opening with inversion of the epoxide at the more electrophilic C6 or S if the opening occured with retention. Nonenzymatic hydrolysis of the tetraene epoxide (broken arrows) would form the all-E tetraene and a diastereomeric mixture of lipoxin B's epimeric at C14 by analogy to the nonenzymatic hydrolysis of LTA$_4$ and the two all-E lipoxin A's.

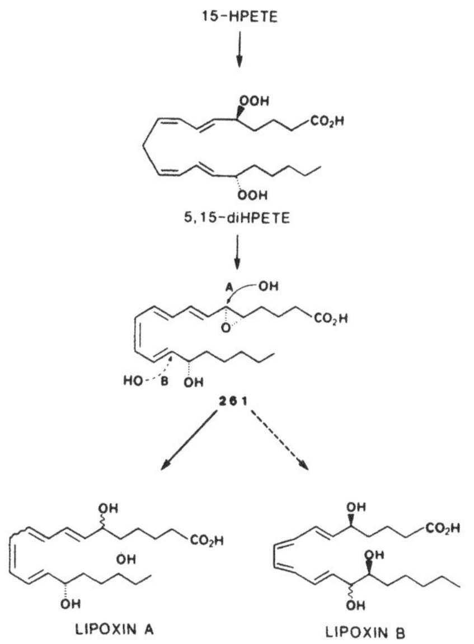

SCHEME 69. Biosynthetic postulate 1.

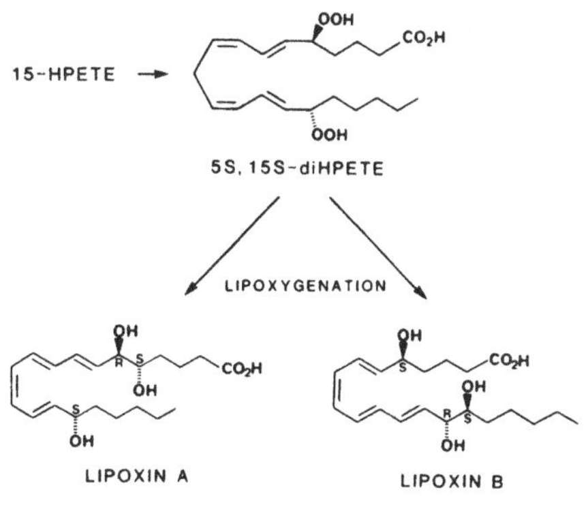

SCHEME 70. Biosynthetic postulate 2.

An alternative biosynthetic route is postulated in Scheme 70 in which 5(S),15(S)-diHPETE is also formed as a common intermediate. However, a third lipoxygenation was then envisaged. This would entail the formation of a carbon-centered radical at C10 (as in the formation of 8- or 12-HPETE), followed by vinylogous trapping of molecular oxygen at C6 to give, after the reduction of the hydroperoxide (s), lipoxin A, or at C14, to give lipoxin B. Based on the stereochemical outcome of similar lipoxygenations,[131] the stereochemistry of the new asymmetric center at C6 in lipoxin A or at C14 in lipoxin B was predicted to be of the R configuration. The tetraene geometry would be as shown in Scheme 70: 7E, 9E, 11Z, 13E for lipoxin A and 6E, 8Z, 10E, 12E for lipoxin B.

The isomers predicted by the above postulates were therefore prepared by total synthesis.

Three separate synthetic approaches to the synthesis of lipoxin A isomers were realized. The first involved the synthesis and the ring opening of the tetraene epoxide **261** (Scheme 69) to give lipoxin A isomers.[132] A second approach made use of LTA$_4$ as a starting material, and this was converted in three steps (two chemical, one enzymatic) to lipoxin A isomers.[133] Finally, total syntheses of four lipoxin A isomers were designed whereby the absolute stereochemistry of the hydroxyl groups was rigorously derived from 2-deoxy-D-ribose and

SHEME 71. Synthesis of the tetraene epoxide and lipoxin A isomers.

L-xylose.[133] Four lipoxin B isomers were also prepared by total synthesis using the same starting materials.[135]

Merck Frosst Synthesis of Lipoxin A Isomers

The tetraene epoxide was prepared[132] as shown in scheme 71. Commencing with 2-deoxy-D-ribose, diene aldehyde trans-5(S)-**177** (Scheme 40), an intermediate used for the synthesis of LTA$_4$, was prepared. The chiral phosphonium bromide **262** was obtained from D-arabinose via the versatile masked dialdehyde synthon **112a** (see Schemes 20 and 26). Coupling these two intermediates using a Wittig reaction to form the conjugated polyene yielded the tetraene epoxide as its protected ethyl ester **263a**.

A 2:1 ratio of pure 11Z and E epoxy tetraenes **263a** and **263b** respectively was obtained in 90% yield. The silyl protecting group was removed to give the free 11-Z and E 15(S) alcohols **264a** and **264b** respectively in 75% yield. The U.V. spectra of the epoxides display a bathochromic shift with respect to the hydrolyzed products (λ max = 309, hexane)

Treatment of epoxide **264a** with KOH in DMSO afforded two diastereomeric triols. The 5(S), 6(R), 15(S) triol **265a** results from S$_{N}$2 opening at C6, and the 5(R), 6(S), 15(S) triol **266a** is produced by an internal displacement at C5 by the Cl carboxylate forming an intermediate δ-lactone followed by saponification. Structural assignment to distinguish the two products was not possible at this stage.

The lack of regiochemical control in the epoxide opening of **264a** prompted us to seek a method to achieve epoxide cleavage at C6 exclusively (the more electrophilic carbon) with the carboxylic acid blocked as its ester. The chemical literature documents the difficulty in epoxide opening using oxygen nucleophiles. In general, polar sovents and high temperatures are required. Heating of the epoxide **263a** in the presence of excess sodium benzoate in anhydrous dimethylsufoxide at 100°C gave no reaction. However, heating the epoxide in dimethylsulfoxide in the presence of 1.2 equivalents of benzoic acid at 60°C for 1 hr produced two benzoates, **267a**, and **268a**, epimeric at C6, in over 80% yield (Scheme 72). Removal of the silyl protecting group and base hydrolysis produced trihydroxyacids **265a** and **269a** epimeric only at C6 (ratio ≃ 1:3).

Although epoxide opening occurs exclusively at C6, it is evident that two mechanisms are operative. One epimer is formed with retention (5(S), 6(S)) at C6 and the second epimer is formed by inversion (5(S), 6(R)) of that center. Again, assignment of which compound corresponds to which relative stereochemistry was not possible at this point.

Conversion of LTA$_4$ to Lipoxin A Isomers

Relying on the availability at Merck Frosst of leukotriene A$_4$ (LTA$_4$) ethyl ester and its

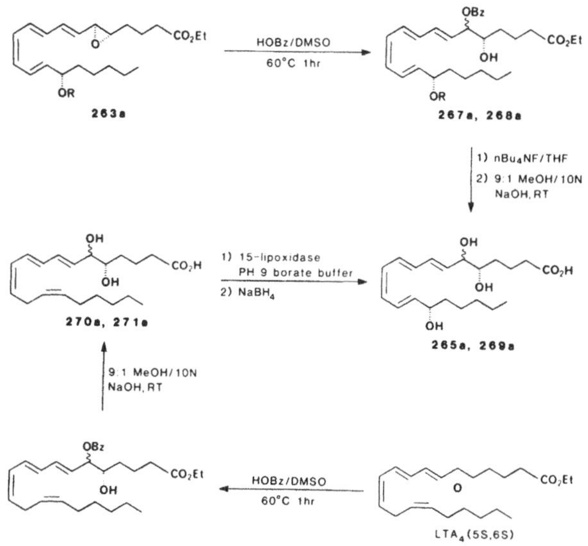

SCHEME 72. Conversion of epoxy tetraene and LTA$_4$ to lipoxin A isomers.

unnatural isomers,[56,57] a three-step synthesis of four lipoxin A isomers from LTA$_4$ ethyl ester was achieved[133] by: (1) opening the epoxide of LTA$_4$ using benzoic acid (Scheme 72); (2) hydrolysis of the benzoate esters to form the diol acids **270a** and **271a** (epimeric at C6); (3) enzymatic conversion of the diols **270a** and **271a** into lipoxin A hydroperoxides using commercially available lipoxidase, and subsequent reduction afforded the lipoxin A isomers **265a** and **269a**. This approach led to the same lipoxin A isomers (5(S), 6(R) and 5(S), 6(S)) as were obtained from opening of epoxide **263a** (Scheme 72, 70% overall yield from LTA$_4$).

The assignment of stereochemistry remained a thorny problem until a correlation scheme was devised using various synthetic isomers of LTA$_4$. Conversion of these isomers into their respective trihydroxy tetraenes, by the method described above, and the correlation of the RP-HPLC analysis of these products and those obtained from treatment of the epoxide **264a** with hydroxide (Scheme 71) was used to determine the relative and absolute stereochemistry of the vicinal diols at C5 and C6, as shown in Scheme 73. Beginning with natural LTA$_4$ (5(S), 6(S)) conversion to the lipoxin A isomers **265a** and **269a** gave a 1:3 ratio of epimers at C6 with the C5 alcohol bearing the S configuration. The same operations performed on 5-epi-6-epi LTA$_4$ (5(R), 6(R)) gave rise to the same ratio of 5(R) diastereomers and **272a**. Turning to the C15-epoxide series, the only change noted was the inversion of the ratio of epimers at C6 ($\simeq$ 3:1) for the corresponding pairs of diastereomers. At this point the stereochemistry is unambiguously established for the C5 alcohol which comes from the LTA$_4$'s and the C15 alcohol which comes from enzymatic oxidation. When the trihydroxy acids obtained from hydroxide openings of the tetraene epoxide **264a** are co-injected, the minor isomer corresponds with peak I in the table and the major isomer coincides with peak II. Since the relative stereochemistry is known (5(S), 6(R) and 5(R), 6(S) respectively) to have come from inversion of only one center of the epoxide (vide supra) identities can be matched based on the established stereochemistry at C5 of compounds derived from LTA$_4$ isomers. The minor isomer from hydroxide opening of **264a** is also the minor isomer (the inversion product) derived from LTA$_4$ as shown in Scheme 72. Thus, the isomer corresponding to peak I has the 5(S), 6(R), 15(S) configuration. Conversely, the minor isomer derived from 5-epi, 6-epi-LTA$_4$ with inversion at C6 is the same product as the major isomer derived from hydroxide opening at **264a**, (inverting C5) and yielding the 5(R), 6(S), 15(S) diastereomer. The identities of the compounds corresponding to peaks I and II are thus

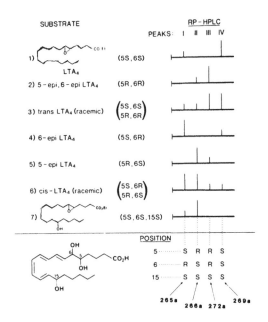

SCHEME 73. Stereochemical correlation of vicinal diols at C_5 and C_6.

established. The remaining peaks III and IV were assigned by default since peaks I and IV, and peaks II and III respectively are representative of epimers at C6.

Considering possible biosynthetic pathways that account for the formation of lipoxins, it is unlikely that a 5(R) isomer exists in nature, based on the reported data about lipoxygenases and other enzymes in arachidonic acid metabolism.

Total Synthesis of Lipoxin A Isomers

In order to obtain sufficient quantities of lipoxin A and its isomers for identification and biochemical evaluation, synthetic schemes were devised which would lead unambiguously to four possible structures of lipoxin A.[133] Since these routes utilize carbohydrate precursors, the absolute and relative stereochemistry of the products are unambiguously predetermined. The totally synthetic compounds also serve to confirm the assignment of structures in the correlation of lipoxin A isomers (scheme 73) from LTA_4's.

The synthesis of the 5(S), 6(R), 15(S) isomers of lipoxin A utilized 2-deoxy-D-ribose as the starting material (Scheme 74). Glycosidation of the sugar followed by formation of the 3,4-cyclic carbonate and then hydrolysis of the glycoside yielded the intermediate 273 in which the future 5,6 diol is protected with a single protecting group and the termini are distinguishable. Condensation of the hydroxyaldehyde **273** with ethyl (triphenylphosphon-ylidene) acetate and hydrogenation of the olefin generated afforded the alcohol **274**. Oxidation of the alcohol, condensation of the aldehyde produced with 4-(triphenylphosphoanyli-dene)but-2-enal, and isomerization provided the diene aldehyde **275** required for the formation of the final carbon-carbon double bond. Condensation of the aldehyde **275** with the ylide **276** yielded the two 5(S), 6(R), 15(S) isomers **277a** and **277b** in a 2:1 ratio. Finally, removal of the silyl protecting group, separation of the 11-Z and the 11-E isomers and basic hydrolysis afforded two 5(S), 6(R) isomers of lipoxin A, **265a** and **265b**.

L-xylose was chosen as the starting material for the synthesis of the 5(S), 6(S), 15(S) isomers of lioxin A. Since it was necessary to excise the C2 hydroxyl group and to protect the C3 and C4 hydroxyl groups, a simple and efficient method to prepare the 5(s) diastereomer **280** of **274** was devised. Pivotal to this scheme was the formation of the cyclic carbonate from an acyclic carbonate upon hydrolysis of an acetonide. Using the method of Wong and Gray,[134] L-xylose was converted into the alcohol **278** in four steps (Scheme 75).

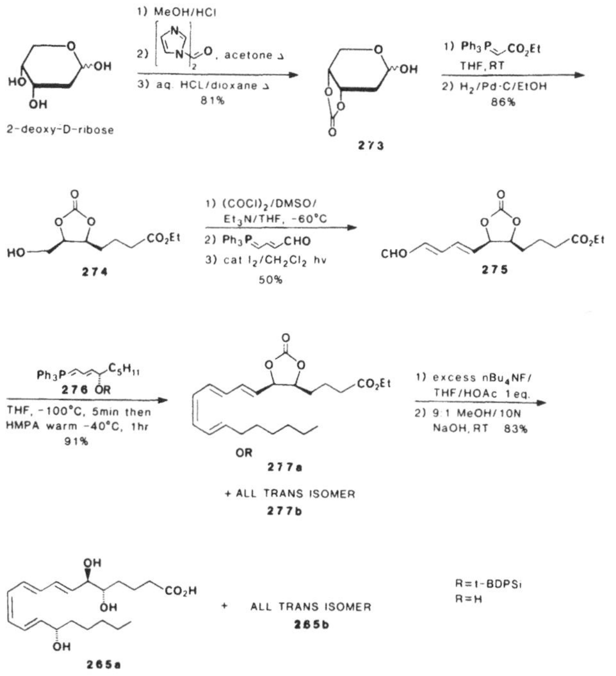

SCHEME 74. Synthesis of the 5(S),6(R),15(S) isomers of lipoxin A. **265a** and **265b**.

Protection of the alcohol **278** as its ethyl carbonate derivative, hydrolysis of the thioacetal and condensation of the resulting aldehyde with ethyl (triphenylphosphoranylidene)acetate gave the acyclic carbonate **279**. Hydrogenation and hydrolysis of the acetonide occurred with concomitant formation of the cyclic carbonate as expected, yielding the alcohol **280**. The alcohol **280** is diastereomeric with the alcohol **274** and was converted to the 5(S), 6(S), 15(S) isomers of lipoxin A, **269a** and **269b**, using the same procedure as for the preparation of the 5(S), 6(R) isomers of lipoxin A from the alcohol **274**.

The two 11-Z isomers of lipoxin A prepared were identical by HPLC and 250 MHz ^{1}HNMR to those prepared from LTA$_4$ by the chemical/enzymatic route described previously. This confirmed the original assignment of the absolute stereochemistry of these compounds.

What remained at this point was to determine which of the isomers prepared was lipoxin A. Therefore, lipoxins A and B were prepared biologically from human leukocytes as described by Samuelsson.[129,130] The natural material thus obtained co-eluted on reverse phase liquid chromatography, both as its free acid (60/40 MeOH-H$_2$O 0.1% HOAc) and as its methyl ester (70/30 MeOH-H$_2$O) with the 5(S), 6(S), 15(S)-11-Z isomer **269a** and its methyl ester respectively. *None* of the other diastereomers or geometric isomers prepared co-eluted with lipoxin A was proposed to possess the structure **269a**.

Total Synthesis of Lipoxin B Isomers

To unambiguously predetermine the absolute stereochemistry of the hydroxyl groups of the lipoxin B isomers, carbohydrate precursors were also utilized.[135] The 5(S) stereocenter was obtained from D-arabinose for all of the lipoxin B isomers. The 14(R), 15(S) and 14(S), 15(S) vicinal diols were transposed from 2 deoxy-D-ribose and L-xylose respectively.

The C1-C8 fragment common to all of the lipoxin B isomers was prepared as shown in Scheme 76. Conversion of the thioacetal **112a**, prepared from D-arabinose, to the phosphonium salt **281** provided the "top half" to the lipoxin B isomers.

The 14(R), 15(S) isomers of lipoxin B **282a** and **282b** were prepared as shown in Scheme

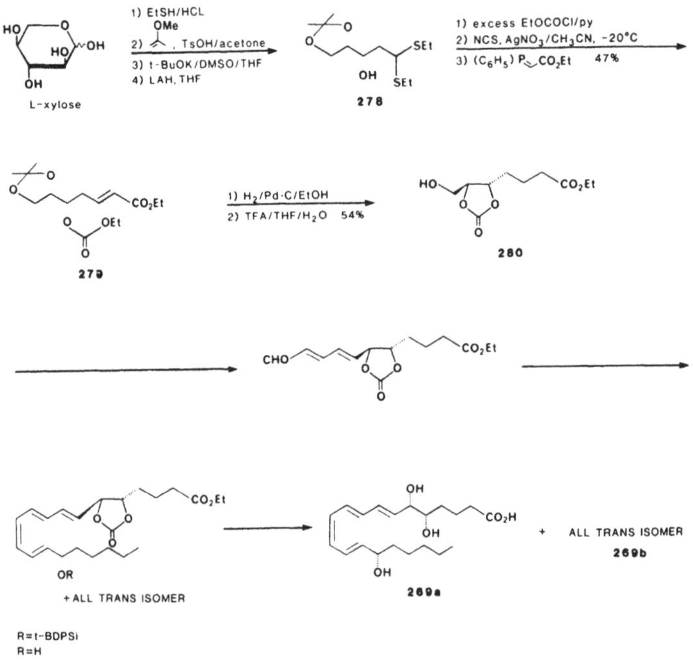

SCHEME 75. Synthesis of the 5(S),6(S),15(S) isomers of lipoxin A. **269a** and **269b**.

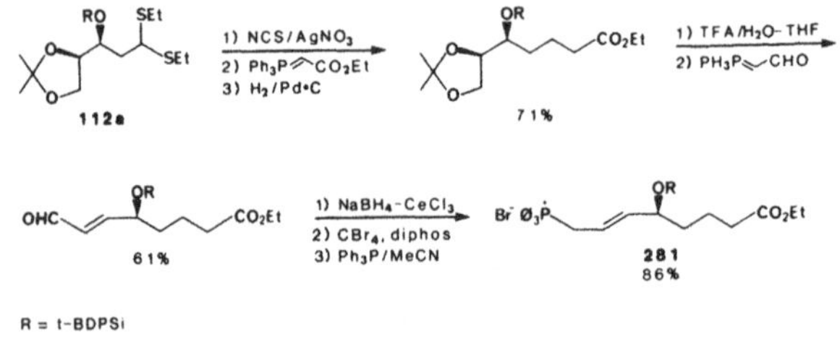

SCHEME 76. Synthesis of the top half of the lipoxin B's.

77. The route parallels the approach taken for lipoxin A total syntheses. The vicinal diol stereochemistry is derived from 2-deoxy-D-ribose and the diol is protected once again as its cyclic carbonate throughout the syntheses. The Wittig reaction with the ylide of fragment **281** proceeds similarly to the final coupling reactions for lipoxin A isomers.

Similarly, the 14(S), 15(S) isomers of lipoxin B, **283a** and **283b**, were prepared as shown in Scheme 78 using L-xylose as the carbohydrate precursor.

Using the procedure outlined by the Samualsson group,[129] authentic samples of lipoxin A and B were prepared from human leukocytes. Reverse phase-HPLC comparison of authentic lipoxin B methyl ester and its free acid with our four synthetic lipoxin B methyl esters and acids were performed in two solvent systems. Using methanol-water (70:30 for the esters; 60:40, 0.05% HOAc for the acids) both the 14(R) and 14(S) all-E compounds **282b** and **283b** were not separable and both co-eluted with authentic lipoxin B as their Me esters and free acids. However in CH_3CH-H_2O (40:60 for the esters; 40:60, 0.05% HOAc for the acids) these isomers were separable. Injection of authentic lipoxin B revealed that it was a mixture of two isomers, with peaks co-eluting with the synthetic standards **282b** and **283b**.

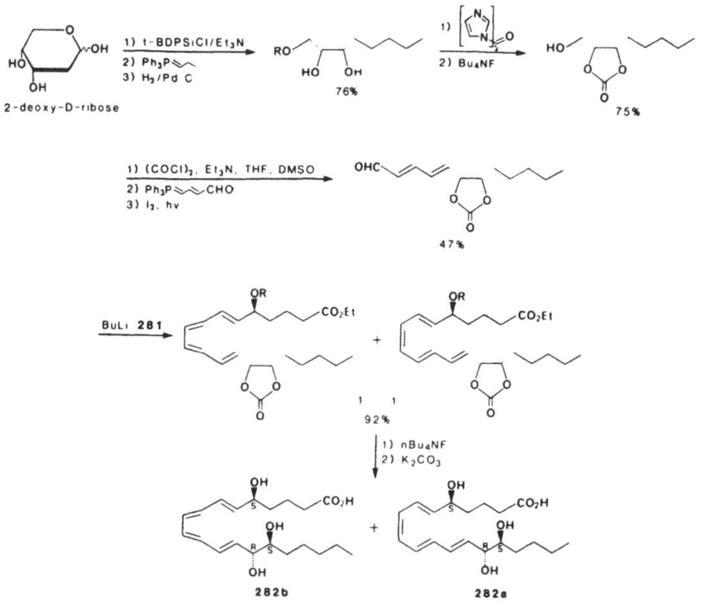

SCHEME 77. Synthesis of the 5(S),14(R),15(S) isomers of lipoxin B.

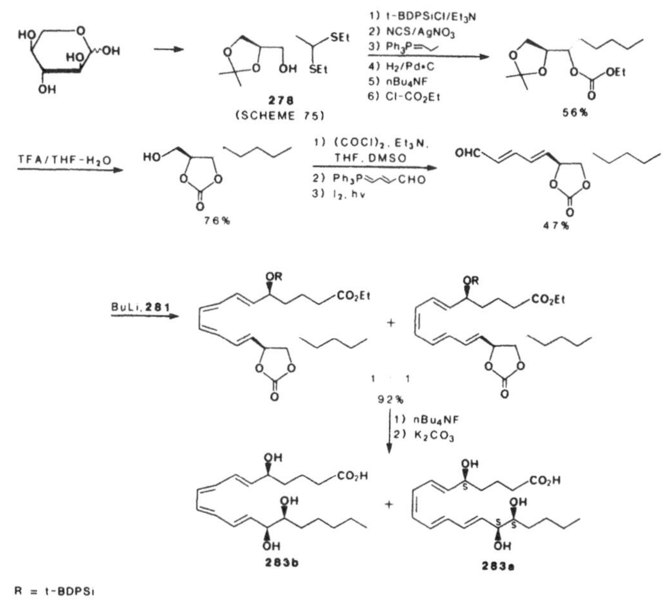

SCHEME 78. Synthesis of the 5(S),14(S),15(S) isomers of lipoxin B.

Revised Stereochemical Assignments of the Lipoxins and Their Biosynthesis

Using the procedure described by Serhan et al.[129] low yields of lipoxin products were obtained derived from leukocyte incubations. Consequently, steps were taken to improve the efficiency of the purification of the biological material.[136] After much experimentation, it was observed that several other lipoxin isomers were also formed during this incubation. Depending on the particular experimental run, as many as 4 lipoxin A isomers and 3 lipoxin B isomers were observed in variable ratios. The original isolation techniques[129] and HPLC

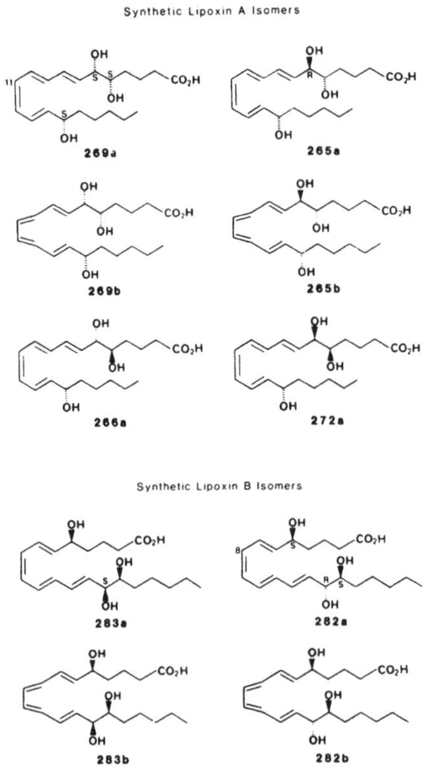

Synthetic Lipoxin A Isomers

Synthetic Lipoxin B Isomers

SCHEME 79. Summary of the synthetic lipoxin A and B isomers (Merck Frosst).

solvent systems revealed only 3 isomers (vide supra), represented by two peaks, in the HPLC chromatogram.

Due to the presence of so many stereoisomers as well as the all-E isomers for lipoxins A and B, the biosynthetic postulate depicted by Scheme 69 is favored. The formation of the unstable 5,6-epoxide followed by hydrolysis (enzymatic or nonenzymatic) accounts for all the observed isomers. To support this hypothesis separate incubations of tetraene epoxide **261** (Scheme 71) with human leukocytes afforded lipoxin A and lipoxin B isomers varying only in relative amounts of products formed. Six lipoxin isomers were formed: 4 lipoxin A isomers and 2 lipoxin B isomers.

On the basis of the biosynthetic pathway postulate (Scheme 69) and the intermediacy of the tetraene epoxide, it was felt that the E-lipoxin isomers were likely to be formed by nonenzymatic hydrolysis of the tetraene epoxide. Therefore, the tetraene epoxide **261** was subjected to the conditions of the leukocyte incubation but in the absence of cells. To our surprise, we found that these conditions produced not only the all-E isomers, but all the products from the incubation of 15-HPETE with leukocytes. The production of the 5(S), 6(S), 15(S), 11-Z lipoxin A isomer **269a** by nonenzymatic hydrolysis of the tetraene epoxide **261** is on the surface suprising. Epoxides are usually opened by either an S_{N_1} (carbocation-mediated) mechanism which could give the 6S isomer but would also lead to the loss of 11-Z double bond geometry, or an S_{N_2} mechanism which usually proceeds with inversion of configuration and therefore should give the 6(R), 11-Z lipoxin A isomer. However, the generation of products which are the result of the opening of similar epoxides with retention of configuration and without loss of double bond integrity was precedented in our previously reported work with LTA$_4$ and the tetraene epoxide **264a**.[132]

The data presented led to the conclusion that the lipoxins are principally nonenzymatic

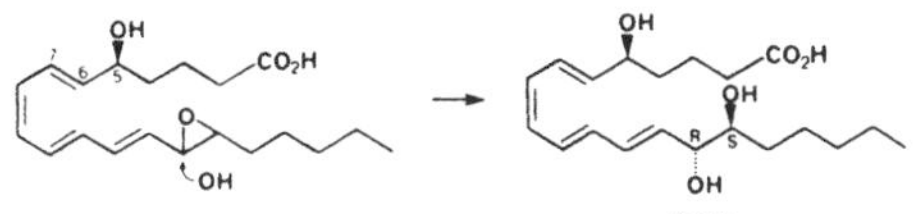

SCHEME 80.

Lipoxin A Isomers

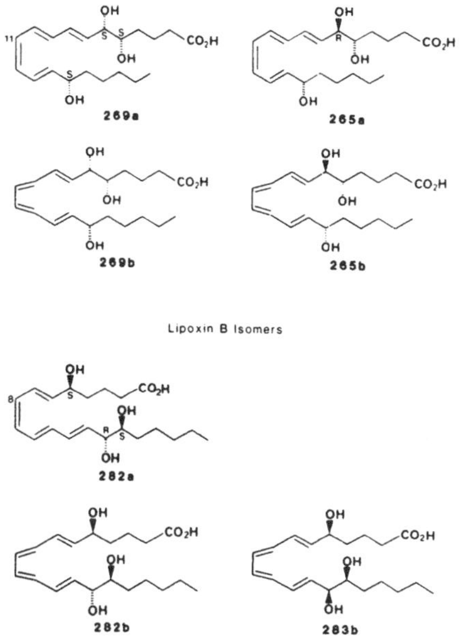

Lipoxin B Isomers

SCHEME 81. Summary of lipoxins A and B isolated
from incubation of 15-HPETE with human leukocytes.

hydrolysis products of the tetraene epoxide **261**. It is not possible to rule out that some of the 11-Z lipoxin A isomers **265a** and **269a** detected may be generated by the action of an epoxide vicinal hydrolase found in leukocytes[137] or by triple lipoxygenation, but clearly all of the lipoxins detected are also generated by nonenzymatic hydrolysis.The tetraene epoxide on the other hand is formed enzymatically.

In one incubation with 15-HPETE the lipoxin B isomer **282a** was formed. It has been suggested that this isomer can arise from hydrolysis of the 14,15 epoxide which may also be formed to a minor extent (Scheme 80).[138]

In summary, it was shown that the lipoxins are a mixture of the 7 isomers, depicted in Scheme 81. The biosynthesis of the lipoxins was also shown to be via an LTA_4-like pathway, with 15-HPETE as the substrate, to give the tetraene epoxide **261**. The lipoxins are then principally formed by nonenzymatic hydrolysis of this epoxide. The incubation of a high concentration of 15-HPETE with calcium ionophore-stimulated leukocytes leads to the production of quantities of the lipoxins which facilitated their structural elucidation and the determination of their biosynthesis. Whether these compounds are produced in significant amounts under physiological conditions remains to be determined.

Corey Synthesis of Lipoxin A Isomers

Following the publication of the Merck Frosst syntheses[132,133] one other synthesis of lipoxin A isomers was reported, and the assignment of the 5(S), 6(S), 15(S), 11-Z, 7, 9, 13E-

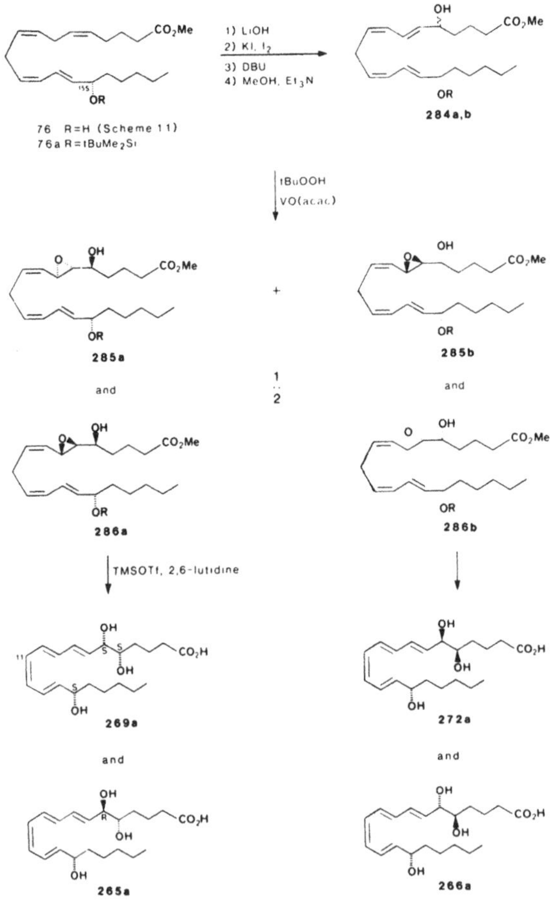

SCHEME 82. Synthesis of lipoxin A isomers (Corey).

trihydroxy eicosatetraenoic acid structure **269a** was confirmed[139] as related to the original isolation procedure described by Serhan et al.[129]

The Corey synthesis of lipoxin A isomers[139] began using 15(S)-hydroxy eicosatetraenoic acid (15(S)-HETE) (Scheme 82). Using standard iodo-lactonization conditions, followed by elimination of HI and methanolysis of the lactone, the diastereomeric alcohols **284a,b** were produced. Directed allylic epoxidation formed the threo epoxy alcohols **285a** and **285b** and erythro epoxy alcohols **286a** and **286b** in a ratio of 1:2 (threo/erythro). The threo epoxy alcohols and the erthryo epoxy alcohols were separated by chromatography and the diastereomeric pairs were treated in parallel sequences. Reaction with trimethylsilyl triflate and of 2,6-lutidine in toluene at −78°C afforded the corresponding tetraene triol silyl ethers. De-silylation and saponification lead to the four diastereomeric lipoxin A isomers **269a**, **265a**, **272a** and **266a**.

Corey and co-workers[140] also published a synthesis of lipoxin B isomers, but a discrepancy exists due to the disagreement of the HPLC relative retention times of the isomers reported, compared to those reported by the Merck Frosst group.[135,141]

Nicolaou Synthesis of Lipoxin B Isomers

Pursuant to the publication by Merck Frosst of the synthesis of lipoxin B isomers, Nicolaou and co-workers reported a synthesis of the same four isomers (Me esters) using the Pd°-Cu[I]-catalysed coupling reaction developed in his laboratories (Scheme 83).[142]

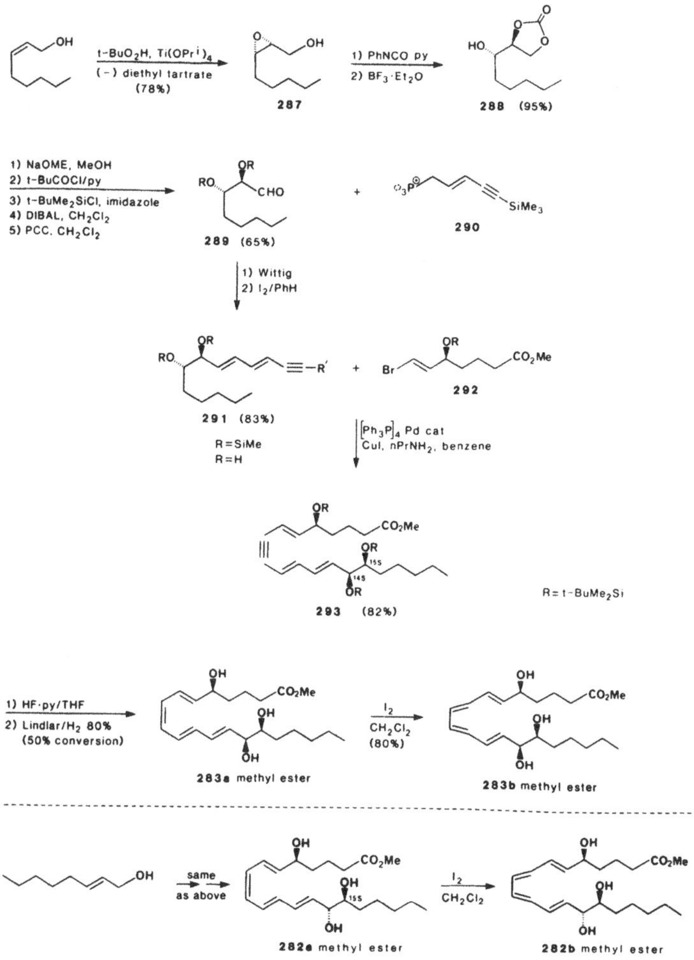

SCHEME 83. Synthesis of lipoxin B isomers (Nicolaou).

Sharpless epoxidation of Z-2-octenol-1-ol gave the epoxy alcohol **287** in high enantiomeric excess. The epoxy alcohol was converted to carbonate **288** with phenylisocyanate in pyridine followed by BF$_3$-Et$_2$O and acidic workup. Carbonate **288** was converted to the aldehyde disilyl ether **289** in 5 steps involving standard operations to protect the 2° alcohols while oxidizing the 1° alcohol to its corresponding aldehyde. Aldehyde **289** was coupled, using a Wittig reaction, with phosphonium salt **290** which gave a 1:1 E/Z mixture which was isomerized exclusively to E with I$_2$ to give **291**. The stage was set for the finally assembly of the C-20 skeleton. The key reaction involved the Pd°-Cu^I coupling reaction of acetylene **291** (R' = H) and optically pure vinyl bromide **292** to produce acetylene **293**. Removal of the silyl protecting groups followed by Lindlar semihydrogenation provided the lipoxin B methyl ester isomer **283a**. The corresponding all-E isomer **283b** Me ester could be obtained by isomerization of the 8-Z double bond using catalytic I$_2$ in CH$_2$Cl$_2$. Similarly the same sequence was performed, beginning with E-2-octen-1-ol to prepare the 14(R), 15(S) isomers **282a** and **282b** as their methyl esters.

Triple Lipoxygenation

While the evidence presented by the Merck group in the section entitled "Merck Frosst Synthesis of Lipoxin A Isomers" favors the biosynthetic pathway requiring tetraene epoxide

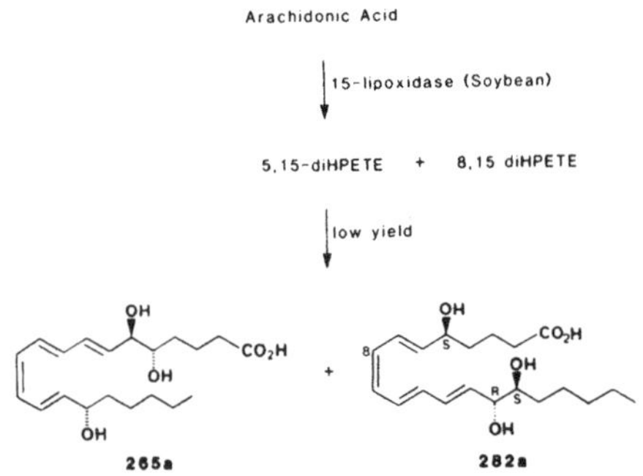

Arachidonic Acid

15-lipoxidase (Soybean)

5,15-diHPETE   +   8,15 diHPETE

low yield

265a   **282a**

SCHEME 84. Triple lipoxygenation of arachidonic acid.

lipoxygenation
12-LOX

5,15 diHPETE

NaBH₄
work up

282a

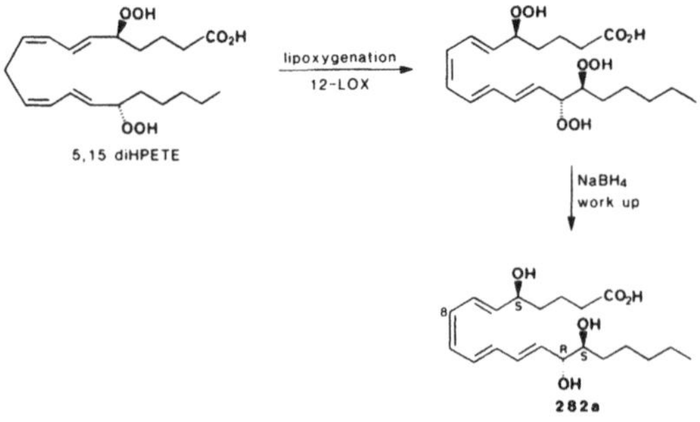

SCHEME 85. Lipoxygenation of 5,15-diHPETE: major pathway.

formation, the possibility of a triple lipoxygenation of arachidonic acid to produce lipoxins could not be ruled out.

In an effort to test this eventuality, arachidonic acid was incubated with commercially available soya bean lipoxidase in pH 9 borate buffer for prolonged reaction times and the reaction was monitored by UV spectroscopy.[143] As expected, the major products consisted of 8,15 diHPETE and 5,15 diHPETE in accordance with previous literature findings.[144] However, a small amount (< 1% yield) of two lipoxin isomers were observed bearing the characteristic tetraene chromophore with λmax = 301 nm. The compounds were isolated and compared to synthetic standards, and were found to be the lipoxin A isomer **265a** and the lipoxin B isomer **282a** in equal proportions. No other isomer was observed. The formation of these isomers uniquely in the presence of the purified enzyme albeit in low yield strongly suggests a triple lipoxygenation mechanism (Scheme 84)

Lipoxins from 12-Lipoxygenase

In a unique collaboration between the Merck Frosst group and Professor Yamamoto of Tokushima University, 12-lipoxygenase, isolated and purified from porcine leukocytes (by the Japanese group), was investigated vis-à-vis its ability to generate lipoxins.[138] When purified 5,15-diHPETE was incubated in the presence of the 12-lipoxygenase enzyme, the major product formed was (50 to 80%) the 5(S), 14(R), 15(S), 8Z-lipoxin B, **282a**, likely resulting from vinylogous lipoxygenation (Scheme 85). The minor components isolated (Scheme 86) included five products which appear to be the result of hydrolytic opening of the 14(S), 15(S)-epoxy tetraene **294**. The lack of appearance of any 11Z isomers of lipoxin A is an indication that no 5(S), 6(S)-tetraene epoxide **261** is formed.

The relative proportions of lipoxygenase and epoxide pathways varies with temperature and pH. The exact ratio is difficult to determine since a portion of the lipoxygenation product

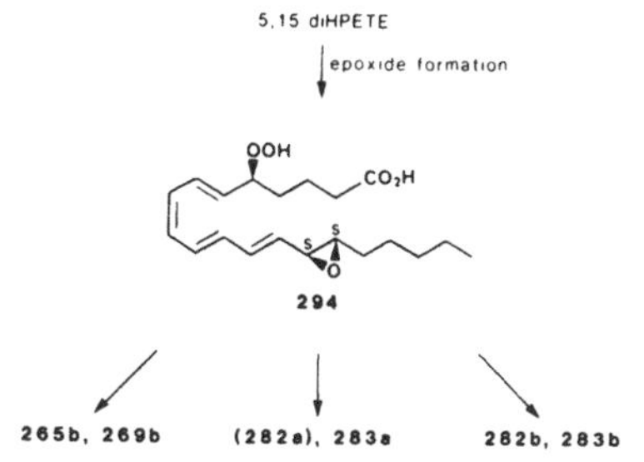

SCHEME 86. Minor products from 5,15-diHPETE.

also comes from the epoxide pathway. However, it is clear that the epoxide pathway is more important at higher temperatures (i.e. 37°C).

Work is in progress to continue to identify biosynthetic pathways to the lipoxins. The elucidation of the biochemical origins of the lipoxins may provide clues as to their cellular sources and ultimately the biochemical and physiological roles for these eicosanoid metabolites.

REFERENCES

1. Various authors, in *Advances in Prostaglandin and Thromboxane Research,* Vol. 1, Samuelsson, B. and Paoletti, R., Eds., Raven Press, New York, 1976.
2. **Hamberg, M. and Samuelsson, B.,** Prostaglandin endoperoxides. Novel transformations of arachidonic acid in human platelets, *Proc. Natl. Acad. Sci. U.S.A.,* 71, 3400—3404, 1972.
3. For the sake of clarity and consistency, the nomenclature system introduced by Samuelsson is used throughout this review: (a) **Samuelsson, B., Borgeat, P., Hammarström, S., and Murphy, R. C.,** Introduction of a nomenclature. Leukotrienes, *Prostaglandins,* 17, 785—787, 1979; (b) **Samuelsson, B. and Hammarström, S.,** Nomenclature for leukotrienes, *Prostaglandins,* 19, 645—648, 1980.
4. For reviews see: (a) related chapters in this monograph; (b) **Samuelsson, B.,** The leukotrienes: a new group of biologically active compounds, *Pure Appl. Chem.,* 53, 1203—1213, 1981; (c) **Marx, J. L.,** The leukotrienes in allergy and inflammation, *Science,* 215, 1380—1383, 1982; (d) **Various authors,** in *Advances in Prostaglandin, Thromboxane and Leukotriene Research,* Vol. 9, *Leukotrienes and other Lipoxygenase Products,* Samuelsson, B. and Paoletti, R., Eds., Raven Press, New York, 1982; (e) **Bailey, D. M. and Casey, F. B.,** Lipoxygenase and the related arachidonic acid metabolites, in *Annual Reports in Medicinal Chemistry,* Vol. 17, Hess, H. J., Ed., Academic Press, New York, 1982, chap. 21.
5. **Murphy, R. C., Hammarström, S., and Samuelsson, B.,** Leukotriene C: a slow-reacting substance from murine mastocytoma cells, *Proc. Natl. Acad. Sci. U.S.A.,* 76, 4275—4279, 1979.
6. (a) **Feldberg, W. and Kellaway, C. H.,** Liberation of histamine and formation of lysocithin-like substances by cobra venom, *J. Physiol. (London),* 94, 187—226, 1938; (b) **Kellaway, C. H. and Trethewie, E. R.,** The liberation of a slow-reacting smooth muscle-stimulating substance in anaphylaxis, *Q. J. Exp. Physiol.,* 30, 121—145, 1940.
7. **Orange, R. P. and Austen, K. F.,** Slow reacting substance of anaphylaxis, *Adv. Immunol.,* 10, 105—144, 1969.
8. **Bailey, D. M. and Chakrin, L. W.,** Arachidonate lipoxygenase, in *Annual Reports in Medicinal Chemistry,* Vol. 16, Hess, H. J., Ed., Academic Press, New York, 1981, chap. 20.
9. **Corey, E. J., Albright, J. D., Barton, A. E., and Hashimoto, S.,** Chemical and enzymic syntheses of 5-HPETE, a key biological precursor of slow-reacting substance of anaphylaxis (SRS), and 5-HETE, *J. Am. Chem. Soc.,* 102, 1435—1436, 1980.
10. **Baldwin, J. E., Reed, N. V., and Thomas, E. J.,** Phenylselenylation of arachidonic acid as a route to intermediates for leukotriene synthesis, *Tetrahedron,* 37(Suppl. 1), 263—267, 1981.
11. **Rokach, J., Guindon, Y., Zamboni, R., Lau, C. K., Girard, Y., Larue, M., Perry, R. A., and Atkinson, J. G.,** Synthesis of leukotrienes. Strategy for the synthesis of lipoxygenase products, in *Abstract Bulletin, 16th Middle Atlantic Regional Meeting,* No. 276, April 21-23, 1982.
12. **Gemal, A. L. and Luche, J. L.,** Lanthanoids in organic synthesis. VII. Selective reductions of carbonyl groups in aqueous ethanol solutions, *Tetrahedron Lett.,* 22, 4077—4080, 1981.

13. **Corey, E. J. and Hashimoto, S.,** A practical process for large-scale synthesis of (S)-5-hydroxy-6-trans-8,11,14-cis-eicosatetraenoic acid(5-HETE), *Tetrahedron Lett.*, 22, 299—302, 1981.

14. **Pirkle, W. H. and Hauske, J. R.,** Trichlorosilane-induced cleavage. A mild method for retrieving carbinols from carbamates, *J. Org. Chem.*, 42, 2781—2782, 1977.

15. **Corey, E. J., Marfat, M., Falck, J. R., and Albright, J. O.,** Controlled chemical synthesis of the enzymatically produced eicosanoids 11-, 12-, and 15-HETE from arachidonic acid and conversion into the corresponding hydroperoxides (HPETE), *J. Am. Chem. Soc.*, 102, 1433—1435, 1980.

16. **Corey, E. J., Niwa, H., and Falck, J. R.,** Selective epoxidation of eicosa-cis-5,8,11,14-tetraenoic (arachidonic) acid and eicosa-cis-8,11,14-trienoic acid, *J. Am. Chem. Soc.*, 101, 1586—1587, 1979.

17. **Just, G. and Luthe, C.,** Total synthesis of 11(R,S)-HETE, *Tetrahedron Lett.*, 23, 1331—1334, 1982.

18. **Corey, E. J. and Kang, J.,** Stereospecific total synthesis of 11(R)-HETE, lipoxygenation product of arachidonic acid via the prostaglandin pathway, *J. Am. Chem. Soc.*, 103, 4618—4619, 1981.

19. **Just, G., Luthe, C., and Potvin, P.,** A method for the systematic resolution of unbranched α-acetoxyalkyl- and aralkylaldehydes: synthesis of 11(R) and 11(S)-HETE, *Tetrahedron Lett.*, 23, 2285—2288, 1982.

20. **Just, G.,** Private communication,

21. **Corey, E. J., Niwa, H., and Knolle, J.,** Total synthesis of (S)-12-hydroxy-5,8,14-cis, -10-trans-eicosatetraenoic acid (Samuelsson's HETE), *J. Am. Chem. Soc.*, 100, 1942—1943, 1978.

22. **Baldwin, J. E., Davies, D. I., Hughes, L., and Gutteridge, N. J. A.,** Synthesis from arachidonic acid of potential prostaglandin precursors, *J. Chem. Soc. Perkin I*, 115—121, 1979.

23. **Porter, N. A., Wolf, F. A., Yarbro, E. M., and Weanen, H.,** The autoxidation of arachidonic acid: formation of the proposed SRS-A intermediate, *Biochem. Biophys. Res. Commun.*, 89, 1058—1064, 1979.

24. **Boeynams, J. M., Brash, A. R., Oates, J. A., and Hubbard, W. C.,** Preparation and assay of monohydroxy-eicosatetraenoic acids, *Anal. Biochem.*, 104, 259—267, 1980.

25. **Porter, N. A.; Logan, J., and Kontoyiannidou, V.,** Preparation and purification of arachidonic acid hydroperoxides of biological importance, *J. Org. Chem.*, 44, 3177—3180, 1979.

26. **Cobern, D., Hobbs, J. S., Lucas, R. A., and MacKenzie, D. J.,** Location of hydroperoxide groups formed by chlorophyll-photosensitised oxidation of unsaturated esters, *J. Chem. Soc. (C)*, 1897—1902, 1966.

27. **Corey, E. J., Marfat, A., Goto, G., and Brion, F.,** Leukotriene B. Total synthesis and assignment of stereochemistry, *J. Am. Chem. Soc.*, 102, 7984—7985, 1980.

28. **Corey, E. J., Marfat, A., Munroe, J., Kim, K. S., Hopkins, P. B., and Brion, F.,** A stereocontrolled and effective synthesis of leukotriene B, *Tetrahedron Lett.*, 22, 1077—1080, 1981.

29. **Guindon, Y., Zamboni, R., Lau, C.-K., and Rokach, J.,** Stereospecific synthesis of leukotriene B_4 (LTB_4), *Tetrahedron Lett.*, 23, 739—742, 1982.

30. **Traverso, G. and Pirillo, D.,** Sintesi del metilestere dell'acido 5-R,S-benzoilossi-5-formil-n-pentanoico, *Farm. Ed. Sci.*, 36, 888, 1981.

31. **Sharpless, K. B. and Katsuki, T.,** The first practical method for asymmetric epoxidation, *J. Am. Chem. Soc.*, 102, 5974—5976, 1980.

32. **Corey, E. J., Hopkins, P. B., Munroe, J. E., Marfat, A., and Hashimoto, S.,** Total synthesis of 6-trans, 10-cis and (±)-6-trans, 8-cis isomers of leukotriene B, *J. Am. Chem. Soc.*, 102, 7986—7987, 1980.

33. **Zamboni, R. and Rokach, J.,** Simple efficient synthesis of LTB_4 and 12-epi-LTB_4, *Tetrahedron Lett.*, 23, 2631-2634, 1982.

34. **Zamboni, R. and Rokach, J.,** Stereospecific synthesis of two metabolites of LTB_4, *Tetrahedron Lett.*, 23, 4751—4754, 1982.

35. **Borgeat, P., Picard, S., Vallerand, P., and Sirois, P.,** Transformation of arachidonic acid in leukocytes. Isolation and structural analysis of a novel dihydroxy derivative, *Prostaglandins Med.*, 6, 557—570, 1981.

36. **Lindgren, J. A., Hansson, G., and Samuelsson, B.,** Formation of novel hydroxylated eicosatetraenoic acids in preparations of human polymorphonuclear leukocytes, *FEBS Lett.*, 128, 329—335, 1981.

37. **Corey, E. J., Marfat, A., and Laguzza, B. C.,** Total synthesis of 5S,12S-dihydroxy-6,10-E,8,14-Z-eicosatetraenoic acid (5S,12S-di-HETE), a new human metabolite of arachidonic acid, *Tetrahedron Lett.*, 22, 3339—3342, 1981.

38. **Corey, E. J., Marfat, A., and Hoover, D. J.,** Stereospecific total synthesis of 12-(R)- and 12-(S)- forms of 6-trans leukotriene B, *Tetrahedron Lett.*, 22, 1587—1590, 1981.

39. **Maycock, A., Anderson, M. S., DeSousa, D. M., and Kuehl, F. A., Jr.,** Leukotriene A_4: preparation and enzymatic conversion in a cell free system to leukotriene B_4, *J. Biol. Chem.*, 257, 13911—13914, 1982.

40. **Rokach, J., Girard, Y., Guindon, Y., Atkinson, J. G., Larue, M., Young, R. N., Masson, P., and Holme, G.,** The synthesis of a leukotriene with SRS-like activity, *Tetrahedron Lett.*, 21, 1485—1488, 1980.

41. **Spangler, C. W.,** Thermal [1,j] sigmatropic rearrangements, *Chem. Rev.*, 76, 187—217, 1976.

42. **Baker, S. R., Jamieson, W. B., McKay, S. W., Morgan, S. E., Rackham, D. M., Ross, W. J., and Shrubsall, P. R.**, Synthesis, separation and N. M. R. spectra of three double bond isomers of leukotriene A methyl ester, *Tetrahedron Lett.*, 21, 4123—4126, 1980.

43. **Atrache, V., Pai, J.-K., Sol, D.-E., and Sih, C. J.**, Biomimetic synthesis of leukotriene A, *Tetrahedron Lett.*, 22, 3443—3446, 1981.

44. **Baker, S. R., Jamieson, W. B., Osborne, D. J., and Ross, W. J.**, Synthesis of the 9Z and 9Z,11E isomers of leukotriene C$_4$, *Tetrahedron Lett.*, 22, 2505—2508, 1981.

45. **Baker, S. R., Boot, J. R., Jamieson, W. B., Osborne, D. J., and Sweatman, W. J. F.**, The comparative *in vitro* pharmacology of leukotriene D$_4$ and its isomers, *Biochem. Biophys. Res. Commun.*, 103, 1258—1264, 1981.

46. **Corey, E. J., Clark, D. A., Goto, G., Marfat, A., Mioskowski, C., Samuelsson, B., and Hammarström, S.**, Stereospecific total synthesis of a "slow-reacting substance of anaphylaxis", leukotriene C-1, *J. Am. Chem. Soc.*, 102, 1436—1439 and 3663, 1980.

47. **Corey, E. J., Arai, Y., and Mioskowski, C.**, Total synthesis of (±)-5,6-oxido- 7,9-trans-11,14-cis-eicosapentaenoic acid, a possible precursor of SRSA, *J. Am. Chem. Soc.*, 101, 6748—6749, 1979.

48. **Schenck, G. O. and Steinmetz, R.**, Photochemische bildungsweisen und umlagerungen von thiopheno- und furano-cyclopropancarbonsäureestern, *Justus Liebigs Ann. Chem.*, 668, 19—30, 1963.

49. **Rokach, J., Girard, Y., Guindon, Y., Atkinson, J. G., Larue, M., and Young, R. N.**, Total synthesis of the leukotrienes from achiral precursors, unpublished results.

50. **Holme, G., Brunet, G., Piechuta, H., Masson, P., Girard, Y., and Rokach, J.**, The activity of synthetic leukotriene C-1 on guinea pig trachea and ileum, *Prostaglandins*, 20, 717—728, 1980.

51. **Rokach, J., Girard, Y., Guindon, Y., Atkinson, J. G., Larue, M., Young, R. N., Masson, P., Hamel, R., Piechuta, H., and Holme, G.**, The synthesis of leukotrienes, in *SRS-A and Leukotrienes*, Piper, P. J., Ed., John Wiley & Sons, New York, 1981, 65—72.

52. **Rosenberger, M. and Neukom, C.**, Total synthesis of (5S, 6R, 7E, 9E, 11Z, 14Z)-5-hydroxy-6-[(2R)-2-amino-2-(carboxyethyl)thio]-7,9,11,14-eicosatetraenoic acid, a potent SRS-A, *J. Am. Chem. Soc.*, 102, 5425—5426, 1980.

53. **Gleason, J. G., Bryan, D. B., and Kinzig, C. M.**, Convergent synthesis of leukotriene A methyl ester, *Tetrahedron Lett.*, 21, 1129—1132, 1980.

54. **Corey, E. J., Park, H., Barton, A., and Nii, Y.**, Synthesis of three potential inhibitors of the biosynthesis of leukotrienes A-E, *Tetrahedron Lett.*, 21, 4243—4246, 1980.

55. **McKay, S. W., Mallen, D. N. B., Shrubsall, P. R., Smith, J. M., Baker, S. R., Jamieson, W. B., Ross, W. J., Morgan, S. E., and Rackham, D. M.**, Semi-preparative high-performance liquid chromatography and spectroscopic characterisation of eight geometric isomers of leukotriene A methyl ester, *J. Chromatogr.*, 214, 249—256, 1981.

56. **Rokach, J., Zamboni, R., Lau, C-K., and Guindon, Y.**, The stereospecific total synthesis of leukotriene A$_4$ (LTA$_4$), 5-epi-LTA$_4$, 6-epi-LTA$_4$ and 5-epi-6-epi-LTA$_4$, *Tetrahedron Lett.*, 22, 2759—2762, 1981.

57. **Rokach, J., Lau, C.-K., Zamboni, R., and Guindon, Y.**, A C-glycoside route to leukotrienes, *Tetrahedron Lett.*, 22, 2763—2766, 1981.

58. **Corey, E. J., Clark, D. A., Marfat, A., and Goto, G.**, Total synthesis of slow-reacting substances (SRS). "Leukotriene C-2" (11-*trans*-leukotriene C) and leukotriene D, *Tetrahedron Lett.*, 21, 3143—3146, 1980.

59. **Ernest, I., Main, A. J., and Menassé, R.**, Synthesis of the 7-*cis* isomer of the natural leukotriene D$_4$, *Tetrahedron Lett.*, 23, 167—170, 1982.

60. **Corey, E. J. and Goto, G.**, Total synthesis of slow reacting substances (SRS's): 6-epi-leukotriene C and 6-epi-leukotriene D, *Tetrahedron Lett.*, 21, 3463—3466, 1980.

61. **Rokach, J., Young, R. N., Kakushima, M., Lau, C.-K., Séguin, R., Frenette, R., and Guindon, Y.**, Synthesis of leukotrienes — new synthesis of natural leukotriene A$_4$, *Tetrahedron Lett.*, 22, 979—982, 1981.

62. **Rossiter, B. E., Katsuki, T., and Sharpless, K. B.**, Asymmetric epoxidation provides shortest routes to four chiral epoxy alcohols which are key intermediates in syntheses of methymycin, erythromycin, leukotriene C$_1$ and disparlure, *J. Am. Chem. Soc.*, 103, 464—465, 1981.

63. **Corey, E. J., Hashimoto, S., and Barton, A. E.**, Chirally directed synthesis of (−)methyl-5(S),6(S)-oxido-7-hydroxyheptanoate, key intermediate for the total synthesis of leukotrienes A, C, D and E, *J. Am. Chem. Soc.*, 103, 721—722, 1981.

64. **Cohen, N., Banner, B. L., and Lopresti, R. J.**, Synthesis of optically active leukotriene (SRS-A) intermediates, *Tetrahedron Lett.*, 21, 4163, 1980.

65. **Marriott, D. P. and Bantick, J. R.**, 5(S),6(R)-5,7-dibenzoyloxy-6-hydroxyheptanoate ester: improved synthesis of a leukotriene intermediate, *Tetrahedron Lett.*, 22, 3657—3658, 1981.

66. **Corey, E. J., Barton, A. E., and Clark, D. A.**, Synthesis of the slow reacting substance of anaphylaxis leukotriene C-1 from arachidonic acid, *J. Am. Chem. Soc.*, 102, 4278—4279, 1980.

67. **Houglum, J., Pai, J.-K., Atrache, V., Sok, D.-E., and Sih, C. J.,** Identification of the slow reacting substances from cat paws, *Proc. Natl. Acad. Sci. U.S.A.,* 77, 5688—5692, 1980.

68. **Corey, E. J. and Barton, A. E.,** Chemical conversion of arachidonic acid to slow reacting substances, *Tetrahedron Lett.,* 23, 2351—2354, 1982.

69. **Corey, E. J., Marfat, A., and Goto, G.,** Simple synthesis of the 11,12-oxido and 14,15-oxido analogues of leukotriene A and the corresponding conjugates with glutathione and cysteinylglycine, analogues of leukotrienes C and D, *J. Am. Chem. Soc.,* 102, 6607—6608, 1980.

70. **Sok, D.-E., Han, C.-O., Shieh, W.-R., Zhou, B.-N. and Sih, C. J.,** Enzymatic formation of 14,15-leukotriene A and C(14)-sulfur-linked peptides, *Biochem. Biophys. Res. Commun.,* 104, 1363—1370, 1982.

71. (a) **Hanessian, S.,** Approaches to the total synthesis of natural products using "chiral templates" derived from carbohydrates, *Acc. Chem. Res.,* 12, 159—165, 1979; (b) **Fraser-Reid, B. and Anderson, R. C.,** Carbohydrate derivatives in the asymmetric synthesis of natural products, in *Progress in the Chemistry of Organic Natural Products,* Vol. 39, Herz, W., Grisebach, H., and Kirby, G. W., Eds., Springer-Verlag, New York, 1980, 1—61.

72. **Hammerström, S., Samuelsson, S., Clark, D. A., Goto, G., Marfat, A., Mioskowski, C., and Corey, E. J.,** Stereochemistry of leukotriene C-1, *Biochem. Biophys. Res. Commun.,* 92, 946—953, 1980.

73. **Lewis, R. A., Austen, K. F., Drazen, J. M., Clark, D. A., Marfat, A., and Corey, E. J.,** Slow reacting substances of anaphylaxis: identification of leukotrienes C-1 and D from human and rat sources, *Proc. Natl. Acad. Sci. U.S.A.,* 77, 3710—3714, 1980.

74. **Murphy, R. C., Mathews, W. R., Rokach, J., and Fenselau, C.,** Comparison of biological-derived and synthetic leukotriene C_4 by fast atom bombardment mass spectrometry, *Prostaglandins,* 23, 201—206, 1982.

75. **Guindon, Y., Young, R. N., and Frenette, R.,** Synthesis of β-trimethylsilyloxythioethers and β-hydroxythioethers by the reaction of epoxides with aryl- and alkylthiotrimethylsilanes, *Synth. Commun.,* 11, 391—398, 1981.

76. **Morris, H. R., Taylor, G. W., Rokach, J., Girard, Y., Piper, P. J., Tippins, J. R., and Samhoun, M. H.,** Slow reacting substance of anaphylaxis, SRS-A: assignment of the stereochemistry, *Prostaglandins,* 20, 601—607, 1980.

77. **Anderson, M. E., Allison, R. D., and Meister, A.,** Interconversion of leukotrienes catalyzed by purified γ-glutamyl transpeptidase: concomitant formation of leukotriene D_4 and γ-glutamyl amino acids, *Proc. Natl. Acad. Sci. U.S.A.,* 79, 1088—1091, 1982.

78. **Bergström, K. and Hammarstöm, S.,** A novel leukotriene formed by transpeptidation of leukotriene E, in *Abstract Book, V International Conference of Prostaglandins,* 667, May 18-21, 1982.

79. **Denis, D., Charleson, S., Rackham, A., Girard, Y., Larue, M., Jones, T. R., Ford-Hutchinson, A. W., Lord, A., Cirino, M., and Rokach, J.,** Synthesis and biological activities of leukotriene F_4 and leukotriene F_4 sulfone, *Prostaglandins,* 24, 801—814, 1982.

80. **Ohnishi, H., Kosozume, H., Kitamura, Y., Yamaguchi, K., Nobuhara, M., Suzuki, Y., Yoshida, S., Tomioka, H., and Kumagai, A.,** Structure of slow-reacting substance of anaphylaxis (SRS-A), *Prostaglandins,* 20, 655—666, 1980.

81. **Girard, Y., Larue, M., Jones, T. R., and Rokach, J.,** Synthesis of the sulfones of leukotriene C_4, D_4, and E_4, *Tetrahedron Lett.,* 23, 1023—1026, 1982.

82. **Jones, T., Masson, P., Hamel, R., Brunet, G., Holme, G., Girard, Y., Larue, M., and Rokach, J.,** Biological activity of leukotriene sulfones on respiratory tissues, *Prostaglandins,* 24, 279—291, 1982.

83. **Lee, C. W., Lewis, R. A., Corey, E. J., Barton, A., Oh, H., Tauber, I., and Austen, K. R.,** Oxidative inactivation of leukotriene C_4 by stimulated human polymorphonuclear leukocytes, *Proc. Natl. Acad. Sci. U.S.A.,* 79, 4166—4170, 1982.

84. **Corey, E. J., Oh, H., and Barton, A. E.,** Pathways for migration and cleavage of the S-peptide unit of the leukotrienes, *Tetrahedron Lett.,* 23, 3467—3470, 1982.

85. **Rokach, J., Guindon, Y., Girard, Y., and Lau, C. K.,** Preparation of the natural and three possible unnatural 5,6-isomers of LTC_4 and LTD_4, unpublished results.

86. **Holme, G., Masson, P., Girard, Y., and Rokach, J.,** Biological activity of the unnatural 5,6-isomers of LTC_4 and LTD_4, unpublished results.

87. **Atrache, V., Sok, D.-E., Pai, J.-K., and Sih, C. J.,** Formation of 11-*trans* slow reacting substances, *Proc. Natl. Acad. Sci. U.S.A.,* 78, 1523—1526, 1981.

88. (a) **Huyser, E. S. and Kellogg, R. M.,** Isomerization of the butenyl methyl sulfides with methanethiol, *J. Org. Chem.,* 30, 2867—2868, 1965; (b) **Hall, D. N., Oswald, A. A., and Griesbaum, K.,** Allene chemistry. IV. Unsymmetrical terminal thiol-allene diadducts. The effect of allylic reversal, *J. Org. Chem.,* 30, 3829—3843, 1965.

89. **Perkins, C. W., Martin, J. C., Arduengo, A. J., Lau, W., Alegria, A., and Kochi, J. K.,** An electrically neutral σ-sulfuranyl radical from the homolysis of a perester with neighboring sulfenyl sulfur: 9-S-3 species, *J. Am. Chem. Soc.,* 102, 7753—7759, 1980.

90. **Drazen, J. M., Lewis, R. A., Austen, K. F., Toda, M., Brion, F., Marfat, A., and Corey, E. J.,** Contractile activities of structural analogs of leukotrienes C and D: necessity of a hydrophobic region, *Proc. Natl. Acad. Sci. U.S.A.,* 78, 3195—3198, 1981.

91. **Young, R. N., Coombs, W., Guindon, Y., Rokach, J., Ethier, D., and Hall, R.,** The preparation of octahydro leukotrienes C, D, and E via a stereoselective sulfenyllactonization reaction, *Tetrahedron Lett.,* 22, 4933—4936, 1981.

92. **Corey, E. J. and Hoover, D. J.,** Total synthesis of 5-desoxyleukotriene D. A new and useful equivalent of the 4-formyl-*trans-trans*-1,3-butadienyl anion, *Tetrahedron Lett.,* 23, 3463—3466, 1982.

93. **Ellis, F., Mills, L. S., and North, P. C.,** A total synthesis of leukotriene F₄ (LTF₄), *Tetrahedron Lett.,* 23, 3735—3736, 1982.

94. **Just, G., Luthe, C., and Phan Viet, M. T.,** The synthesis of 11R and 11S-HETE and of 11-R,S-HPETE methyl esters, *Can. J. Chem.,* 61, 712—717, 1983.

95. **Clark, D. A. and Marfat, A.,** Structure elucidation and the total synthesis of the leukotrienes, in *Annual Reports in Medicinal Chemistry,* Vol. 17, Hess, H. J., Ed., Academic Press, New York, 1982, chap. 29.

96. **Zamboni, R. and Rokach, J.,** Stereospecific synthesis of 5S-HETE, 5R-HETE and their transformation to 5(±) HPETE, *Tetrahedron Lett.,* 24, 999—1002, 1983.

97. **Okuyama, S., Miyamoto, S., Shimoji, K., Konishi, Y., Fukushima, D., Niwa, H., Arai, Y., Toda, M., and Hayashi, M.,** Structural analogs of leukotrienes C and D and their contractile activities, *Chem. Pharm. Bull.,* 30, 2453—2462, 1982.

98. **Buck, J. C., Ellis, F., and North, P. C.,** A novel stereospecific synthesis of (±)-leukotriene A₄ (LTA₄) methyl ester, *Tetrahedron Lett.,* 23, 4161—4162, 1982.

99. **Rokach, J., Adams, J., and Perry, R.,** A new general method for the synthesis of lipoxygenase products: preparation of (±)-5-HETE, *Tetrahedron Lett.,* 24, 5185—5188, 1983.

100. **Gunn, B. P.,** An efficient and general methodology for the synthesis of the HETES: synthesis of (±)-5-hydroxy-6-trans-8,11,14-cis-eicosatetraenoic acid (5-HETE), *Tetrahedron Lett.,* 26, 2869—2872, 1985.

101. **Adams, J. and Rokach, J.,** Synthesis of (±)-8- and (±)-9-HETES, *Tetrahedron Lett.,* 25, 35—38, 1984.

102. **Corey, E. J., Kyler, K., and Raju, N.,** A short, three component total synthesis of 12-hydroxyeicosa-5,8,14(Z), 10(E)-tetraenoic acid (12-HETE) via the corresponding ketone, *Tetrahedron Lett.,* 25, 5115—5118, 1984.

103. **Just, G. and Wang, Z. Y.,** A simple synthesis of methyl 11(S)- and 12(S)-HETE, *Tetrahedron Lett.,* 26, 2993—2996, 1985.

104. (a) **Leblanc, Y., Fitzsimmons, B. J., Adams, J., Perez, F., and Rokach, J.,** The total synthesis of 12-HETE (12-hydroxyeicosatetraenoic acid) and 12,20-diHETE, *J. Org. Chem.,* 51, 789—793, 1986; (b) **Manna, S., Viala, J., Yadagiri, P., and Falck, J. R.,** Synthesis of 12(S),20-, 12(S),19(R)-, and 12(S),19(S)-dihydroxy-eicosa-cis-5,8,14-trans-10-tetraenoic acids, metabolites of 12(S)-HETE, *Tetrahedron Lett.,* 27, 2679—2682, 1986.

105. **Nicolaou, K. C., Ladduwahetty, T., and Elisseou, E. M.,** Stereocontrolled total synthesis of (5Z,8Z,11Z,13E)(15S)-15-hydroxyeicosa-5,8,11,13-tetraenoic acid (15S-HETE) and analogs, *J. Chem. Soc., Chem. Commun.,* 1580—1581, 1985.

106. **Fuganti, C., Servi, S., and Zirotti, C.,** Non-carbohydrate based synthesis of natural LTB₄, *Tetrahedron Lett.,* 24, 5285—5288, 1983.

107. **Nicolaou, K. C., Zipkin, R. E., Dolle, R. E., and Harris, B. D.,** A general and stereocontrolled total synthesis of leukotriene B₄ and analogs, *J. Am. Chem. Soc.,* 106, 3548—3551, 1984.

108. **Davies, H., Roberts, G., Wakefield, S. M., and Winders, J. A.,** Synthesis of (3S,4R)-eldanolide and (5S,12R)-leukotriene B₄ through photolysis of optically active hydroxy-7,7-dimethylbicyclo[3.2.0]heptones, *J. Chem. Soc., Chem. Commun.,* 1166—1168, 1985.

109. **Han, C. Q., DiTullio, D., Wang, Y. F., and Sih, C. J.,** A chemoenzymatic synthesis of leukotriene B₄, *J. Org. Chem.,* 51, 1253—1258, 1986.

110. **Young, R. N., Zamboni, R., and Rokach, J.,** Studies on the conjugation of leukotriene B₄ with proteins— for development of a radioimmunoassay for leukotriene B₄, *Prostaglandins,* 26, 605—613, 1983.

111. **Adams, J., LeBlanc, Y., and Rokach, J.,** Synthesis of 5S,12S-diHETE (LTBₓ), *Tetrahedron Lett.,* 25, 1227—1230, 1984.

112. **Fitzsimmons, B. J. and Rokach, J.,** The total syntheses of several 8,15-dihydroxyarachidonic acid derivatives (8,15-LTB's), *Tetrahedron Lett.,* 25, 3043—3046, 1984.

113. **Nicolaou, K. C. and Webber, S. E.,** Total synthesis of 5(S),15(S)-dihydroxy-6,13-trans-8,15-cis-eicosatetraenoic acid (5,15-diHETE) and 8(S),15(S)-dihydroxy-5,11-cis-9,13-trans-eicosatetraenoic acid (8,15-DiHETE), two novel metabolites of arachidonic acid, *J. Am. Chem. Soc.,* 106, 5734—5736, 1984.

114. **Radmark, O., Serhan, C., Hamberg, M., Lundberg, U., Ennis, M. D., Bundy, G. L., Oglesby, T. D., Aristoff, P. A., Harrison, A. W., et al.,** Stereochemistry, total synthesis, and biological activity of 14,15-dihydroxy-5,8,10,12-eicosatetraenoic acid, *J. Biol. Chem.,* 259, 13011—13016, 1984.

115. **Spur, B., Crea, A., Peters, W., and Koenig, W.,** Synthesis of leukotriene B₃, *Arch. Pharm. (Weinheim),* 318, 225—228, 1985.

116. **Corey, E. J., Pyne, S. G., and Su, W. G.,** Total synthesis of leukotriene B$_5$, *Tetrahedron Lett.*, 24, 4883—4886, 1983.

117. **Pirillo, D., Gazzaniga, A., and Traverso, G.,** The synthesis methyl (RS)-5-benzoyloxy-5-formylpentanoate, a leukotriene B$_4$ synthon, *J. Chem. Res., Synop.*, 1, 1983.

118. **Mills, S. and North, P. C.,** A short synthesis of a key leukotriene B$_4$ synthon, *Tetrahedron Lett.*, 24, 409—410, 1983.

119. **Merrer, Y. L., Dureault, A., Gravier, C., Languin, D., and Depezay, J. C.,** Synthesis of chiral α-hydroxy aldehydes from D-mannitol. Intermediates for the synthesis of arachidonic acid metabolites, *Tetrahedron Lett.*, 26, 319—322, 1985.

120. (a) **Tolstikov, G. A., Miftakhov, M. S., and Tolstikov, A. G.,** Synthesis of methyl (±)-5,6-epoxy-7E,9E,11Z,14Z-eicosatetraenoate, *Izv. Akad. Nauk SSSR, Ser. Khim.*, 475, 1984; (b) **Tsuboi, S., Masuda, T., and Takeda, A.,** A new approach to (±)-leukotriene A$_4$ methyl ester, *Chem. Lett.*, 12, 1829—1832, 1983.

121. **Wang, Y., Li, J., Wu, Y., Huang, Y., Shi, L., and Yang, J.,** A facile stereoselective synthesis of leukotriene A$_4$ (LTA$_4$) methyl ester, *Tetrahedron Lett.*, 27, 4583—4584, 1986.

122. **Corey, E. J., Guo Su, W., and Mehrotra, M. M.,** An efficient and simple method for the conversion of 15-HPETE to 14,15-EPETE (lipotriene A) and 5-HPETE to leukotriene A as the methyl esters, *Tetrahedron Lett.*, 25, 5123—5126, 1984.

123. **Rosenberger, M., Newkom, C., and Aig, E. R.,** Total synthesis of leukotriene E$_4$, a member of the SRS-A family, *J. Am. Chem. Soc.*, 103, 3656—3661, 1983.

124. **Zamboni, R., Milette, S., and Rokach, J.,** Stereospecific synthesis of 11S,12S-oxido-5Z,7E,9E,14Z-eicosatetraenoic acid, *Tetrahedron Lett.*, 25, 5835—5838, 1984.

125. **Zamboni, R., Milette, Z., and Rokach, J.,** The stereospecific synthesis of 14S,15S-oxido-5Z,8Z,10E,12E-eicosatetraenoic acid, *Tetrahedron Lett.*, 24, 4899—4902, 1983.

126. **Miftakhov, M. S., Tolstikov, A. G., and Tolstikov, G. A.,** Prostanoids. VIII. Preparation of C$_7$-synthons used in the synthesis of leukotrienes, *Zh. Org. Khim.*, 20, 678—683, 1984.

127. **Cohen, N., Banner, B. L., Lopresti, R. J., Wong, F., Rosenberger, M., Liu, Y. Y., Thom, E., and Liebman, A. A.,** Enantiospecific syntheses of leukotrienes C$_4$, D$_4$, and E$_4$, and (14,15-^{3}H$_2$) leukotriene E$_4$ dimethyl ester, *J. Am. Chem. Soc.*, 105, 3661—3672, 1983.

128. **Adams, J., Millette, S., Rokach, J., and Zamboni, R.,** Oxidized leukotrienes; synthesis of 20-OH and 20-COOH LTD$_4$. Possible metabolites in the lipoxygenase pathway, *Tetrahedron Lett.*, 25, 2179—2182, 1984.

129. **Serhan, C. N., Hamberg, M., and Samuelsson, B.,** Trihydroxytetraenes: A novel series of compounds formed from arachidonic acid in human leukocytes, *Biochem. Biophys. Res. Commun.*, 118, 943—949, 1984.

130. **Serhan, C. N., Hamberg, M., and Samuelsson, B.,** Lipoxins: Novel series of biologically active compounds formed from arachidonic acid in human leukocytes, *Proc. Natl. Acad. Sci. U.S.A.*, 81, 5335—5339, 1984.

131. **Maas, R. L. and Brash, A. R.,** Evidence for a lipoxygenase mechanism in the biosynthesis of epoxide and dihydroxy leukotrienes from 15(S)-hydroperoxyeicosatetraenoic acid by human platelets and porcine leukocytes, *Proc. Natl. Acad. Sci. U.S.A.*, 80, 2884—2888, 1983.

132. **Adams, J., Fitzsimmons, B. J., and Rokach, J.,** Synthesis of lipoxins: Total synthesis of conjugated trihydroxy eicosatetraenoic acids, *Tetrahedron Lett.*, 25, 4713—4716, 1984.

133. **Adams, J., Fitzsimmons, B. J., Girard, Y., LeBlanc, Y., Evans, J. F., and Rokach, J.,** Enantiospecific and stereospecific synthesis of lipoxin A. Stereochemical assignment of the natural lipoxin A and its possible biosynthesis, *J. Am. Chem. Soc.*, 107, 464—469, 1985.

134. **Wong, M. Y. H., and Gray, G. R.,** 2-Deoxypentoses. Stereoselective reduction of ketene dithioacetals, *J. Am. Chem. Soc.*, 100, 3548—3553, 1978.

135. **LeBlanc, Y., Fitzsimmons, B. J., Adams, J., and Rokach, J.,** Total Synthesis of lipoxin B: Assignment of stereochemistry, *Tetrahedron Lett.*, 26, 1399—1402, 1985.

136. **Fitzsimmons, B. J., Adams, J., Evans, J. F., LeBlanc, Y., and Rokach, J.,** The lipoxins. Stereochemical identification and determination of their biosynthesis, *J. Biol. Chem.*, 260, 13008—13012, 1985.

137. **Seidegard, J., DePierre, J. W., and Pero, R. W.,** Measurement and characterization of membrane-bound and soluble epoxide hydrolase activities in resting mononuclear leukocytes from human blood, *Cancer Res.*, 44, 3654—3660, 1984.

138. **Rokach, J., Fitzsimmons, B. J., LeBlanc, Y., Ueda, N., and Yamamoto, S.,** unpublished results.

139. **Corey, E. J. and Su, W.,** Simple synthesis and assignment of stereochemistry of lipoxin A, *Tetrahedron Lett.*, 26, 281—284, 1985.

140. **Corey, E. J., Mehrotra, M. M., and Su, W.,** On the synthesis and structure of lipoxin B., *Tetrahedron Lett.*, 26, 1919—1922, 1985.

141. **Fitzsimmons, B. J. and Rokach, J.,** On the biosynthesis and the structure of lipoxin B., *Tetrahedron Lett.*, 26, 3939—3942, 1985.

142. **Nicolaou, K. C., Veale, C. A., Webber, S. E., and Katerinopoulos, H.,** Stereocontrolled total synthesis of lipoxins A., *J. Am. Chem. Soc.,* 107, 7515—7518, 1985.

143. **Fitzsimmons, B. J. and Rokach, J.,** unpublished results.

144. **Van Os, C. P. A., Rijke-Schilder, G. P. M., Van Halbeek, H., Verhagen, J., and Vliegenthart, J. F. G.,** Double dioxygenation of arachidonic acid by soybean lipoxygenase-1. Kinetics and regiostereospecificities of the reaction steps, *Biochim. Biophys. Acta,* 663, 177—193, 1981.

INDEX

I

K

Printed and bound by CPI Group (UK) Ltd, Croydon, CR0 4YY

18/10/2024

01776251-0008